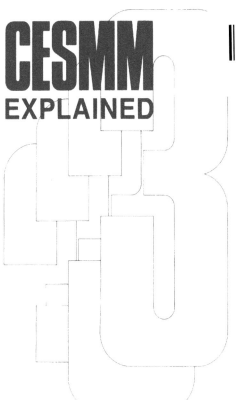

CESMM
EXPLAINED

Other titles from E & FN Spon

Spon's Architects' and Builders' Price Book
Davis Langdon & Everest

Spon's Civil Engineering and Highway Works Price Book
Davis Langdon & Everest

Spon's Mechanical and Electrical Services Price Book
Davis Langdon & Everest

Spon's Landscape and External Works Price Book
Davis Langdon & Everest and Derek Lovejoy Partners

Spon's European Construction Costs Handbook
Davis Langdon & Everest

Spon's Contractors' Handbook Series
Spain and Partners
> **Minor Works, Alterations, Repairs and Maintenance**
> **Painting, Decorating and Glazing**
> **Roofing**
> **Floor, Wall and Ceiling Finishings**
> **Electrical Installation**
> **Plumbing and Domestic Heating**

Spon's Budget Estimating Handbook
Spain and Partners

A Concise Introduction to Engineering Economics
P. Cassimatis

Design and Construction of Engineering Foundations
F.D.C. Henry

Planning and Design of Engineering Systems
G. Dandy and R. Warner

Spon's Fabrication Norms for Offshore Structures
Franklin & Andrews

The Piping Guide
D. Sherwood

Standard Method of Specifying for Minor Works
L. Gardiner

Worked Examples in Quantity Surveying Measurement
P.E. Goodacre and W. Crosbie-Hill

Estimating Checklist for Capital Projects
Association of Cost Engineers

Spon's Construction Cost and Price Indices Handbook
M.C. Fleming and B.A. Tysoe

For more information on these and other titles please contact:
The Promotion Department, E & FN Spon, 2–6 Boundary Row, London,
SE1 8HN. Telephone 071-865 0066

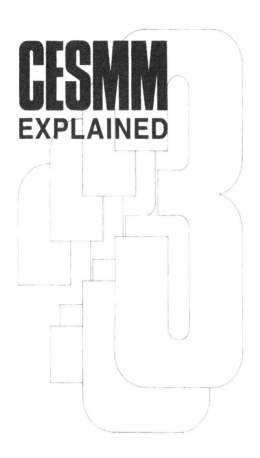

CESMM
EXPLAINED

Routledge
Taylor & Francis Group

LONDON AND NEW YORK

First published 1992 by Routledge

2 Park Square, Milton Park, Abingdon, Oxfordshire OX14 4RN
52 Vanderbilt Avenue, New York, NY 10017

Routledge is an imprint of the Taylor & Francis Group, an informa business

First issued in paperback 2019

A catalogue record for this book is available from the British Library

Library of Congress Cataloging-in-Publication data available

ISBN 978-0-419-17700-5 (hbk)
ISBN 978-0-367-86554-2 (pbk)

CONTENTS

CONTENTS

PREFACE

The first edition of this book was published in 1986 shortly after CESMM 2 was produced. CESMM 2 has served the civil engineering industry remarkably well in the last few years and the number of measurement disputes has declined.

The revisions that have been made are stated at the beginning of each chapter (changes from CESMM 1 to CESMM 2 have also been listed). The most important changes are:

(a) bringing CESMM in line with changed BS numbers and linking them to 'equivalent national standards' of EEC countries

(b) making the method compatible with the new 6th edition of the ICE Conditions of Contract

(c) enlarging Class Y to include repairs and renovations to water mains

(d) including a new Class Z to cover building works 'incidental to civil engineering works'

(e) the re-defining of materials in Class I and J in line with common usage

(f) sundry changes in various classes.

In the first edition of this book, most of the technical input was provided by Len Morley, who is now the partner in charge of civil engineering work at Tweeds, and this volume is soundly based on his work and research subject to changes listed above.

I am also indebted to Gil Nicholls, Paul Spain, Dorothy Spain, Elizabeth Young and Nikki Lark for their help in the production of this volume. The thanks expressed in the first edition to those who helped at that stage are still relevant.

> Bryan J.D. Spain FInstCES MACostE
> Tweeds (incorporating Spain and Partners)
> Cavern Walks
> 8 Matthew Street
> Liverpool L2 6RE

This book is intended as a guide to the use of CESMM 2 and has been presented in a style which it is hoped will be helpful both to students and to working quantity surveyors or engineers.

It has not been written as a comparison between CESMM 1 and CESMM 2 although references to some of the more significant changes have been made.

Throughout the book reference is made to CESMM 1 and CESMM 2 and these should be taken as the Civil Engineering Standard Method of Measurement published in 1976 and 1985 respectively.

The first chapters deal with general matters connected with the measurement of civil engineering works and each class of CESMM 2 is dealt with individually in the subsequent chapters, together with a library of the most commonly used item descriptions. Examples of taking off have been included in Appendices A to E and Appendix F contains useful data.

Because of the very nature of civil engineering work there are generally more unknown factors at the document preparation stage of civil projects than in building. CESMM 2 therefore is much more flexible than the Standard Method of Measurement for Building Works. This flexibility is most readily observed in the authority given by paragraph 5.10 to the taker-off to add additional descriptive matter if he wishes in certain circumstances.

The flexibility sometimes extends to the interpretation of the rules and the authors have set out their views and have sometimes recommended alternative courses of action. It should be noted that these are personal opinions based on general matters which may not be appropriate in all circumstances.

Each chapter dealing with the individual classes has been set out thus:-

(a) Principal changes from CESMM 1

(b) Measurement Rules

(c) Definition Rules

(d) Coverage Rules

(e) Additional Description Rules

(f) Item Measurement

(g) Library of Standard Descriptions

PREFACE TO THE FIRST EDITION

Section A above has been included to assist those surveyors and engineers who are already familiar with CESMM 1. Sections b) to e) provide a detailed examination of the most pertinent rules together with an explanation of their significance. Section f) is a checklist of the items which require consideration when undertaking the measurement of any particular item. The rules which are itemised separately are those which directly affect the quantity or item description of the work being measured and must be taken into account. Subsidiary rules which do not directly affect the quantity or item description appear at the end of the 'Generally' section of each item or group of items.

The reader should refer to a copy of Civil Engineering Standard Method of Measurement and The CESMM 2 Handbook by Martin Barnes. Both books are published by Thomas Telford Ltd., 1-7 Great George Street, London SW1P 3AA.

Acknowledgements are due to the following individuals and firms for their comments and assistance.

Institution of Civil Engineering Surveyors
Stephen Booth
Cementation Ltd
Paula Wood
Gil Nicholls
B.T. Rathmell
John Taylor and Sons
Rona Harper
Stuart Mackrell

and finally to John McGee for his valuable
contribution throughout the preparation of this book.

INTRODUCTION

The latest revision to the Civil Engineering Standard Method of Measurement (originally published in November 1985) is part of a continuous policy of appraisal of measurement techniques of civil engineering works undertaken by the Institution of Civil Engineers for over half a century.

In 1933 a report was published by the Council of the Institution which laid down 'uniform principles for drafting bills of quantities for civil engineering work'. This report served as a guide to the industry until after the Second World War when a committee was appointed in 1950 to update and amend the document in the light of 17 years use.

In 1951 a revised report 'Standard Method of Measurement of Civil Engineering Quantities' was published and this document (with metric addendum added later) survived until the mid-seventies.

Martin Barnes and Peter Thompson carried out a research project for CIRIA (Construction Industry Research and Information Association) in 1971 on 'Civil Engineering Quantities' and it was this work that led to the publication of the Civil Engineering Standard Method of Measurement in 1976 by the ICE.

The CESMM was quite different from its predecessors and indeed from any other method of measurement in use at that time. It attempted, and research has shown that it has succeeded, to produce bills of quantities which had a more realistic relationship with both the methods of working and to the cost of carrying out the work. The performance of CESMM 1 was closely monitored and in 1983 the ICE appointed Martin Barnes and Partners to commence work on the revision which led to the publication of CESMM 2 in 1985.

Apart from correcting the minor ambiguities and uncertainties which had been revealed during seven years of use, it was also intended to produce a closer link to actual construction costs brought about by advances in construction techniques.

The major change in CESMM 2 in presentation was the upgrading of the notes which appeared with the Work Classification tables. It was felt that the attention of the 'taker-off' should be more forcibly directed to the provisions of the notes.

To achieve this the information previously listed under the heading of notes was altered, placed under the heading of rules, divided into four sections and presented in a tabular form to match the Work Classification tables opposite.

The other significant change was the introduction of a new class for the measurement of sewer renovation work. This area of work was increasing rapidly and there was a well defined need for uniformity in its measurement.

There were a number of factors which led the ICE to believe that there was a need to update CESMM2. These included the necessity to make the method compatible with the ICE Conditions of Contract 6th Edition which was published in January 1991. It was also felt that recent changes in BS numbers should be recognised but clause 1.15 now links the British Standards to 'equivalent national standards' of member countries of the European

Community.

The two main changes in CESMM3, however, lie in the extension of the scope of work to be measured. Class Y which previously dealt with renovation to sewers has been broadened to include water main repairs and upgrading. A new Class Z has been included which covers work described as 'simple building works incidental to civil engineering works'. This is intended for pump house superstructures, small administration offices and the like which occur on large civil contracts. Large building projects would still be needed to be measured on SMM7 of course, but Class Z offers a welcome framework for the measurement of small buildings.

PART ONE

GENERAL

Chapter 1

GENERAL PRINCIPLES OF MEASUREMENT

Until comparatively recently, the person preparing the Bill of Quantities - the 'taker-off' - had a limited choice of how to convert the information on the drawings into a Bill of Quantities.

Traditionally the systems followed a procedure of:

Taking off - measuring from the drawings and entering the dimensions on to specially ruled dimension paper

Squaring - calculating and totalling the lengths, areas and volumes of the dimensions

Abstracting - collecting the totals from the dimension paper on to an abstract to produce a final total for each individual description.

Billing - reproducing the items from the abstract on to bill paper in draft form ready for typing.

It may be that some offices still adopt this system of taking off and working up as they are commonly called but they cannot be in the majority. In any case there is less need for the preparation of the abstract in civil engineering work as in building. For example, in a school all the painting dimensions for every room are added together on the abstract and stated in the bill as the total for the whole project. In a sewage treatment works, however, the work will usually be presented as a series of locational sub-bills each containing similar items e.g. the Inlet Works, Primary Treatment Tanks will each be billed separately and will contain similar items. The adoption of this system greatly increases the efficiency of the post contract administration.

An experienced civil engineering taker-off can usually take off in bill order and if he adopts a system of allocating only one item to each dimension sheet it removes the need for abstracting.

Conversely, some practices have adopted a system of writing full descriptions on the abstract sheets in bill order (a skill possessed by an experienced worker-up) and typing the bill direct from the abstract.

In the last thirty years most quantity surveying practices have adopted the cut-and-shuffle method. This comprises of the writing of item descriptions and dimensions on to sensitised paper to produce two copies. When the taking off and squaring is complete the copies are split or 'cut' and one copy 'shuffled' into bill order with all sheets for the same item pinned together and their totals collected to produce a final quantity. More recently other systems have come into use where the taker-off enters dimensions into a computer (sometimes by using a digitiser) which will then perform the squaring, abstracting, billing and printing functions.

Dimension paper

The ruling of dimension paper should conform to the requirements of BS3327 - Stationery for Quantity Surveying and the paper is vertically separated into two parts by a double line each with four columns (Figure 1).

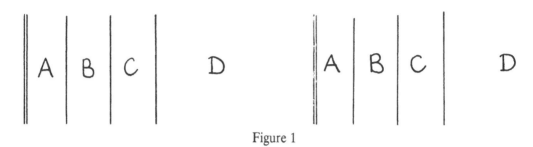

Figure 1

Column A is the timesing and 'dotting on' column where multiplication and addition of the dimensions can be recorded (Figure 2).

A	B	C	D	A	B	C	D
5/	2	10	2 multiplied by 5	π/2.00	2.00	12.57	a circle with a 2 metre radius multiplied by pi (3.142)
3.5/	2	16	2 multiplied by the sum of 3 and 5	4/½/4.00	2.00	16.00	Four triangles with base 4 metres and height 2 metres multiplied by ½ to produce the area.
5/	6.00 7.00	210.00	5 superficial areas with sides 6m and 7m long				
2/3.5/	6.00 7.00	672.00	a superficial area with sides 6m and 7m long multiplied by the sum of 3 and 5 and further multiplied by 2				

Figure 2

2

GENERAL PRINCIPLES OF MEASUREMENT

The practice of 'dotting-on' should be used only where absolutely necessary because of the dangers of mistaking the dot for a decimal point.

Column B is the dimension column and receives the measurements taken off from the drawings. The dimensions are normally expressed to two decimal points (Figure 3).

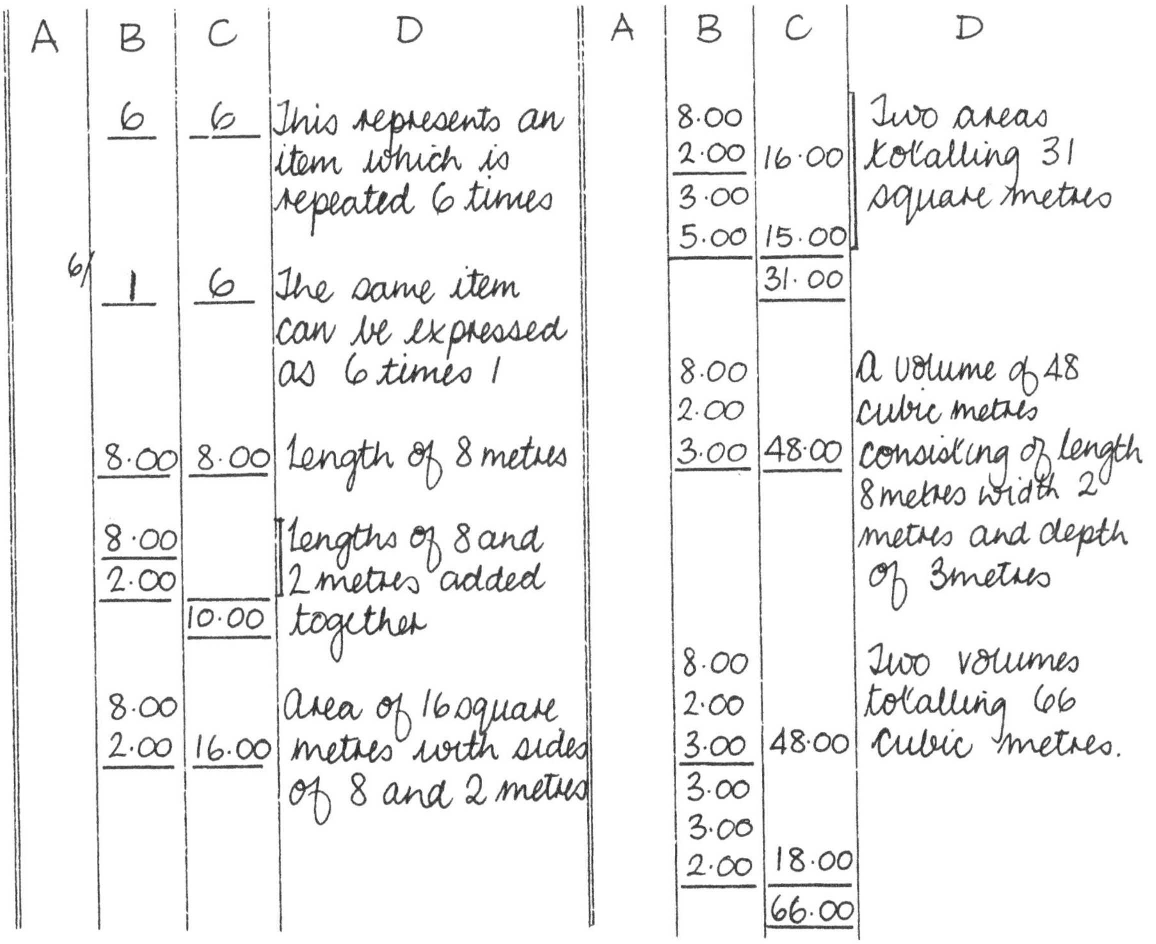

A	B	C	D	A	B	C	D
	6	6	This represents an item which is repeated 6 times		8.00		Two areas totalling 31 square metres
					2.00	16.00	
					3.00		
6/	1	6	The same item can be expressed as 6 times 1		5.00	15.00	
						31.00	
					8.00		A volume of 48 cubic metres consisting of length 8 metres width 2 metres and depth of 3 metres
	8.00	8.00	Length of 8 metres		2.00		
					3.00	48.00	
	8.00		Lengths of 8 and 2 metres added together				
	2.00						
		10.00			8.00		Two volumes totalling 66 cubic metres.
					2.00		
	8.00	16.00	Area of 16 square metres with sides of 8 and 2 metres		3.00	48.00	
	2.00				3.00		
					3.00		
					2.00	18.00	
						66.00	

Figure 3

It is important to note that it is the insertion of the horizontal line which determines whether the dimension is intended as a linear, superficial or cubic measurement (Figure 4).

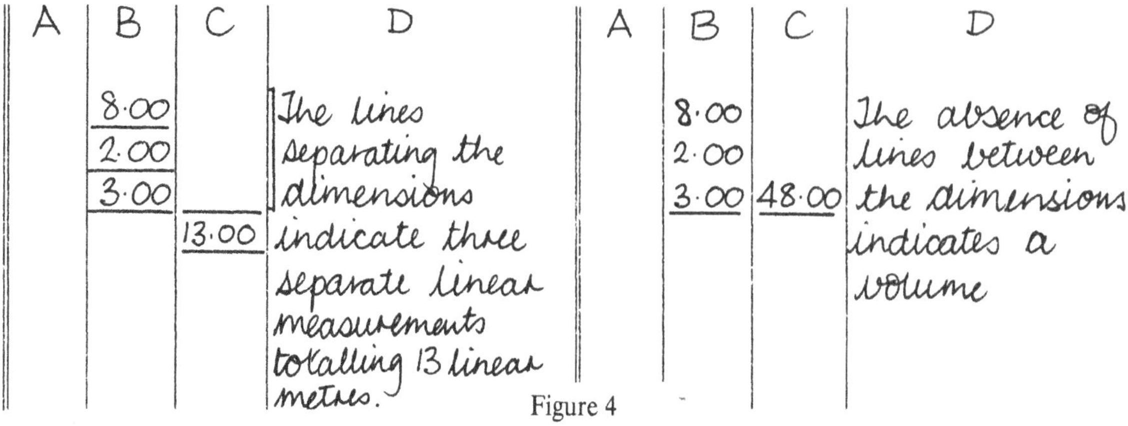

A	B	C	D	A	B	C	D
	8·00		The lines		8·00		The absence of
	2·00		separating the		2·00		lines between
	3·00		dimensions		3·00	48·00	the dimensions
		13·00	indicate three				indicates a
			separate linear				volume
			measurements				
			totalling 13 linear				
			metres.				

Figure 4

The dimensions should always be recorded in the order of length, width and height. Column C is the squaring column where the result of the addition, subtraction or multiplication of the entries in the dimension column is recorded. Figures which are to be added or subtracted are bracketed together in the manner shown.

Deductions are sometimes necessary where it is easier to take an overall measurement and deduct the parts not required (Figure 5).

Column D is the description column where the item being measured is described. This is done by using a form of standard abbreviations which have been listed separately. This column also contains annotations giving the location of the dimensions and waste calculations which show the build up of the figures entered in the dimension column. (Figure 6).

Quite often two item descriptions share the same measurement and this is indicated by linking the descriptions with an ampersand.

It may be considered desirable to insert the appropriate CESMM 3 code in the description column as shown in Figure 6 but the value of doing this will depend upon the subsequent method of processing the dimensions and descriptions that is adopted.

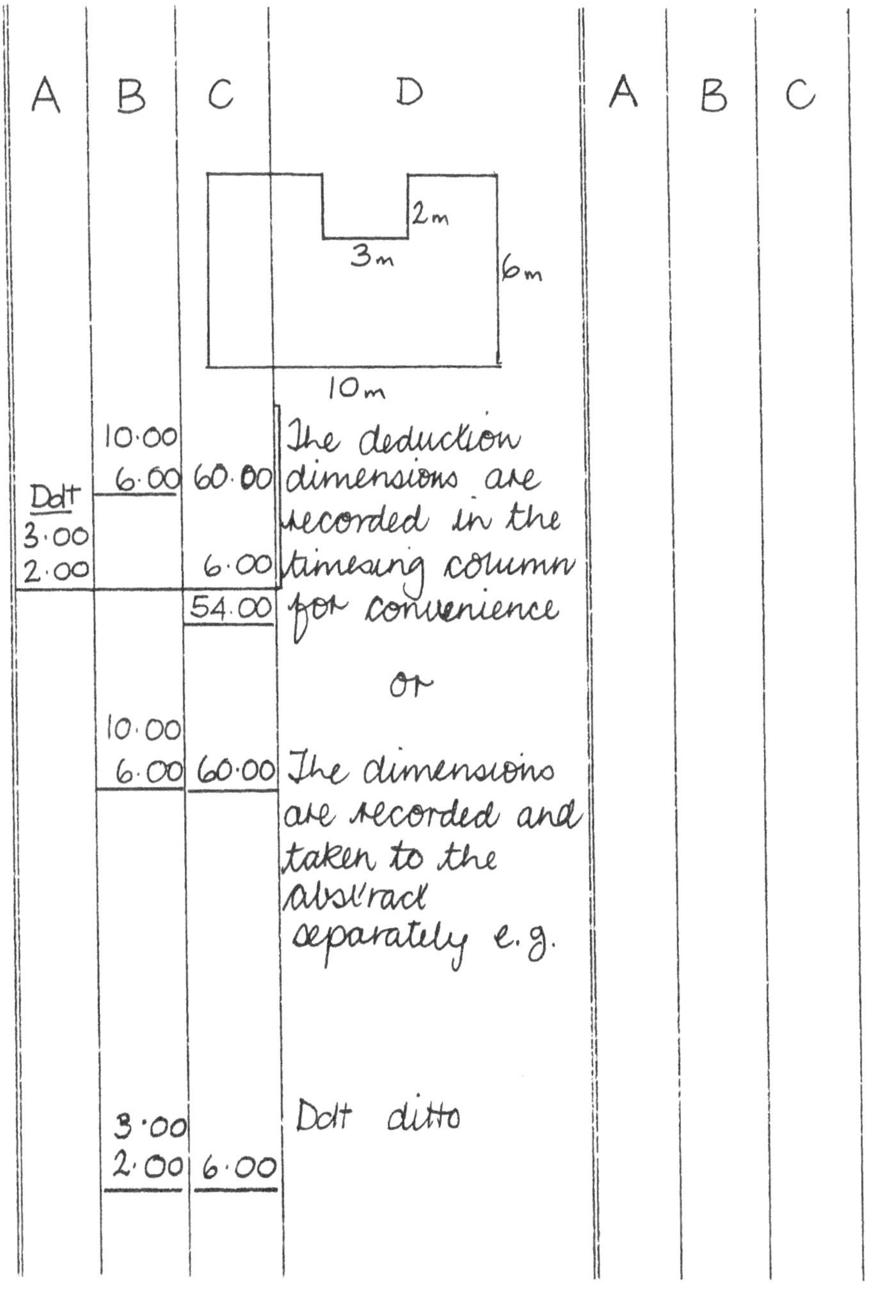
A
B
C
D
A
B
C
D
2m
3m
6m
10m
10.00
6.00
60.00
The deduction dimensions are recorded in the timesing column for convenience
Ddt
3.00
2.00
6.00
54.00
or
10.00
6.00
60.00
The dimensions are recorded and taken to the abstract separately e.g.
3.00
2.00
6.00
Ddt ditto

Figure 5

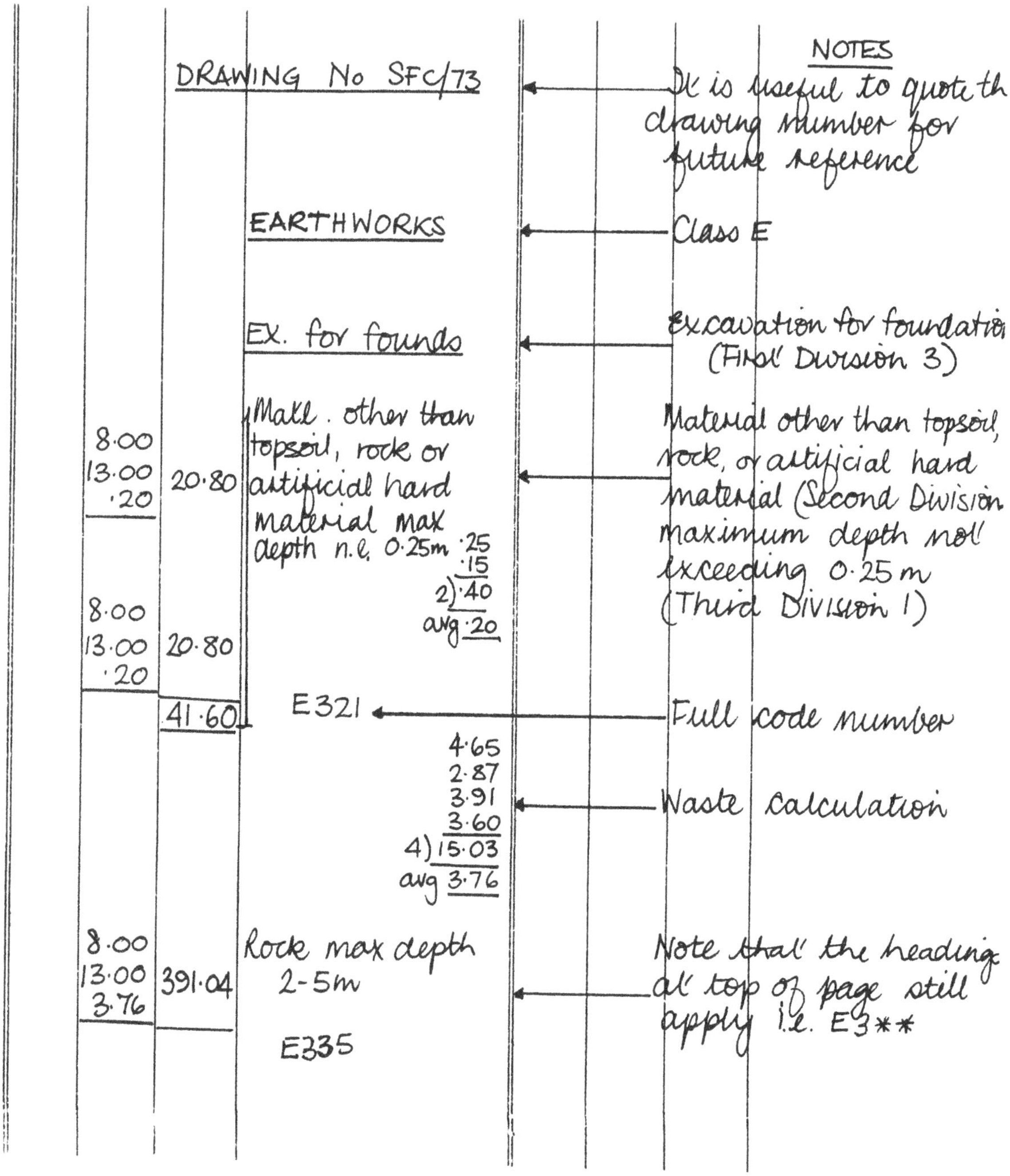

				NOTES
		DRAWING No SFC/73	←	It is useful to quote the drawing number for future reference
		EARTHWORKS	←	Class E
		EX. for founds	←	Excavation for foundation (First Division 3)
8·00 13·00 ·20	20·80	Matl. other than topsoil, rock or artificial hard material max depth n.e. 0.25m	←	Material other than topsoil, rock, or artificial hard material (Second Division maximum depth not exceeding 0·25 m (Third Division 1)
			·25 ·15 2)·40 avg ·20	
8·00 13·00 ·20	20·80			
	41·60	E 321	←	Full code number
			4·65 2·87 3·91 3·60 4)15·03 avg 3·76	Waste calculation
8·00 13·00 3·76	391·04	Rock max depth 2-5m	←	Note that the heading at top of page still apply i.e. E3**
		E335		

Figure 6

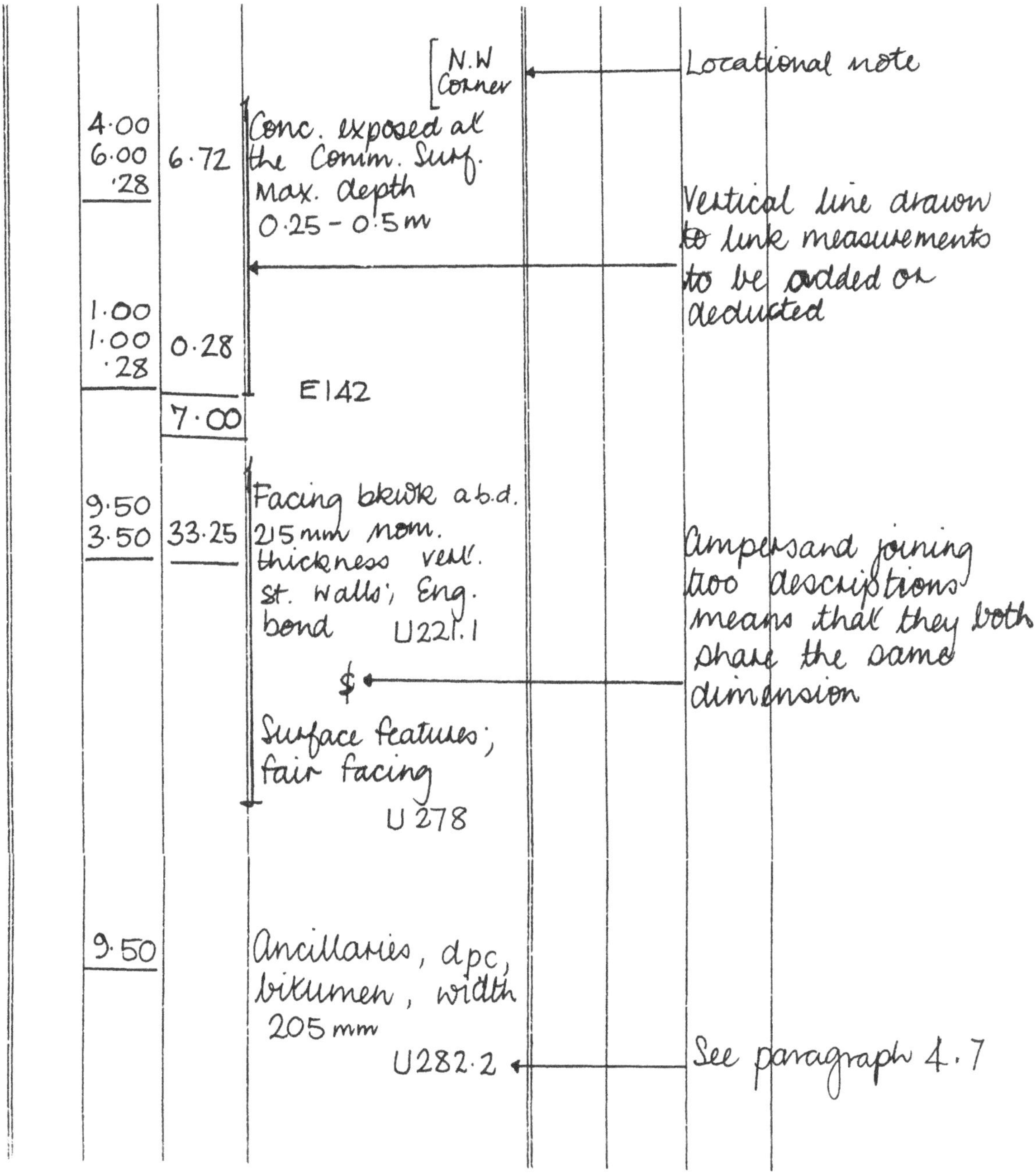

4.00 6.00 .28	6.72	[N.W Corner Conc. exposed at the Comm. Surf. Max. depth 0.25 - 0.5 m
1.00 1.00 .28	0.28	E142
	7.00	
9.50 3.50	33.25	Facing bkwk a.b.d. 215 mm nom. thickness vert. st. walls; Eng. bond U221.1 $ Surface features; fair facing U 278
9.50		Ancillaries, dpc, bitumen, width 205 mm U282.2

Locational note

Vertical line drawn to link measurements to be added or deducted

Ampersand joining two descriptions means that they both share the same dimension

See paragraph 4.7

Figure 6 (continued)

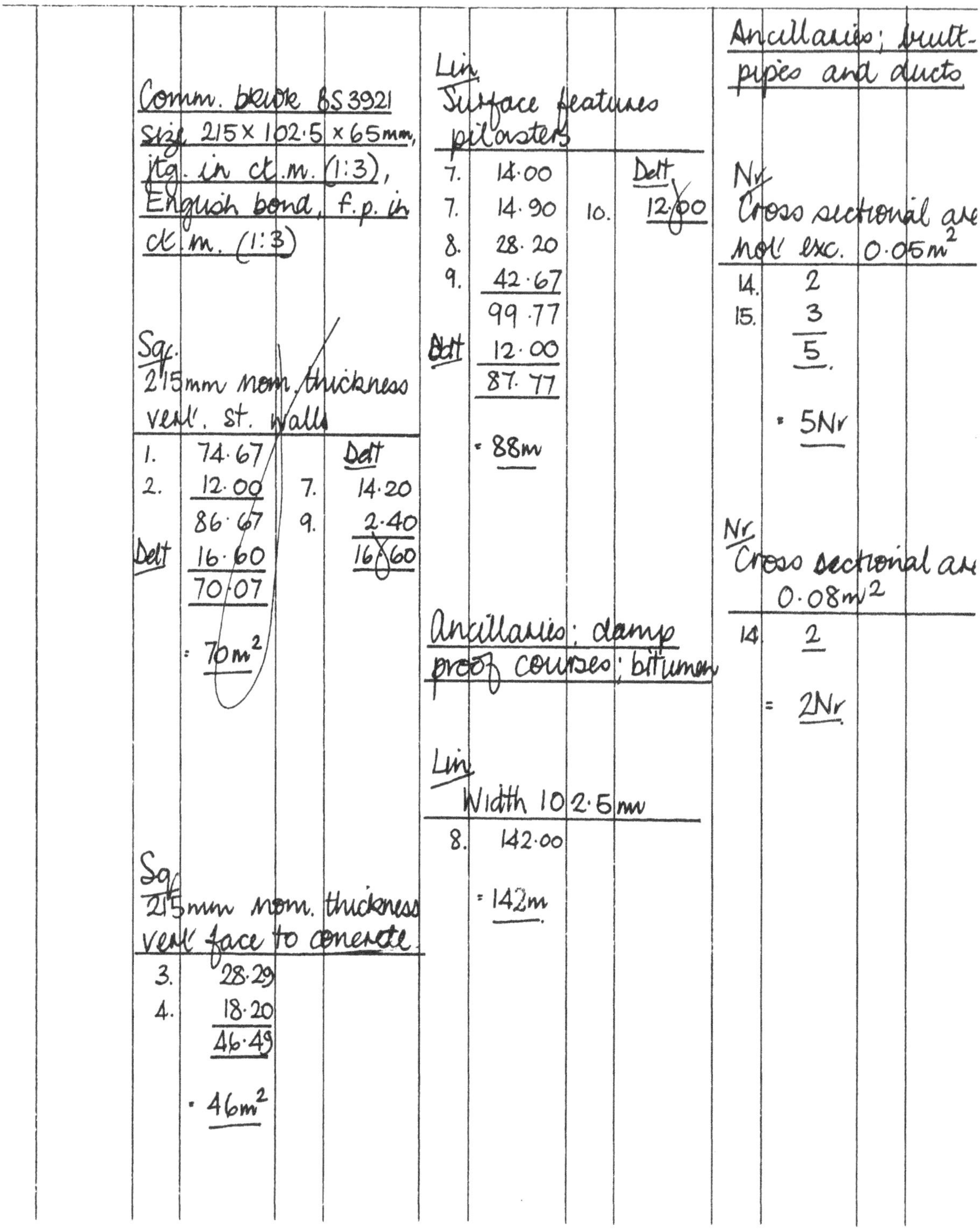

Comm. brick BS 3921
Size 215 × 102.5 × 65mm,
jtg. in ct.m. (1:3),
English bond, f.p. in
ct.m. (1:3)

Sq.
215mm nom. thickness
vert. st. walls

1.	74.67		Dedt	
2.	12.00	7.	14.20	
	86.67	9.	2.40	
Dedt	16.60		16.60	
	70.07			

= 70m²

Sq.
215mm nom. thickness
vert. face to concrete

3.	28.29
4.	18.20
	46.49

= 46m²

Lin.
Surface features
pilasters

7.	14.00		Dedt	
7.	14.90	10.	12.00	
8.	28.20			
9.	42.67			
	99.77			
Dedt	12.00			
	87.77			

= 88m

Ancillaries: damp
proof courses: bitumen

Lin.
Width 102.5mm

8.	142.00

= 142m

Ancillaries: built-
pipes and ducts

Nr.
Cross sectional are
not exc. 0.05m²

14.	2
15.	3
	5

= 5Nr

Nr.
Cross sectional are
0.08m²

14	2

= 2Nr

Figure 7

Abstracting

The skill of preparing an abstract lies in the ability of the worker-up to arrange the items abstracted from the dimension sheets in bill order. This may not seem too difficult a task to anyone who has not tried it but when tender documents are being prepared in a rush against a tight deadline (which must be 99% of the time!) the worker-up may be handed the dimension sheets in small lots but must lay out his abstract to accommodate items he has not yet seen.

A typical abstract is set out in Figure 7. The figures on the left hand side are the column numbers of the dimension sheets and the first item has been stroked through to indicate that it has been transferred to the draft bill.

Chapter 2

CESMM 3 - HOW IT WORKS

SECTION 1: DEFINITIONS

Reference should be made to the Method of Measurement when considering the following notes:

1.1 All the words and expressions used in the Method of Measurement and in the Bills of Quantities are deemed to have the meaning that this section assigns to them.

1.2 Where reference is made to the Conditions of Contract, it means the ICE (6th Edition) Conditions of Contract issued January 1991.

1.3 Where words and expressions from the Conditions of Contract are used in CESMM 3 they shall have the same meaning as they have in the Contract.

1.4 Where the word 'clause' is used it is referring to a clause in the Conditions of Contract. The word 'Paragraph' refers to the numbered paragraphs in Sections 1 to 7 inclusive of CESMM 3.

1.5 The word 'work' is defined in a broader sense than that in common usage to include not only the work to be carried out but also the labour materials and services to achieve that objective and to cover the liabilities, obligations and risks that are the Contractor's contracted responsibility as defined in the Contract.

1.6 The term 'expressly required' refers to a specific stated need for a course of action to be followed. This would normally take the form of a note on the drawings, a statement in the specification or an order by the Engineer in accordance with the appropriate clause in the Conditions of Contract.

1.7 The 'Bill of Quantities' is defined as a list of brief descriptions and estimated quantities. The quantities are defined as estimated because they are subject to admeasurement and are not expected to be totally accurate due to the unknown factors which occur in civil engineering work.

1.8 'Daywork' refers to the practice of carrying out and paying for work which it is difficult to measure and value by normal measurement conventions.

1.9 'Work Classification' is the list of the classes of work under which the work is to be measured in Section 8, e.g. Class E Earthworks.

1.10   'Original Surface' is defined as the ground before any work has been carried out. It should be stressed that this definition refers not to virgin untouched ground but ground on which no work has been carried out on the contract being measured.

1.11   'Final Surface' is the level as defined on the drawings where the excavation is completed. Pier or stanchion bases, soft spots or any other excavation below this level would be described as 'below the Final Surface'.

1.12   The definitions of 'Commencing Surface' and 'Excavated Surface' in CESMM 1
and   were amended significantly in CESMM 2.
1.13   These amendments were made because of the confusion surrounding their definitions. It is important when considering these changes that Rules M5, D4 and A4 of Class E are also considered.

Where only one type of material is encountered in an excavation the Commencing Surface is always the top surface prior to excavation and the Excavated Surface is always the bottom surface after excavation unless separate stages of excavation are expressly required. Complications arise however when the excavation penetrates through more than one type of material.

Previously, many takers-off were incorrectly stating the maximum depth of each *layer* of material instead of the maximum depth of the excavation itself.

The additional sentences added to Paragraphs 1.12 and 1.13 clarified what was always intended i.e. that the top of the excavation is the Commencing Surface and the bottom is the Excavated Surface no matter how many layers of different materials lie between the two.

The maximum depth of the excavation of an individual layer of material in accordance with Paragraph 5.21 is the maximum depth of the excavation even though the depth of the layer is significantly less. This rule is acceptable when the excavation is a regular shaped hole in the ground but problems arise when it is applied to a large area of land-forming or reduced level excavation.

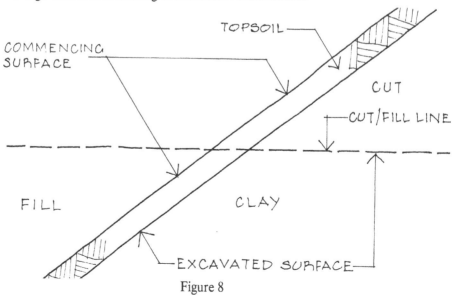

Figure 8

Figure 8 illustrates a sloping site which is being cut and filled to produce a level surface with the topsoil removed separately. The first operation to remove the topsoil is where the anomaly occurs. The information which is of most interest to the estimator is the depth of the topsoil itself. Over the filled area the Commencing Surface and the Excavated Surface is the top and bottom of the topsoil layer and the maximum depth in accordance with Paragraph 5.21 is the difference between the two.

Unfortunately this is not the case over the cut area. The Excavated Surface is the cut and fill line so the maximum depth is far greater in this area than over the filled section. This creates two separate items for identical work. This is obviously undesirable because the estimator only requires one item to cover all the work to be carried out under the scope of the description. Careful thought should be given to including either an additional preamble or enlarging the description to cover this set of circumstances.

1.14 A 'hyphen between two dimensions' means the range of dimensions between the figures quoted in excess of the first number but including the second e.g. '150 - 300mm' means 151 to 300mm inclusive.

Other words commonly used in CESMM 3 are defined as follows:-

'shall'   - mandatory
'should'   - optional
'may'   - optional
'given'   - stated in the Tender Documents
'inserted' - included by the Contractor

1.15 Problems arose in the last few years over the reference to BS numbers where the documents were intended for use in countries belonging the European community, particularly Ireland. This has been overcome by the widening of the clauses' meaning to include equivalent standards of other countries.

SECTION 2: GENERAL PRINCIPLES

2.1 The formal title of the document is 'Civil Engineering Standard Method of Measurement' which can be referred to in an abbreviated form as CESMM. Although not mentioned it is inevitable that the revised edition will be called CESMM 3 and the previous versions CESMM 1 and 2.

2.2 Building (with the exception of work covered by Class Z), mechanical, electrical or any other work which is not civil engineering, but which is part of a civil contract should be measured in accordance with the appropriate method of measurement for the particular work involved. An item should be included in the Preamble to the Bill of Quantities stating which work is affected and how it has been measured.

2.3 Although this paragraph states that the defined procedures of the CESMM 3 shall be observed in the preparation and pricing of the Bill of Quantities and the description and measurement of the quantities and items of work, it is possible to depart from the rules where thought appropriate. The authority for the departure lies in Paragraphs 5.4 and 5.10.

2.4 The object of preparing the Bill of Quantities is stated as twofold. First, to assist estimators produce an accurate tender efficiently. It should be borne in mind, however that the quality of the drawings plays a major part in achieving this aim by enabling the taker-off to produce an accurate bill and also by allowing the estimator to make sound engineering judgements on methods of working. Second, the Bill of Quantites should be prepared in such a style (within the framework of the CESMM 3) to assist the post contract administration to be carried out in an efficient and cost-effective manner.

2.5 This paragraph defines the need to present the measured items in the Bill of Quantities in sufficent detail so that items covering separate classes of work can be easily distinguished. It also requires that work of the same nature carried out in different locations is kept separate. This is a direct result of complying with the requirements of the preceding Paragraph 2.4 and is intended to assist the site surveyor or measurement engineer in the admeasurement and valuing of the work.

   For example if the Bill of Quantities is being prepared for a water treatment works, it is desirable that separate parts are given for say the raw water storage reservoir, the slow sand filters and pump houses. The alternative of adding together similar items from each structure would produce a Bill of Quantities which would have limited post contract value.

2.6 This paragraph states that all work (as defined in Paragraph 1.5) that is expressly required (as defined in Paragraph 1.6) should be covered in the Bill of Quantities. This paragraph is intended to remove any doubt about the status of formwork or any other temporary works which are required but are not left on site on completion to become the property of the Employer.

2.7 It is the proper application of the Work Classification tables that enables the aims of Paragraphs 2.4, 2.5 and 2.6 to be achieved. The tables and rules state how the work is to be divided, the scope of the item descriptions, the measurement unit for each individual item and the method (with a small 'm') to be adopted to produce the quantities.

SECTION 3: APPLICATION OF THE WORK CLASSIFICATION

This section deals with the details of how to use the Work Classification and the Rules. There are now 26 classes in the Work Classification and each class comprises:

   a) up to three divisions each containing up to a maximum of eight descriptive features of the work

   (b) units of measurement

   (c) Measurement Rules

   (d) Definition Rules

   (e) Coverage Rules

   f) Additional Description Rules.

3.1 The Work Classification is prefaced by an 'Includes' and 'Excludes' section which defines in general terms the nature and scope of the work contained in each individual class. This is particularly useful where there would appear to be a choice of which class to use. For example in Class R (Roads and Pavings) it clearly states that associated Earthworks and Drainage are to be measured separately in Class E and Classes I, J, K and L respectively.

There are three divisions or levels of description and up to eight part descriptions or descriptive features in each division. These descriptive features are intended to cover the broad range of activities in civil engineering but they are not exhaustive. At its simplest, it should be possible to take a descriptive feature from each division and produce an item description, e.g. in Class C - Concrete Ancillaries the following item description could be assembled.

First Division   2   Formwork: fair finish

Second Division   3   Plane battered

Third Division   1   Width: not exceeding 0.1m

This item would then appear in the bill as 'Formwork fair finish, plane battered, width not exceeding 0.1m' with a coding of C. 2 3 1.

This example is straightforward and it is not typical of the compilation of most item descriptions. It is essential that the rules are carefully studied before attempting to build up a description in case there is a restriction which would not be apparent by merely assembling descriptive features from the divisions. The use of the exact form of wording in the divisions is not mandatory but it would be unwise to depart from the printed descriptive features without a sound reason.

It is not essential that the punctuation is adhered to rigidly. The Work Classification is intended as a foundation upon which the 'taker-off' can base the needs of the individual project he is working on.

In the selection process from the three divisions it is important to note the need to observe the function of the horizontal lines. The selection must be made horizontally and be contained within the lines joining adjacent divisions. It would not be possible for example to select features in Class G to produce an item coded G. 1 1 8 because of the line preventing the Second Division Code 1 being linked with any of the void depth descriptive features.

3.2 This paragraph states that there is a basic assumption built into the descriptive features which removes the need for a comprehensive list of activities which the Contractor must perform in order to achieve the fixing of the material described. For example:

Item U. 1 2 1 does not require a preface stating that the rate set against the item must include for unloading the bricks, transporting them to a suitable place, carrying them to the Bricklayers' side, lifting them by the Bricklayers' hand and laying to the correct line and level. That is all assumed in the item 'Common brickwork, thickness 230mm, vertical straight wall'.

3.3 If the scope of the work to be carried out is less than that normally covered by item descriptions in similar circumstances then that limitation must be clearly defined.

For example it is not uncommon for the Employer to purchase special pipes or fittings in advance of the main civil contract where there is a long delay between order and delivery.

These materials would be handed to the Contractor on a 'free issue' basis and the item description would be headed 'Fix only'.

The scope of the term 'Fix only' should be clearly defined in the Preamble and could include such activities as taking charge of, handling, storing, transporting, multiple handling, laying and jointing including all necessary cutting.

The item or items involved must be unequivocally stated to avoid any ambiguities and if a number of items are covered by the limitation it would be prudent to insert 'End of fix only' after the last item.

3.4 The taker-off must not take more than one descriptive feature from any one division when compiling an item description.

3.5 The units of measurement are stated within the Work Classification and apply to all the items to which the descriptive features relate.

3.6 The Measurement Rules are defined as the circumstances existing for the implementation of the rules listed under M1 et seq. Reference must be made to Paragraph 5.18 for general additional information on measurement conventions.

3.7 Definition Rules lay down the parameters of the class of work covered by words or phrases in the Work Classification or the Bill of Quantities.

3.8 Coverage Rules describe the scope of work that is included in an item description although part of the required action may not be specifically mentioned. The rule will not necessarily cover all the work required and does not cover any work included under the Method Related Charges section.

3.9 The Additional Description Rules are included to provide a facility for the inclusion of extra descriptive features where those listed in the Work Classification are not considered comprehensive enough. The authority for this comes from Paragraph 3.1.

3.10 This note clears any confusion between the Work Classification and the Additional Description Rules. The example quoted refers to Class I. If item I. 5 2 3 was assembled by reprinting the descriptive features, the items would read:-

'Clay pipes, nominal bore 200 - 300mm, in trenches depth 1.5 - 2m.'

Rule A2 requires, among other things, that the nominal bore is stated. Paragraph 3.10 states that where these circumstances occur the provision of the note should override that of the descriptive feature so this item would read:

'Clay pipes, nominal bore 250mm, in trenches depth 1.5 - 2m.'

N.B. The other requirements of A2 would probably be included in a heading or in an enlargement of the First Division entry "Clay pipes".

3.11 It should be noted that on the rules side of the page in many classes there is a horizontal double line near the top of the page. All the rules above this line refer to all the work in the class. Rules below the double line refer only to the work contained within the same horizontal ruling.

SECTION 4: CODING AND NUMBERING OF ITEMS

This section deals with coding of items and it is important to remember that the provisions of the section are not mandatory. The value of applying the CESMM system of coding must be judged by the engineer or surveyor preparing the Bill of Quantities. If by following the recommendations of this section a series of unwieldly codings is produced it may be better not to apply them.

    The aim of the coding is to produce a uniformity of presentation to assist the needs of the estimator and the post contract administration.

4.1 The structure of the coding system is simple and easy to apply. Each item is allocated four basic symbols to produce a four unit code. The class is represented by the class letter (e.g. G - Concrete Ancillaries) followed by the numbers taken from the First, Second and Third Division respectively.

    Item G. 1 2 3 therefore refers to a item description of 'Formwork, rough finish plane sloping, width 0.2 - 0.4m'.

4.2 Where the symbol * appears it denotes all the numbers in the appropriate division. For example item G. 1 * 3 refers to items G. 1 1 3, G. 1 2 3, G. 1 3 3, G. 1 4 3 and G. 1 5 3. This symbol would never appear in the coding of an item description in a Bill of Quantities because by definition it refers to more than one item. Its main use is to assist, in an abbreviated form, in the application of the rules (e.g. See Rule M2 in Class P).

4.3 The option is given whether to apply the provisions of this section in the Bill of Quantities or not. The authors' reservations stated at the beginning of this section about the use of the codes only apply to their appearance in the Bill of Quantities. It is desirable that they are used in the taking off and working up stages as an aid to the presentation of the items in a regular and uniform bill order.

4.4 This paragraph states that if the code numbers are to appear in the Bill of Quantities they must be placed in the item number column and not become part of the item description. The code numbers have no contractual significance.

4.5 The highest number listed in the Work Classification is 8 and if a completely new descriptive feature is to be added it should be given the digit 9 in the appropriate division.

4.6 Conversely the digit 0 should be used if no descriptive feature in the Work Classification applies or if there are no entries in the division itself.

4.7 The code numbers only refer to the descriptive features in the three divisions. If additional descriptive features are required (See Paragraph 3.9) by the implementation of the Additional Description Rules they shall be identified by the addition of a further digit at the end of the code number. The example quoted in Paragraph 4.7 quotes item H. 1 3 6 but when Rule A1 is applied the additional information generates a code number of H. 1 3 6.1. If there was a need for more than one item they would appear as H. 1 3 6.2, H. 1 3 6.3, H. 1 3 6.4, etc.

What is not explained is the technique for dealing with this situation when the number of items containing additional descriptive feature exceeds 9. The choice lies between H. 1 3 6.10, H. 1 3 6.9.1 or H. 01 03 06.01 but the authors feel that the latter nomenclature is probably the most suitable.

SECTION 5: PREPARATION OF THE BILL OF QUANTITIES

It should be noted that the term used in this section heading i.e. 'Bill of Quantities' is correct. The phrase 'Bills of Quantities' is more appropriate to a building contract where the General Summary contains a list of individual Bills. In civil engineering documents the equivalent Bills are called Parts (Paragraph 5.23) so the overall document is a Bill of Quantities.

5.1 This paragraph states that the rules and provisions used in the pre- contract excercise of measuring the work also apply to the post-contract task of measurement. The correct term for this task is re-measurement where the work is physically measured on site or admeasurement where the actual quantities are calculated from records.

5.2 There are five sections in the Bill of Quantities

       A List of principal quantities

       B Preamble

       C Daywork Schedule

       D Work items (divided into parts)

       E Grand Summary

The Daywork Schedule can be omitted from the Bill of Quantities if required. The above sections should be allocated the letters A to E and the parts of the Bill contained within section D are enumerated, e.g.

Section A List of Principal Quantities

       B Preamble

       C Dayworks Schedule

    D Work Items
        Part 1   General Items
        Part 2   Boldon Sewers
        Part 3   Cleadon Sewers
        Part 4   Rising Mains
        Part 5   Shackleton Pumping Station
        Part 6   Roker Pumping Station

    E Grand Summary

5.3 It should be noted that the list of principal quantities is prepared by the taker-off or the person assembling the Bill of Quantities and should satisfy two requirements. First, to give the estimator an early feel for the scope of the work before he commences pricing and second, to assist the participants at the Contractor's pre-tender meeting with regard to the type and size of the contract when considering the application of the adjustment item (Paragraph 5.26). The list has no contractual significance. A notional list of principal quantities for the job mentioned in Paragraph 5.2 could be as follows:-

Part 1 General Items
        Provisional Sum   75,000
        Prime Cost Sums   100,000

Part 2 Boldon Sewers
        Pipelines   1200m
        Manholes   22nr

Part 3 Cleadon Sewers
        Pipelines   1500m
        Manholes   28nr

Part 4 Rising Mains
        Pipelines   2000m
        Valve Chambers   12nr

Part 5 Shackleton Pumping Station
        Excavation   600m3
        Concrete   50m3
        Brickwork   200m2

Part 6 Roker Pumping Station
        Excavation   650m3
        Concrete   80m3
        Brickwork   240m3

5.4 The Preamble is an extremely important section of the Bill of Quantities and is the potentially vital source of information to the estimator. If any other Methods of Measurement have been used in the preparation of the Bill of Quantities, the fact should be recorded here. This is not uncommon where say the Administration Building of a Sewage

Treatment Works or even the superstructure of a large pumping station has been measured in accordance with the current Method of Measurement for Building Works, although the inclusion of Class Z should reduce the need for this.

The Preamble should also contain information on work to be designed by the Contractor or where the Contractor is involved in alternative forms of construction. The style of measurement to deal with these events will involve a departure from the rules laid down so warrants an insertion in the Preamble.

The Preamble will also contain a list of departures from the rules and conventions of CESMM 3 if the taker-off considers it desirable. Because the preamble note will usually commence 'Notwithstanding the provisions of.....' these notes have become known as 'notwithstanding' clauses. A common example affects Paragraph 5.9. Many surveyors and engineers do not wish to adopt the lining out system as set out in this paragraph and would insert the following clause in the Preamble:

'Notwithstanding the provisions of Paragraph 5.9, lines have not been drawn across each bill page to separate headings and sub-headings'.

It should be noted that the Preamble can also be used as a vehicle to extend the Rules. The Item Coverage Rules are the most likely to be enlarged and the taker-off should not hesitate to use this facility in order to improve the quality of the information provided to the estimator.

Where the word 'Preamble' is used in this book it refers to the section of the Bill of Quantities as defined in this paragraph. The term 'preamble' (with a small p) has been used to mean a clause or note.

5.5 It is also necessary to include in the Preamble a definition of rock. On first consideration it may seem odd that what is primarily an engineering matter should find its place in the Preamble of the Bill of Quantities. The reason of course is that it is the definition of what rock is, that will determine to what extent it is measured. It is the measurement of rock which is the main consideration in the Preamble and the definition should clearly state in geological terms what materials will be defined and paid for as rock. If any borehole information is available it would be useful to make reference to the logs and use the same terms wherever possible.

The practice of defining rock as 'material which in the opinion of the Engineer can only be removed by blasting, pneumatic tools, or wedges' is not recommended because it creates doubt in the minds of the taker-off, the estimator and, most importantly, of the people engaged upon the post contract work.

5.6 It is not mandatory that a Daywork Schedule is included in the Bill of Quantities. If it was omitted, either by design or error, any daywork that occurred would be measured in accordance with Clause 52(3) of the Conditions of Contract and valued at the rates applicable to the FCEC schedules without any increase or decrease to the current percentages.

The other two methods of including dayworks in the Bill of Quantities are fully described in Chapter 3 General Items.

5.7 Where the method set out in Paragraph 5.6(6) is adopted for dayworks it is usual to include separate provisional sums for the Labour, Materials, Plant and Supplementary

Charges. The Contractor would be given the opportunity to insert his adjustment percentages after each item. (Chapter 4 General Items).

5.8 It is important that careful thought is given to layout of the Bill of Quantities. Almost the first task of the taker-off should be to consult the Engineer and draw up the Grand Summary to identify the various parts. In the example given in Paragraph 5.2 the various parts are easily identified. The work in Part 2 headed Boldon Sewers should be presented in a style which locates the work in more detail e.g. Manhole 1 to Manhole 2 etc.

In sewage disposal works and water treatment works it is usually quite straightforward to prepare a list of parts based on individual structures in the same order in which they are involved in the treatment process. It is more difficult in major bridge contracts and it is usual for the parts to be related more to CESMM 3 work classes than the locations of the work.

Whatever decisions are taken regarding the arrangements of the parts, the order of billing within each part should conform to the order of classes and items created by the Classifications within each class.

5.9 This paragraph provides for the placing of headings and sub-headings above item descriptions to prevent the repetition of material common to each item. These headings and sub-headings should be repeated at the top of each new page (perhaps in an abbreviated form) to assist the estimator in appreciating the full content of the item he is pricing.

A more controversial part of this paragraph deals with the procedure of what has come to be known as 'lining out'. This is the arrangement by which lines are drawn across the Item Description column of the bill page to end the influence of a previous heading or sub-heading.

Figure 9 shows how the lining out is done.

5.10 This paragraph gives the authority to the taker-off to add additional descriptive material to a description constructed from the three divisions if the work being measured has special characteristics which 'give rise to special methods of construction or consideration of cost'.

The implications of this paragraph are far reaching for the taker-off. He can impose his own judgement on the measurement of any item and depart from the format provided he believes it is a special case. It would be unwise however for the taker-off to abuse the power entrusted to him by this paragraph. The larger the number of items in the Bill of Quantities that conform to the preferred style of the CESMM 3 the more uniformity will be achieved which will benefit all parties. The taker-off should use the powers of this paragraph sparingly but on the occasions where it is felt that a new form of item description or additional descriptive material is necessary, the opportunity should be taken with the needs of the estimator and post contract administration overriding those of the generalities of CESMM 3.

If the new form of item description conflicts with the rules of the method, a 'notwithstanding' clause should be raised in the Preamble (Paragraph 5.4).

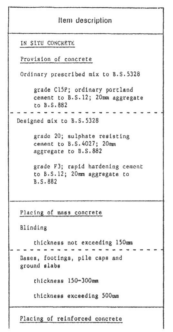

Figure 9

5.11 This paragraph reinforces the secondary role of the Bill of Quantities. The estimator is actively discouraged from relying on the item descriptions as a sole source for the information he requires to build up his rates. The 'exact nature and extent of the work' (or as near as it is possible to define it) must be determined from the Drawings, the Specification and the Contract. The item descriptions should not be held to be comprehensive but used to 'identify' the work being measured. This downgrading of the descriptions does not in any way relieve the taker-off from his responsibility of producing the most lucid descriptions he can within the framework of the method of measurement.

5.12 Where an unusual feature occurs in the work it is sometimes easier and more accurate to direct the estimator to a clause in the Specification or a detail of a drawing rather than produce a clumsy description which does not fully cover the work to be measured. Although this paragraph gives the authority for this form of referencing, it also contains an important proviso that the reference must be precise. A general reference to a drawing containing standard details would be unacceptable unless it identified the exact detail in the drawing being referred to.

5.13 In civil engineering contracts the work is subject to admeasurement. The quantities that are measured in the contract document are approximate because of the uncertainties inherent in civil engineering. The method of measurement has by necessity great flexibility and affords the taker-off opportunities to use his professional skill and judgement denied to his opposite number in the building side of the industry. This paragraph demonstrates this freedom. There are many situations where the choice of the style of measurement and placing of the items in the bill is entirely at the discretion of the taker-off. One example of this concerns thrust blocks in Class L. Where a block is large, say 10m3, the taker-off may feel it more helpful to measure it in detail using Classes E, F and G rather than item L. 7 8 0. A major consideration in this decision would be the knowledge that if the item was enumerated and the drawing reference given (as Paragraph 5.12) each tendering estimator would need to measure the excavation, concrete and formwork so it is less wasteful if the taker-off prepares these. Whatever decision the taker-off makes in these matters it should be clear and unequivocal so that the estimators are not confronted by ambiguities and uncertainties.

The comments made earlier about the quantities being approximate should not give the impression that anything less than the highest professional standards should be employed in the preparation of the tender documents. The quantities are described as approximate because in many cases the scope of the work is not known but the measurements should be as accurate as possible even in the knowledge that they will be taken again on completion of the works.

5.14 Where a range of dimensions is given in the Work Classification tables but the measured items have an identical thickness it is permissible to state the thickness instead of the range. For example item K. 1 1 3 describes brick manholes in a depth range of 2 to 2.5m. If there were three manholes all 2.2m deep the item description should read "Manholes, brick, depth 2.2m" and would carry the same code number K. 1 1 3.

5.15 Where work is to be carried out by a Nominated Sub-Contractor the estimated cost of the work should be given as a Prime Cost Item. Items to cover what used to be called general and special attendances follow this sum and are dealt with in Chapter 3 General Items.

The scope of the facilities to be available to the Nominated Sub-Contractor include for temporary roads, hoists and disposing of rubbish.

5.16 Any goods, materials or services supplied by a Nominated Sub-Contractor which are to be used by the Contractor must be referenced to the Prime Cost Item involved by a heading or mention made in the item itself.

5.17 The use of provisional quantities is discouraged by CESMM. Prior to 1976, items frequently appeared in Bills of Quantities under a heading of Provisional. This procedure was usually adopted because the design Engineer either did not know the scope of the work or did not have enough time to design it. The assumption that the Contractor had better knowledge at tender stage than the Engineer, and was able to price the work was completely unacceptable. On occasions, if the provisional quantities included were small, the Contractor would insert high rates which would hardly affect his tender total but could lead to a windfall if the quantities increased on admeasurement.

This paragraph states how the cost of uncertainties in design should be treated. If there are specific areas of work where the design has not advanced far enough to allow accurate quantities to be prepared, the work should be placed in the General Items against a Provisional Sum. It is also usual to include a Provisional Sum in the Grand Summary for general contingencies.

Recently, however, some Employers are resisting the inclusion of this general contingency allowance in the spurious belief that the Contractor will somehow regard that sum as 'spendable' and attempt to recover it through claims. This notion shows little confidence in the skills of post-contract management team acting on the Employer's behalf.

5.18 This paragraph confirms the long standing convention that measurements are taken net unless there is a specific requirement to the contrary. Ideally, the quantities are computed from dimensions on the drawings. Common sense must be applied in the matter of rounding off quantities. The total quantity and the effect on it of rounding off must be considered.

5.19 The units of measurement are set out in this paragraph and the abbreviations must be used in the Bill of Quantities. Care should be taken when using the abbreviation for Number because the handwritten 'nr' is very similar to 'm' and mistakes can be made when documents are produced at speed by confusing the two abbreviations.

5.20 It is a requirement that where a body of open water is either on the site or bounds the site, it shall be identified in the Preamble to the Bill of Quantities stating its boundaries and levels or fluctuating levels. This requirement should not be taken too literally. If a power station was to be constructed on the West Cornish coast it would be sufficient to state in the Preamble that the Atlantic Ocean was adjacent to the site together with tidal information. It would be unnecessary and foolish to attempt to define the bounds of the Atlantic!

It is interesting to note that Rule A2 in Class E provides a further requirement for the body of water to be identified in the item description for work which is below the feature, this requirement is not thought necessary in other Classes such as F, I or P where similar situations could occur.

5.21 This paragraph deals with the definition of the terms Commencing Surface and Excavated Surface. This matter has been deal with under paragraphs 1.12 and 1.13. See also Class E.

5.22 A sample of the ruling and headings of bill paper is shown in Figure 10.

Number	Item description	Unit	Quantity	Rate	Amount	

Figure 10

5.23 The summary of each Part would be printed on standard bill paper but the Part total would be styled 'Carried to Grand Summary'. (See Figure 11).

PART 4 – RISING MAINS

Number	Item description	Unit	Quantity	Rate	Amount	
	COLLECTION					
	Page 6/1					
	Page 6/2					
	Page 6/3					
	Page 6/4					
	Page 6/5					
	Page 6/6					
	Total Carried to Grand Summary £					

Figure 11

5.24 The Grand Summary collects the totals from the parts of the Bill of Quantities and is usually printed on plain paper (Figure 12).

GRAND SUMMARY

	£	p
PART 1 GENERAL ITEMS		
PART 2 BOLDON SEWERS		
PART 3 CLEADON SEWERS		
PART 4 RISING MAINS		
PART 5 SHACKLETON PUMPING STATION		
PART 6 ROKER PUMPING STATION		
	£	
GENERAL CONTINGENCY ALLOWANCE	50,000	00
ADJUSTMENT ITEM ADD/DEDUCT*		
TENDER TOTAL	£	

* Delete as required.

Figure 12

5.25 The General Contingency Allowance is discussed in Paragraph 5.17.

5.26 The Adjustment Item is to be placed at the end of the Grand Summary and its significance and purpose is discussed in Paragraph 6.3, 6.4 and 6.5.

5.27 The Grand Summary must contain a provision for the addition of the individual bill parts, the General Contingency Allowance and the addition or subtraction of the Adjustment Item. This total is often called the Tender Total but it should not strictly receive that title until the acceptance of the Contractors Tender for the Works in accordance with Clause 1(i)(h).

SECTION 6: COMPLETION, PRICING AND USE OF THE BILL OF QUANTITIES

6.1 The rates to be inserted in the rates column shall be expressed in pounds sterling with the pence given as a decimal fraction. Thus 6.47 denotes 6 pounds 47 pence. It is important that the amount is written clearly with the decimal point well defined to avoid subsequent misunderstandings and disputes. If 647 was entered in the rate column and it was intended to be 647 pounds it should be expressed as 647.00. Careful inspection of the presentation of the rates together with their values should be part of the tender appraisal process. Where rates are not inserted the other priced items are deemed to carry the price of the unpriced items.

6.2 This paragraph confirms the requirement made in Paragraph 5.22 that each part must be totalled and then carried to the Grand Summary.

6.3 The introduction of the Adjustment Item was warmly welcomed by the industry and in 1976 when CESMM 1 was published and its use is now well established.

6.4 Most Contractors contend that they are rarely allowed sufficient time to prepare their tenders. Each job needs careful scrutiny and the application of sound engineering judgements to determine how the construction work should be tackled. Enquiries for material prices and subcontractors quotations must be sent out and it frequently happens they do not arrive until quite late in the tender period.

　　If for example a quote for ready mixed concrete was obtained on the day before a tender was due to be submitted which was substantially below other quotations the Contractor would be keen to include the effect of the offer in his tender.

　　Pre-1976 he would have probably deducted the difference from a convenient sum in the General Items and thus created an imbalance in the pricing structure. By using the Adjustment Item the Contractor can now increase or decrease his Tender Total at a stroke yet still present a well balanced bid.

　　Another reason for using the device could arise from the Contractor winning or losing other contracts during the tender period which would lessen or increase his determination to put in a keen bid. This decision would normally be taken at a tender appraisal meeting before the signing of the offer.

　　The sum inserted should be regarded as a lump sum and will be paid or deducted in instalments in the same proportion that the amount being certified bears to the Tender Total before the application of the Adjustment Item in the Grand Summary.

It is a requirement of CESMM that this should be stated in the Preamble to the Bill of Quantities.

The amount involved shall be calculated before the deduction of retentions and the aggregate total must not exceed the amount inserted in the Grand Summary. When the Certificate of Substantial Completion (Clause 48) is issued the difference (if any) between the aggregate total and the amount in the Grand Summary should be paid or deducted in the next certificate to be issued.

6.5 This new paragraph clears up any misunderstandings over the position of applying the Adjustment Item when the Contract is subject to a Contracts Price Fluctuation (CPF) clause. When the Effective Value is calculated it should take into account the effect of deducting or adding the Adjustment Item as appropriate in assessing the amount due to the Contractor under Clause 60.

SECTION 7: METHOD-RELATED CHARGES

Method-Related Charges were first introduced in CESMM 1 in 1976. It was felt that a different approach was required in the valuation of items where quantities were increased or decreased from those in the tender document. Research had shown that modern construction techniques had substantially increased the proportion of the non-quantity related part of a Contractor's costs to a level where it was becoming inequitable both to the Employer and the Contractor that changes in the quantities should be valued merely by multiplying the admeasured quantity by the bill rate.

The unit rates are made up of quantity related costs - the labour, material and that part of the plant and overheads directly related to the item of work being constructed, and the non-quantity related items such as the transporting to site, erection, maintenance, dismantling of plant, cabins and other consumables which may have no direct link with the quantity of the permanent works being constructed.

It is sensible therefore to give the Contractor the opportunity to declare the cost of those items which he does not wish to be subject to the admeasurement process so that his real costs are recovered without being affected by changes in quantity.

7.1 A Method-Related Charge is the sum inserted in a Bill of Quantities in the space provided (Class A) and is either a Time-Related Charge or a Fixed Charge.

A Time-Related Charge is a sum which is directly proportional to the time taken to carry out the work which is described.

A Fixed Charge is a sum which is neither quantity-related nor time-related but is a set cost regardless of changes in the admeasured work or the time taken to execute it, e.g. the cost of bringing a batching plant on to site.

7.2 The Contractor has the opportunity to insert the cost of Time-Related and Fixed Charges in the Bill of Quantities (see General Items Class A).

7.3 The Contractor should enter the item description for his Method-Related Charges in the same order as the order of classification in Class A. He must also list the Time-Related Charges separately from the Fixed Charges and insert a sum against each item. He has the freedom, of course, to enter other items which are not listed or do not have a direct counterpart in Class A.

7.4 The Contractor should unambiguously describe the scope of the work that is covered by each sum. He should also list the labour, plant and materials involved and, where applicable, state the parts of the Permanent or Temporary Works that are linked to the sum inserted.

7.5 The Contractor is not obliged to follow the method he has set out in the tender document when he carries out the work on site.

7.6 This paragraph states that the Method-Related Charges are not to be admeasured. The wording was expanded in CESMM 2 to include the words '...but shall be deemed to be prices for the purposes of Clauses 52(1), 52(2) and 56(2)'.

The addition of these words confirms what was always inferred in CESMM 1. It is sometimes difficult for students to understand the true meaning of what this paragraph covers. An unequivocal statement that Method-Related Charges are not to be admeasured seems to sit uneasily beside the assertion that they are subject to the provisions of Clause 56(2).

If the items for Time-Related Charges and Fixed Charges have been set out by the Contractor in a sensible fashion it should be a straightforward task of apportionment each month to arrive at the amount due. One complication may arise if the time being expended on a Time-Related Charge looks like increasing or decreasing from that shown in the Bill of Quantities. If, for example, an operation was scheduled to occupy 6 months so after 1 month the Contractor would rightly ask for 1/6 of the sum. If his progress increased dramatically and the work looked like being completed in only 4 months, he would be fully entitled to ask for 1/2 the sum at the end of the second month. It can be seen therefore that the numerator in the fraction will increase each month by 1 but the denominator could vary as the Contractor and the Engineer determine the likely length of time the event will last.

It should be noted that a statement must be included in the Preamble to the Bill of Quantities confirming that payment must be made in accordance with Clauses 60(1)(d) and 60(2)(a). This apparent confliction is explained if one remembers that the charges will be paid in full whether they were incurred ten fold or not at all providing the risk the Contractor undertook and priced did not vary.

If there was a significant change in quantity or in the time an item of plant was required which was substantially different from that envisaged when the Contractor prepared his tender, then an adjustment to the Method-Related Charges would be in order and the provisions of Clauses 52(1), 52(2), and 56(2) would be implemented.

7.7 Method-Related Charges are to be certified and paid for in exactly the same manner as other parts of the work and this should be stated in the Preamble.

7.8 It may be that the method of working stated by the Contractor is not adopted (Paragraph 7.5) but in the absence of a variation (see Paragraph 7.6) the sum inserted must be paid in full. It is obviously desirable that the Contractor and the Engineer agree a method of apportioning the sum each month for payment by linking it to progress of a relevant part of the works or indeed the whole works. If agreement cannot be reached the sum would then be added to the Adjustment Item (which would increase a positive Adjustment Item and decrease a negative one) and would be treated as described in Paragraph 6.4.

SECTION 8: WORK CLASSIFICATION

This section lists the twenty six classes in CESMM 3. Each class consists of the Classification Tables containing three divisions of descriptive features and four types of rules. See Section 3 for details on the application of the tables and rules.

PART TWO

CLASSIFICATION TABLES

Chapter 3

CLASS A: GENERAL ITEMS

GENERAL TEXT

Principal changes from CESMM 1

Supplementary charges were added to the list of Daywork items (A. 4 1 7 and 8).

Note A4 of the CESMM 1 was deleted. It stated that any plant measured as specified requirements had to distinguish between plant operating and standing.

Additional Description Rule A1 requires that items for specified requirements which are carried out after the Certificate of Completion has been issued must be stated.

Principal changes from CESMM 2

The item descriptions in the 2nd Division of Contractual requirements have been amended to tie in with the ICE Form of Contract 6th Edition.

Measurement Rules

M1 The specified requirements allow the taker-off freedom in the use of units and of
& measurement for work done under this classification. However, the units used should
M2 always be related to the work involved, e.g. it would not be appropriate to measure the establishment of cabins in weeks. The following are examples of appropriate units for various types of work.

Contractual requirements	- sum
Items of establishment and removal	- sum
Items of continuing operation or maintenance	- day/wk/mth
Testing of materials	- sum/nr/set
Provisional sums	- sum

M2 For time related items such as maintenance of Engineer's accommodation, the quantity given may have to be determined in conjunction with the Engineer. Quantities should only be given for those items which are likely to vary or for which the pricing of quantities would assist in the valuing of interim applications. The principal items measured in this way are those which are time related such as continuing operation or maintenance items, and those such as testing of materials (concrete cube tests and the like). For time related items which in theory extend for the entire duration of the contract e.g. maintenance of Engineer's accommodation, the full contract time should be given although it would rarely all be used in practice. This is because it would be impractical for the Contractor to establish all required accommodation from day one. The actual establishment of the date from which the remeasurement should start is often not clear because the Contractor may take several weeks to complete any one item, and in cases such as these the date would have to be negotiated. (See also Definition Rule D1).

Time related specified requirements should only be remeasured up to the date for completion, be it the original date or any properly extended date (See also Additional Description Rule A1).

M3 It is necessary to read the whole of the specification and drawings to ascertain what testing is required. If the testing is not set out in the relevant class then it should be measured here or the coverage rules in the class should be extended to cover them in accordance with common practice.

M4 It is usual to give examples of the most common Method-Related Charges in the Bill but no units or quantities should be inserted. It is then up to the Contractor to decide which items he wishes to price and to insert any further items which he wants to be included. Although this is the usual practice it is not strictly necessary to list any examples of Method-Related Charges at all, and in fact doing so can often confuse the Contractor because he may only price those items listed and not realise that he has to read all the documentation and insert anything further which he requires. It is therefore recommended that CESMM is strictly adhered to and that the completion of the Method-Related Charges is left solely to the Contractor.

Method-Related Charges are generally priced by the Contractor for items which do not vary proportionally with either an increase or decrease in quantities. For example the cost of bringing to site piling plant and equipment will not vary irrespective of how many piles it installs. In this instance the cost of mobilisation could be included in the Method-Related Charges, which protects the Contractor if the number of piles decreases, because he is still recouping the fixed cost of the plant mobilisation and the Employer is protected if the quantity increases.

Blank pages should be left in the Bill for the insertion of Method-Related Charges, under a general heading which includes this rule and also Additional Description Rule A4. (See also Definition Rule D1).

See also Section 7.

M5 There are normally two methods of dealing with Dayworks in civil engineering bills.

The first method, which is not commonly used nowadays, is to schedule the various classes of labour, plant and materials which will be used on the contract; labour and plant are usually given in units of one hour and materials in the unit most appropriate to their nature e.g. cement/kg, bricks/thousand, steel/tonne. The tenderer is required to insert the prices he requires against these items for services provided on a Daywork basis and a provisional sum is then included for the work to be carried out. However, this is not satisfactory because the tenderer does not have any incentive to keep his Daywork rates competitive. It is more satisfactory to provide provisional quantities against the unit rates and have the extensions incorporated as part of the Tender Total because this introduces the competitive element. Purists will say that tenders should be assessed on the rates anyway and not the Tender Total, but tenders are invariably assessed on the total cost.

The second method which is commonly used is based on the Federation of Civil Engineering Contractor's Schedule of Dayworks, which details how to value Daywork. The Schedules are divided into four sections, Labour, Plant, Materials and Supplementary Charges. The taker-off inserts provisional sums for each of these categories as provided for in CESMM 3, together with an item which enables the tenderer to adjust the sums either up or down by any percentage he wishes to insert, thus providing the competitive element. Any subsequent Daywork which is done during the contract is valued and paid for in accordance with the Schedules taking account of the percentage adjustment.

It is felt that percentage adjustment to Supplementary Charges is not adequately treated. The Supplementary Charges section of the Schedules covers several Daywork items which are charged at basic cost plus a percentage addition. This percentage allowance varies depending upon the item involved from 10% for ordinary subcontractors and up to 64% for plant operatives and the 'norm' is 10-12.5%. Whilst a Contractor may be willing to put a minus figure on the plant operatives percentage of say 20% this would mean that because the percentage adjustment applies to all Supplementary Charges, he would be negating his percentage addition on the other items as stated in the Schedules. The Contractor would rarely insert a negative percentage on the Supplementary Charges Section and the Employer would lose any potential benefit.

The answer is to either subdivide the Supplementary Charges into their individual components, or it is recommended that a preamble be incorporated stating that plant operatives are excluded from the Supplementary Charges Section and included in the Labour Section where they would be treated in the same way as labour only subcontractors.

Additional Description Rule A1 states that descriptions for specified requirements carried out after the Completion Certificate shall so state, and it is felt that this rule should apply equally to Daywork. Prior to the maintenance period, the Contractor will begin winding down his site operations and any Daywork items instructed during the maintenance period would not be incidental to the main contract works. It is quite probable that Daywork will be instructed after the Certificate of Substantial Completion has been issued and it is recommended that Daywork operations distinguish between those carried out before and after the issuing of the Certificate of Substantial Completion (See also Additional Description Rule A1).

M6 General labours include general facilities which the Contractor has already provided for himself. This is something of an anomaly in that strictly speaking only the Contractor himself knows exactly what he is providing. If the Nominated Subcontractor was pricing for the subcontract when the main contract was actually on site then there should not be a problem, because he could liaise with the Contractor on these matters. If, however, and this is the case more often than not, the nominated subcontracts are let before the main contract, then it is not possible for the Subcontractor to know exactly what he is likely to be provided with, and he may elect to include in his price for items which are eventually provided for him, which means that the Employer would be paying twice for the same item. Great care must always be taken regarding general labours and it is recommended that the Contractor is specifically told to provide anything which the Subcontractor is informed will be provided to him under general labours and that it is not left to chance.

Special labours would include items such as three-phase power supplies, special scaffolding etc.

Definition Rules

D1 It is necessary for the taker-off to examine all the tender documentation - Specification, Drawings, Letter of Instruction etc., in order to ascertain the items to be given under specified requirements. There is sometimes a very fine line between whether an item should be included here or left for the Contractor to include under his Method-Related Charges. For example the Specification may instruct the Contractor to keep adjacent roads clean by sweeping on a regular basis with a suitable road sweeper. In general terms, an item should be included in the specified requirements only if the Contractor is required to do something in a particular way or to a particular extent. On inspection, this example would appear to fulfil these criteria. However, it would be wrong to include this example in the specified requirements because in this case it is not possible for the Engineer to define precisely what work is involved e.g. it would depend upon how dirty the roads were as to how often they need to be cleaned, i.e. it is at the Contractor's risk. If however, the Specification said that adjacent areas of roads for a radius of 5 miles were to be cleaned once a week, then this is quite clearly defined and could be included in the specified requirements.

Other criteria which should be considered are whether the item under consideration requires remeasurement or is likely to be varied or even deleted. If any of these are applicable then the item should be included in the specified requirements because it is generally not possible to tamper with the Method-Related Charges.

Coverage Rules

C1 The Contractor is only to price for the insurances specifically required under the contract. Any additional cover he requires would be included by him as a Method-Related Charge.

Additional Description Rules

A1 Careful consideration needs to be given to this rule, and the Engineer's cabins are examined to highlight the problem. Although not specifically stated, items should only be given for specified requirements after the Certificate of Substantial Completion where it is an express requirement that work should be executed after that date. The Engineer may specifically require the cabins to remain on site after the completion date in order to supervise the work of other contractors as defined in Clause 31 of the Contract. In this case it would be proper and correct for the continuing operation, maintenance and removal of the cabins to be measured in accordance with this rule. However, on a 'normal' contract more often than not, the Contractor will apply for a

Certificate of Substantial Completion when the actual date for completion is due, together with an undertaking that outstanding items of work will be carried out during the Maintenance Period (all this is perfectly acceptable and provided for in Clause 48 of the Contract), and is done to obviate the possible imposition of liquidated damages. In this instance there should not be any items included in the remeasurement for the cabins beyond the Certificate of Substantial Completion, be it the original date for completion or any properly extended date. Items for specified requirements after the Certificate of Substantial Completion would only be included in the Bill therefore, when the Engineer has a specific requirement for them and not when they are caused by actions taken by the Contractor.

A2 It is common practice for the establishment and removal of specified requirements to be given as a single sum. However, it assists in the post contract administration if these are separated and given as two distinct items.

A3 More often than not, reference will be made in the item description to the relevant specification clause. Testing should be further described by reference to the individual component concerned, e.g. testing of concrete, testing of pipelines etc.

A4 There is no specific requirement by CESMM for the Contractor to insert the actual length of time applicable to any particular Method-Related Charge, but merely to state whether a charge is fixed or time related. However, for ease of post contract administration it may be desirable to request the Contractor to identify the periods of time over which the charge applies.

ITEM MEASUREMENT

A. 1 1-4 0 Contractual Requirements

Divisions Contractual requirements are given as a sum stating the type of requirement as classified.

Rules -

Generally The Second Division gives the four items from the Contract most commonly priced but that is not to say that there are no other contract conditions to which the Contractor may wish to ascribe a value and there are several ways of solving this problem. The first would be to add a further contractual requirement, 'Complying with other conditions of contract'. If the Contractor priced this item, it would not be possible to identify the Contract Clause to which it is related. The second would be for the taker-off to list all the relevant contract clauses which he thinks should be priced, but the danger here would be in missing one out. The third would be to leave a blank space and ask the Contractor to insert any further clauses himself together with their relevant value (in a similar way to Method-Related Charges). It would be of more benefit if the headings of all the contract clauses were scheduled in the item description column together with the full text of any amendments or additional conditions. Since this is traditionally always done in section one of the Specification no additional work would be created in adopting this approach.

See also Rule C1.

A.2 * * Specified requirements generally

Applicable to all items

Divisions   -

Rules   M2 A unit and quantity shall be given for all items to be remeasured.

A1 State when specified requirements are to be carried out after the issue of the Completion Certificate.

A2 Distinguish between establishment and removal, and continuing operation or maintenance.

Generally   The items given in CESMM under specified requirements should be taken as being indicative only and should be added to or further sub-divided as necessary. It should also be made perfectly clear in the Preamble or description what exactly the item given is to include. For example, does an item for continuing operation and maintenance of cabins include for the cleaning and servicing of them or is a separate item given under services for the Engineer's staff?

See also Rule D1.

A. 2 1-3 *   Accommodation, services and equipment for the Engineer's staff

Divisions   State the nature of the accommodation, service or equipment to be provided. Establishment and removal are usually given as a sum and continuing operation in weeks, although other units of time could be used.

Rules   -

Generally -

A. 2 4 *   Attendance upon the Engineer's Staff

Divisions   State the nature of the attendance to be provided. Attendances are generally time related and given in weeks, although other units of time could be used.

Rules   -

Generally -

CLASS A: GENERAL ITEMS

A. 2 5&6 0   Testing

Divisions   State the nature of the testing either to the works or to materials. Testing is given in the unit most appropriate to the nature of the testing concerned e.g. testing of pipelines - sum; testing of concrete cubes - nr.

Rules   M3 Only give items for testing here when there is no provision for the testing under any of the other classes.

A3 Give particulars of the samples and methods of testing.

Generally   Also state the nature of the component to be tested, i.e. pipelines, water tanks, concrete, timber, etc. Testing should not be measured again for additional tests necessitated due to failure of the initial test.

A. 2 7 *   Temporary works

Divisions   State the nature of the temporary works to be provided. Establishment and removal are usually given as a sum, and continuing operation in weeks, although other units of time could be used.

Rules   -

Generally   It should be stressed that items should only be measured under this classification when they are designed and expressly required by the Engineer.

A. 3 * *   Method-Related Charges

Divisions   The items listed in the divisions are given as examples only of Method-Related Charges which the Contractor may wish to include. The list is not exhaustive or definitive. The unit is usually the sum.

Rules   M4 Method-Related Charges are inserted by the tenderer in accordance with Section 7.

A4 Distinguish between Time-Related and Fixed Charges.

Generally   It is recommended that after the specified requirements section of the Bill, a heading stating Rules M4 and A4 is given and two or three blank pages are left for insertion of Method-Related Charges by the tenderer.

See also Section 7.

A. 4 1 *   Daywork

Divisions   State that the Contractor is to be paid for work executed on a Daywork basis at rates according to the Schedules of Dayworks carried out incidental to Contract

Work issued by the Federation of Civil Engineering Contractors.

Give provisional sums in words and figures for Daywork labour, materials, plant and supplementary charges and items to allow for the percentage adjustment of same.

Rules   -

Generally   Alternatively, give schedules of the various classes of labour, materials and plant and include a provisional sum or give provisional quantities (see Measurement Rule M5).

Distinguish between Daywork carried out before and after the issue of the Completion Certificate.

See also Paragraphs 5.6 and 5.7.

A. 4 2 0   Other provisional sums

Divisions   -

Rules   -

Generally   Although not a specific requirement, state briefly the work which the provisional sum is to cover and give the amount of the sum in words and figures. It is common to include at the beginning of the provisional sums a statement to the effect that the tenderer should allow for the work covered by provisional sums (but not contingencies) in his programme and that they may or may not be expended only by instruction from the Engineer, and that no adjustment of General Items will be allowed.

CESMM makes no reference in its text to work done or services provided by Local Authorities, Statutory Undertakers or other Public Bodies. Public Bodies may be paid direct by the Employer or through the Contract. If paid through the Contract they would be included as provisional sums and the Contractor may require and be entitled to a payment in respect of overheads and profit and provision should be made for this (generally expressed as a percentage of the sum). In either case there may be attendances or facilities which the Contractor has to provide and these can either be dealt with in a similar way to A. 5&6 2&3 0 or measured in detail in accordance with the various classes of CESMM.

A. 5&6 * * Prime Cost Items

Divisions   Prime cost items are given as a sum distinguishing between those which do and do not include for work on-site (effectively Nominated Subcontractors and Nominated Suppliers). Items are given to allow the tenderer to insert his charges for labours, special labours (both expressed as sums) and other charges and profit (expressed as a percentage).

Rules   A5 Identify the work involved.

           A6 Describe the nature of special labours.

Generally   State the amount in words and figures. Although not specifically stated prime costs items should make it clear whether or not the sums include for any discount to the Contractor for prompt payment to the Nominated Subcontractor, and if so how much, because this will affect the percentage addition the Contractor inserts for other charges and profit. Needless to say this should conform to whatever is contained within the Subcontract.

STANDARD DESCRIPTION LIBRARY

Contractual requirements

A. 1 1 0	Performance bond	sum
A. 1 2 0	Insurance of the Works	sum
A. 1 3 0	Third party insurance	sum

Specified requirements

Accommodation for the Engineer's staff

A. 2 1 1	offices; establishment	sum
A. 2 1 1.1	offices; continuing operation or maintenance	wk
A. 2 1 1.2	offices; continuing operation or maintenance; after the Completion Certificate	wk
A. 2 1 1.3	offices; removal	sum

Services for the Engineer

A. 2 2 2	telephones; establishment	sum
A. 2 2 2.1	telephones; continuing operation or maintenance	wk
A. 2 2 2.2	telephones; continuing operation or maintenance; after the Completion Certificate	wk
A. 2 2 2.3	removal	sum

Specified requirements (cont'd)

Equipment for the Engineer's staff

A. 2 3 1	office equipment; establishment	sum
A. 2 3 1.1	office equipment; continuing operation or maintenance	wk
A. 2 3 1.2	office equipment; continuing operation or maintenance; after the Completion Certificate	wk
A. 2 3 1.3	office equipment; removal	sum

Attendance upon the Engineer's staff

A. 2 4 1	drivers	wk
A. 2 4 2	chainmen	wk

Testing of materials

A. 2 5 0	concrete cube tests; compressive strength; as specification clause 4.1	sum

Testing of the Works

A. 2 6 0	potable water pipelines; by smoke; as specification clause 4.2	sum

Temporary Works

A. 2 7 1	traffic diversions; establishment	sum
A. 2 7 1.1	traffic diversions; continuing operation or maintenance	wk
A. 2 7 1.2	traffic diversions; removal	sum
A. 2 7 3	access roads; establishment	sum
A. 2 7 3.1	access roads; continuing operation or maintenance	wk
A. 2 7 3.2	access roads; removal	sum
A. 2 7 8	compressed air for tunnelling; establishment	sum
A. 2 7 8.1	compressed air for tunnelling; continuing operation or maintenance	wk

Specified requirements (cont'd)

Temporary Works (cont'd)

A. 2 7 8.2   compressed air for tunnelling; removal   sum

Method-Related Charges

Items for Method-Related Charges, if any, shall be inserted hereinafter by the tenderer in accordance with Section 7. Item descriptions for Method-Related Charges shall distinguish between Fixed and Time-Related Charges.
     Leave 2-3 blank pages to enable the Contractor to insert the Method-Related Charges he requires.

Provisional Sums

Daywork - The Contractor shall be paid for work executed on a Daywork basis at rates and prices calculated by adding the percentage additions stated in the Schedule of Dayworks carried out incidental to Contract Work issued by the Federation of Civil Engineering Contractors to the rates and prices contained in the aforementioned Schedules and by making further adjustments as follows:

Schedule 1   Labour addition/deduction* of + percent

Schedule 2   Materials addition/deduction* of + percent

Schedule 3   Plant addition/deduction* of + percent

Schedule 4   Supplementary charges ** addition/deduction* of + percent

     * appropriate deletion to be made by Contractor when tendering

     + percentage to be entered by the Contractor when tendering

     ** supplementary charges shall not include the charges referred to in notes and conditions 2(ii), 3 and 6 of schedule 4.

Include the following amounts for work to be executed or services to be provided on a Daywork basis

A. 4 1 1   Labour (amount in words)   sum

A. 4 1 2   percentage adjustment to Provisional sum for Daywork labour -
                    addition/deduction*   %

Provisional Sums (cont'd)

A. 4 1 3	Plant (amount in words)	sum
A. 4 1 4	percentage adjustment to Provisional sum for Daywork plant - addition/deduction*	%
A. 4 1 5	Materials (amount in words)	sum
A. 4 1 6	percentage adjustment to Provisional sum for Daywork materials - addition/deduction*	%
A. 4 1 7	Supplementary charges (amount in words)	sum
A. 4 1 8	percentage adjustment to Provisional sum for Daywork supplementary charges - addition/deduction*	%
	* delete as appropriate	

Other Provisional sums

A. 4 2 0.1	site notice board (amount in words)	sum
A. 4 2 0.2	telephone charges on behalf of the Employer (amount in words)	sum

Nominated Subcontracts which include work on the site

A. 5 1 0.1	Electrical installation (amount in words)	sum
A. 5 2 0.1	labours	sum
A. 5 3 0.1	special labours, labour in unloading materials; supplying special scaffolding	sum
A. 5 4 0.1	other charges and profit	%
A. 5 1 0.2	Lightning protection installation (amount in words)	sum
A. 5 2 0.2	labours	sum
A. 5 3 0.2	special labours, labour and plant in unloading materials; providing 3-phase power supply	sum
A. 5 4 0.2	other charges and profit	%

Chapter 4

CLASS B: GROUND INVESTIGATION

GENERAL TEXT

Principal changes from CESMM 1

There were extensive changes to this class.

The trial hole classification were amended completely and pumping was added as a further item (B. 1 8 0).

The classifications for boreholes and pumping test wells were replaced with new classifications for light cable percussion boreholes and rotary drilled boreholes (B. 2&3 * *).

Additional types of samples were added (B. 4 * *).

The classification for site tests and observations had additional tests added and some of the old tests deleted (B. 5 * *). Additional items were added to the instrumental observations classification (B. 6 * *).

The laboratory tests classification were changed considerably and the tests given were much more specific (B. 7 * *).

A completely new classification was added for professional services to be provided (B. 8 * *).

Principal changes from CESMM 2

The term trial holes has been changed to 'trial pits and trenches' in 1 * *, 4.1 * and in A1 of the Additional Description Rules.

A new Coverage Rule C2 has been inserted which deals with the disposal of surplus excavated material and dead services in connection with trial pits and trenches. The old Coverage Rules C2, C3, C4 and C5 are renumbered C3, C4, C5 and C6 to accommodate the insertion.

Coverage Rule A2 has also been amended slightly and states that the excavations which must be done by hand, shall be identified separately.

The reference to BS1377 in items B.5 2 5 and B. 7 1-6 has been deleted.

The units of measurement in items B. 6 3-8 * which were missing or incorrect in CESMM 2 but were included in the corrigenda have now been inserted.

Item B.7 6 * which referred to strength tests generally has been enlarged to cover both soil and rock strengths. B.7 6 * becomes soil strength and a new B.7 7 * covers rock strengths. The creation of this new B.7 7 * item generates five new insertions in the third division and five new Additional Description Rules coded A23 to A27.

The type of protective fences to be used must be stated in Additional Description Rules A15.

CLASS B: GROUND INVESTIGATION

Measurement Rules

M1 It follows that unless the specification or drawings instruct him to, the Contractor can only chisel to prove rock or penetrate obstructions when specifically instructed to do so by the Engineer. Therefore in practice, should the Contractor meet an obstruction, he must stop work and seek the Engineer's instructions. The Contractor will, of course, wish to be paid for any standing time whilst awaiting instructions, and it follows that the Engineer must be expeditious in his decision.

M2 The supply of professional services for any other reason than analysis of records and results is not measurable.

    The Contractor's liability for items measured for samples, tests and observations ends when those samples, tests and observations have been made. The classification for professional services should only be used where the Engineer specifically requires professional services to be supplied by the Contractor for the analysis of the records and results of samples, tests and observations.

    Services measured under this classification would be for intermittent services only. If services are required for long continuous periods then it would be more appropriate to deal with them under Class A - specified requirements.

Definition Rules

D3 In other words the depth shall be measured from the top of the borehole to the bottom of the borehole irrespective of how the borehole is formed.

Coverage Rules

C2 The item description for trial pits and trenches is deemed to include the cost of disposal of all surplus excavated material and dead services. If the value of dead services is thought to be significant, this should be drawn to the Contractor's attention in the Preamble in the expectation of receiving a credit.

C5 The professional services' classification is potentially open to abuse by the Contractor in that hours for services which are provided off site cannot be easily checked by the Engineer, and there is no simple way of overcoming this problem.

Additional Description Rules

A1 The maximum length in this context applies to the length of a trench and must be kept separate from trial pits. The size of the plan area at the bottom of the hole must be stated for trial pits.

A2 In certain instances, such as in the location of services, the excavation of the bottom part of the hole would always be done by hand and this is required to be stated.

A3 Class B is the only class in CESMM 3 where the disposal of water (by pumping) is measured separately. CESMM 3 does not specifically state what type of water is included and it is therefore construed that it applies to water from all sources.

CLASS B: GROUND INVESTIGATION

Paragraph 5.21

It is necessary to state the Commencing Surface for trial holes and boreholes for which the Commencing Surface is not also the Original Surface, and also to state the Excavated Surface when it is not the Final Surface. This particularly applies to rotary drilled boreholes which are continuations of light cable percussion boreholes (see Additional Description Rule A6).

ITEM MEASUREMENT

B. 1 * *   Trial pits and trenches

Divisions   Trial pits and trenches are classified by several items. Items must be given for:

a) the number of pits and trenches, differentiating between those which include and do not include rock, stating the maximum depth as classified. The actual depth must be stated where this exceeds 20m. The maximum depth is the maximum depth of each pit and trench and not the maximum depth of several pits and trenches contained within a group. Therefore two pits and trenches which have depths which fall into different classifications would have to be measured separately

b) the total depth of the pits and trenches in material other than rock

c) the total depth of the pits and trenches in rock

d) the total depth which is to be supported and the total depth backfilled stating the nature of the backfilling material. Although not specifically stated, these items should only be measured where they are expressly required

e) the total number of hours which is required for obstruction removal

f) The total number of hours which is required for pumping continued operations.

Rules   A1 State the minimum plan area at the bottom of the pit and trench or the maximum length at the bottom where it is used for locating services.

A2 State when the excavation is expressly required to be carried out by hand.

A3 State any special de-watering techniques which are expressly required.

Paragraph 5.21

State the Commencing Surface if it is not also the Original Surface. State the Excavated Surface if it is not also the Final Surface.

CLASS B: GROUND INVESTIGATION

Generally   There are several confusing aspects in the measurement of trial holes.

The first is in the classification of the number of trial holes. CESMM 3 requires that numbers of holes are separated into those which include and do not include rock. The item for the number of holes is generally to cover the cost of moving the plant from one hole to another, and the fact that the hole may or may not contain rock has little bearing on this cost. B. 1 2 * is also open to abuse by the Contractor in that there is no minimum piece size stated (see Measurement Rule M8 of Class E). This means that if a piece of rock appears in a trial hole, irrespective of size then theoretically that hole should be classified as including rock. The Contractor who has a higher rate in for the number of holes which includes rock will obviously try and use this argument. B. 1 2 * should therefore be clarified or possibly even deleted.

The second is whether trial holes in rock include for other artificial hard materials and one must assume that it does in the absence of any other classification. It should also be made clear whether removal of obstructions is measured in addition to, instead of, or not at all when measuring trial holes in rock, and it is recommended that the latter course of action is adopted. The preamble should explain exactly the measurement philosophy used.

The third is concerned with the measurement of rock itself. B. 1 4 0 adequately deals with the situation where the rock extends across the entire hole, but would not be applicable where isolated volumes of rock are encountered. A suitable preamble should be included which has a similar content to Measurement Rule M8 in Class E, which defines when an isolated volume of rock should be measured separately. Secondly, the method in which the isolated volume is measured should be made clear. This could either be by volume, in which case a separate classification should be created, or by hours in accordance with B. 1 7 0.

See also Rule C1.

B. 2 * *   Light cable percussion boreholes

Divisions   Light cable percussion boreholes are classified by several items. Items must be given for:

a) the number of boreholes

b) the total depth stating the maximum depth as classified. The maximum depth in instance would be the maximum depth which is not exceeded by any borehole included in the item (see Definition Rule D1 of Class P)

c) the total depth which is expressly required to be backfilled stating the nature of the backfilling material

d) the total number of hours which is required for proving rock or penetrating obstructions.

Rules   M1 Measure chiselling to prove rock or penetrate obstructions only where it is expressly required.

A4 State the nominal diameter of the bases of the boreholes.

Paragraph 5.21

State the Commencing Surface if it is not also the Original Surface. State the Excavated Surface if it is not also the Final Surface.

Generally   See also Rules C1, C3.

B. 3 * *   Rotary drilled boreholes

Divisions   Rotary drilled boreholes are classified by several items. Items must be given for:

a)   the number of boreholes.

b)   the total depth bored, differentiating between holes with and without core recovery. The maximum depth in this instance would be the maximum depth which is not exceeded by any borehole included in the item (see Definition Rule D1 of Class P)

c)   the total depth which is to be cased, and the total depth to be backfilled stating the nature of the backfilling material. Although not specifically stated, these would only be measured where expressly required.

d)   the total number of core boxes to be supplied stating the length of the core.

Rules   D2 State when core boxes are to remain the property of the Contractor.

D3 Classify the length rotary drilled boreholes which are extensions of light cable percussion boreholes from the Commencing Surface of the latter.

A5 State the nominal minimum core diameter.

A6 State when continuations of light cable percussion boreholes.

A7 State the angle of inclination for inclined boreholes.

Paragraph 5.21

State the Commencing Surface if it is not also the Original Surface. State the Excavated Surface if it is not also the Final Surface.

Generally   See also Rules D1, C1.

B. 4 * *   Samples

Divisions   Samples are enumerated stating whether they are taken from the surface, from trial holes or from boreholes. The type of sample is stated in accordance with the Third Division.

Rules   A8 State the size, type and class in accordance with B.S.5930.

Generally   See also Rule C1.

B. 5 1&2 *   Site tests and observations

Divisions   Site tests and observations are enumerated except for permeability tests which are given in hours. State the test or observation to be covered.

Rules   A9 State the type and particulars of permeability tests.

A10 State when the measurements are to be taken for ground water level tests.

A11 State whether standard penetration tests are in light cable percussion boreholes or rotary drilled boreholes.

A12 State whether plate bearing tests are in trial pits, trenches or boreholes.

A13 State the maximum depth of the cone for static cone sounding tests. State the maximum capacity of the machine where electric cones are used.

A14 State the minimum diameter and maximum depth of hand auger borehole tests.

Generally   See also Rule C1.

B. 6 * *   Instrumental observations

Divisions   Instrumental observations are classified according to their type in the Second Division. The Third Division separates the observations into installation of equipment and readings. Readings are enumerated and installations given in linear metres or by number as appropriate.

Rules   A15 State types of both the observations and protective fences.

A16 State whether inclinometers and settlement gauges are in special boreholes.

Generally   See also Rules C1, C4.

B. 7 * *   Laboratory tests

Divisions   Laboratory tests are enumerated stating the type of test as classified including strength tests for both soil and rock.

Rules   A17 State the standards required and contaminants to be analysed for tests for contaminants.

A18 State the number of increments and effective pressures for triaxial cell and Rowe cell tests.

A19 State the diameter and whether single, multi-stage or set of three specimens is required for quick undrained triaxial tests.

A20 State the diameter and effective pressures for consolidated triaxial tests. Identify multi-stage tests.

A21 State the normal pressure and size of shearbox for shearbox tests.

A22 State the compactive effort, surcharge and whether soaking is required for California bearing ratio tests.

A23 State the diameter and height of the samples of unconfined compressive strength tests.

A24 State the diameter, height and effective pressures of consolidated triaxial tests.

A25 State the diameter and length of the Brazilian test samples.

A26 State the diameters and normal pressures of the samples at the ring shear tests.

A27 State the type of test required and the minimum dimensions of the samples of the point load test.

Generally   See also Rule C1.

CLASS B: GROUND INVESTIGATION

B. 8 * *   **Professional services**

Divisions   Professionals time is given in hours stating the type and level of the person or persons involved. Site visits and overnight stays are enumerated.

Rules   M2 Measure only where expressly required for analysis of records and results.

M3 Exclude hours occupied in travel, meals, etc.

Generally   Other professionals required to be supplied should also be given such as chemists, etc.
See also Rules C1, C5, C6.

STANDARD DESCRIPTION LIBRARY

Trial pits; minimum plan size 750 x 750mm

Number in material other than rock

B. 1 1 1	maximum depth not exceeding 1m	nr
B. 1 1 5	maximum depth 5-10m	nr
B. 1 1 8	maximum depth 22.5m	nr

Depth

B. 1 3 0	in material other than rock	m
B. 1 5 0	supported	m
B. 1 6 0	backfilled with excavated material	m

Removal of obstructions

B. 1 7 0	generally	h

Pumping

B. 1 8 0	40 litres/hr	h
B. 1 8 0.1	200 gallons/day	h

Trial trenches to locate services; maximum length 3m; by hand

Number in material other than rock

B. 1 1 2	maximum depth 1-2m	nr

Trail trenches (cont'd)

Depth

B. 1 3 0.1	in material other than rock	m

Light cable percussion boreholes

Nominal diameter 300mm at base of hole

B. 2 1 0	number	nr
B. 2 3 3	depth in holes of maximum depth 10-20m	m
B. 2 3 6	depth in holes of maximum depth 45m	m
B. 2 6 0	depth backfilled with excavated material	m
B. 2 7 0	chiselling to prove rock or penetrate obstructions	h

Rotary drilled boreholes

Nominal minimum core diameter 300mm

B. 3 1 0	number	nr

Nominal minimum core diameter 300mm continued

B. 3 3 2	depth without core recovery; in holes of maximum depth 5-10m	m
B. 3 4 6	depth with core recovery; in holes of maximum depth 50m	m
B. 3 5 0	depth cased	m
B. 3 6 0	depth backfilled with grout as specification clause 5.12	m
B. 3 7 0	core boxes; core length 500mm	nr

Nominal minimum core diameter 350mm; inclined at an angle of 5 degrees to the vertical; Commencing Surface 2m below Original Surface

B. 3 1 0.1	number	nr

Rotary drilled boreholes (cont'd)

B. 3 4 4	depth with core recovery; in holes of maximum depth 20-30m	m
B. 3 5 0.1	depth cased	m
B. 3 7 0.1	core boxes; core length 750mm; to remain the property of the Contractor	nr

Samples

From the surface or trial holes

B. 4 1 3	rock; type_____; size_____; class_____	nr

From boreholes

B. 4 2 1	open tube; type_____; size_____; class_____	nr

Site tests and observations

Standard penetration

B. 5 1 3.1	in light cable percussion boreholes	nr
B. 5 1 3.2	in rotary drilled boreholes	nr

Plate bearing

B. 5 2 1	in pits	nr
B. 5 2 1.1	in boreholes	nr

Instrumental observations

Pressure head

B. 6 1 1	standpipes	m
B. 6 1 2	piezometers	m
B. 6 1 3	install covers	nr
B. 6 1 4	readings	nr

Instrumental observations (cont'd)

Settlement gauges

B. 6 3 1	installations	nr
B. 6 3 4	readings	nr

Settlement gauges; in special boreholes

B. 6 3 1.1	installations	nr
B. 6 3 4.1	readings	nr

Laboratory tests

Classification

B. 7 1 1	moisture content	nr

Chemical content

B. 7 2 1	organic matter	nr

Compaction

B. 7 3 1	standard	nr

Consolidation

B. 7 4 1	oedometer cell	nr

Permeability

B. 7 5 1	constant head	nr

Soil strength

B. 7 6 1	quick undrained triaxial; 100mm diameter; single stage	nr

Rock strength

B. 7 7 1	unconfirmed compressive strength of core samples	nr

CLASS B: GROUND INVESTIGATION

Professional services

Engineer

B. 8 3 1   graduate   h

Geologist

B. 8 3 2   chartered   h

Chapter 5

CLASS C: GEOTECHNICAL AND OTHER SPECIALIST PROCESSES

GENERAL TEXT

Principal changes from CESMM 1

There were been several changes to this Class.

The classification for injection of grout was rationalised. There are now five separate First Division headings instead of three, and the Second and Third Divisions were amended accordingly (C. 1-5 * *).

Diaphragm walling had only minor amendments. The classification for reinforcement was extended (C. 6 5&6 *) and the depth of excavation was classified as the maximum depth (C. 6 1-3 *).

The classification of the materials in which ground anchors were placed was improved and it was necessary to state the type of corrosion protection in accordance with the Third Division (C. 7 1-4 *).

The classification for sand drains was extended to include band and wick drains (C. 8 * *). The other principal changes were the addition of a classification for predrilling through overlying material, and the changing of the Third Division classification from diameter to cross-sectional dimension.

Principal changes from CESMM 2

The item description in 6. 5 * has been altered to 'Plain round bar reinforcement'.

The item description in 6. 6 * has been altered to 'Deformed high yield steel bar reinforcement to BS4449'.

A printing error in the lines on page 27 has been corrected (this fault appears in the corrigenda). Coverage Rule C1 has changed to mean what was always intended i.e. the item description 1-3 * * includes for the removal of dead services not live.

Measurement Rules

M1 This rule prevents any argument about the Commencing Surface to be used for the remeasurement of the completed work if the Contractor should, due to his modus operandi, carry out the work from a level other than that adopted in the Bill of Quantities.

M2 Although this rule states, 'holes for ground anchorages and drains' it should be read as being the ground anchors and drains themselves because there are no actual holes measured separately for ground anchors.

M3 Upon first reading of this rule the two sentences appear to be contradictory but they are not.

The first sentence refers to holes which are required to be done by the technique known as stage grouting. This is a technique which is commonly used in soils which are liable to collapse as the drill is withdrawn. The first operation is to drill the first stage and grout, effectively stabilising the surrounding area of ground around that stage. The second operation is to then drill though the first stage grout in order to drill the second stage and then grout the second stage and so on for the third, fourth, fifth stages etc., until all the required stages are completed. The first sentence of this rule means that for a hole drilled out in this manner, the drilling is only measured once irrespective of the number of stages.

The second sentence of this rule applies only to the situation on site where the Contractor is specifically requested by the Engineer to go back to a completed hole and redrill through the grout to extend the length of the hole. This only applies to holes which it was envisaged would not be required to be extended. Obviously, had the extension been envisaged then the hole would have been described as being in stages anyway, and the second sentence would therefore not apply (see Figure C1 for clarification of this point). It follows then that the second sentence only applies to remeasurement and items for hole extensions would not appear in the tender documentation itself because they would by definition be variations instructed under Clause 51 of the Conditions of Contract.

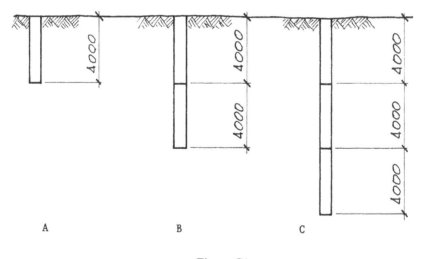

Figure C1

In Figure C1 the hole was originally drilled in 'normal' materials and considered complete (A). The Engineer subsequently ordered the extension as in (B) and then ordered a further extension (C). The ground throughout was a coarse granular material. The remeasurement of the hole would be as follows:

CLASS C: GEOTECHNICAL AND OTHER SPECIALIST PROCESSES

For (A) Drilling for holes vertically downwards in material other than
rock or artificial hard material in holes of depth not exceeding
5m.   4m

Number of holes.   1nr

Number of stages.   1nr

For (B) Drilling for holes vertically downwards in rock or artificial
hard material in holes of depth 5-10m (previously injected
in (A)).   4m

Drilling for holes vertically downwards in material other than
rock or artificial hard material in holes of depth 5-10m
(extension).   4m

Number of holes.   1nr

Number of stages.   1nr

For (C) Drilling for holes vertically downwards in rock or artifical
hard material in holes of depth 10-15m (previously injected
in (A) and (B)).   8m

Drilling for holes vertically downwards in material other than
rock or artificial hard material in holes of depth 10-15m
(extension).   4m

Number of holes.   1nr

Number of stages.   1nr

If, at tender stage, it was known that the hole was to be done in stages as in (C) the
measured items for the hole would have been:

Drilling for holes vertically downwards in material other than
rock or artificial hard material in holes of depth 10-15m.   12m

Number of holes.   1nr

Number of stages.   3nr

Problems may occur in the post contract stage in valuing holes which have been extended. As previously stated, extensions to holes would not normally be measured separately in the Bill and therefore the valuation of them would be dependent upon applying other rates in the Bill to the remeasured items for the extension in accordance with Clause 52 of the Conditions of Contract. Take the example shown in Figure C2.

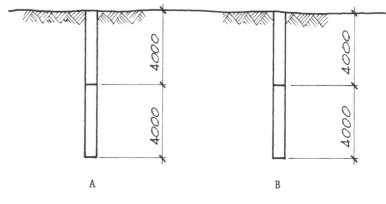

Figure C2

(A) is a hole which was originally envisaged as being 4m deep and was drilled and grouted as such. The Engineer subsequently instructed a 4m extension to the hole and this was done after the original hole was completed. (B) is a hole which was always envisaged as being done in two stages and was measured as such in the Bill. As can be seen from the diagram there is no physical difference in the two holes. The question arises whether the rate in the Bill for (B) should be applied to the total value of the work done in (A), or whether the value of (A) should be based on the original bill rates for the first four metres plus the value of the extension based on its remeasurement and other applicable rates in the Bill. The Contractor would of course choose to value the work in the manner which gives him the greatest financial return but in fact the correct valuation would be the latter because this would more accurately reflect the work involved.

M4 In order to save confusion and give the estimator a better chance to price the work with a degree of accuracy it is recommended that where two holes have different stage lengths, they are measured quite separately and that the length of the stage is stated where this would otherwise be unclear.

The situation is also confused where some holes are to be done by stage grouting and others are not. Obviously to combine the two into one group of measured items would give an artificially high number of stages because each single hole with no extensions would have to be classed as a one stage hole whereas in actual fact it does not have any stages at all in the context of this rule. It is recommended that 'single stage' holes ie. those without any stages are also kept separate.

M5 The mass of grout materials and injection should include the mass of all M7. other materials except for mixing water.

M6 When read in conjunction with Additional Description Rule A3 it is not clear what constitutes an injection, particularly in the case of stage grouting, (Figure C3).

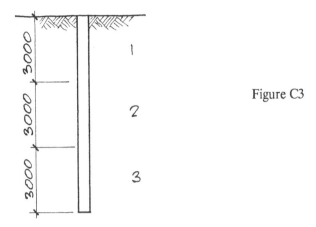

Figure C3

Figure C3 is a nine metre hole grouted in three stages. These could be in ascending stages or descending stages. If in ascending stages it could be argued that the injection is one number only, described as being in ascending stages i.e. starting from the bottom and rising to the top. The rig would only be required once at the location and this argument would seem to be reasonable.

However, if the hole was to be grouted in decending stages then the rig may have to be brought to the hole on three separate occasions. Additional Description Rule A3 obviates this problem to some extent by requiring that the grouting is described as being in descending stages but this would still not enable the estimator to ascertain the number of rig moves which is dependent on the number of injections.

It is recommended that in stage grouting, whether ascending or descending, an injection is measured for each stage. Apart from the above, this is for the following reasons:

1. whilst the specification may give specific stage lengths, there is a strong possibility that these may be varied because the ground conditions encountered on site could vary from those envisaged. Measuring the injections in the manner recommended deals more easily with the valuation of the varied work

2 each stage of injection has its own distinct work content i.e. raising of the probe, isolating the stage for grouting and injection of materials.

M8 Voids are dealt with under G. 1-4 7 *. Rebates and fillets are dealt with under G. 1-4 8 5&6 where they do not exceed 0.01m2 in cross-sectional area, and under the formwork classification generally where the cross-sectional area exceeds this size, (see Class G). Because of the specialist nature of this class, it is recommended that the items for formwork identify those which are in diaphragm walls and that they are not 'lost' in the general formwork measurement. This is most easily done by grouping them together under a separate heading.

It is not clear whether any formwork which is required to the ends of the panels should be measured separately. They are not mentioned at all so it is considered that they are deemed to be included although a suitable preamble should make this clear.

One problem with the measurement of formwork to voids, rebates and fillets concerns the striking. Whilst the formwork will always be put in by the specialist subcontractor, more often than not the excavation to expose it to enable it to be struck will not take place until some time after the subcontractor has left the site. This means that the subcontractor must make a return visit to the site just to remove the formwork and this is obviously expensive. Usually the Main Contractor will strike the formwork after the excavation work is completed. One method to overcome this is to remeasure the installation and striking of the formwork separately. In his subcontract enquiry the Contractor can then ask the subcontractor to price the installation only, whilst allowing for the striking himself.

M9 This refers to the levels which the Engineer designs them to be cast to. See comments on Measurement Rules M1 and M2 in Class G.

The volume should be based on the designed wall thickness and the Contractor should allow for the wastage in casting against the excavated face in his rates.

M10 It follows that the Engineer must design and include on his schedules for the stiffening, lifting and supporting steelwork. Where the Engineer wishes to leave this to the discretion of the Contractor then this rule must be omitted and the situation made clear to the Contractor.

M12 It is considered that guide walls are not adequately dealt with. Guide walls are low walls, mostly constructed in reinforced concrete which are accurately aligned along the front and back face of diaphragm walls and which serve to guide the grab in the excavation of the trench. They are sometimes left to the specialist subcontractor to design, but are more often designed by the Engineer after consultation with a specialist. They are almost always used. Note that this Rule does not contain the words 'expressly required'.

If the walls are left to the Contractor's discretion, then the simple way in which CESMM 3 deals with them is adequate. However, if the guide walls are designed by the Engineer then far more information must be provided to the estimator, either in the form of additional description or by reference to the relevant drawing. Such information would include, but not necessarily be limited to, specification of concrete, reinforcement details, joint details, finishes, etc.

It is also not clear exactly what the item for guide walls is to include for because there is no coverage rule. The item is supposed to include for all work associated with the item including excavation although it would be prudent to make this clear in the Preamble. One point to note is that whereas the inside guide wall is usually removed, the outside guide wall is often left in place and it should be made clear whether the guide walls are temporary or permanent.

Guide walls are measured to both sides of the diaphragm wall.

M13 The outside face of anchorage is the outside face of the anchorage itself and not the end of the tendon.

M14 Predrilling is the term given to the drilling through embankments or fill material overlying the strata to be consolidated. Predrilling should only be measured where it is an express requirement that the drains be installed from the top of the embankment or overlying fill.

Definition Rules

D1 It is not necessary to state the nature of artificial hard material for items measured under C. 2 * * - drilling for grout holes, although it is necessary for excavation for diaphragm walls measured under C. 6 3 * (Additional Description Rule A5).

D2 Whilst this rule is correct in its definition, the classification of diaphragm walls in CESMM 3 is dedicated to walls constructed in in situ concrete whereas nowadays they can also be constructed in precast concrete. The technique involved is to construct the trench as normal with a bentonite slurry and then introduce a bentonite/cement slurry immediately prior to the placing of the precast unit. The units usually have dovetailed ends and the next unit slots on to it. Several panels can be laid in a day and the following day the excavation to the outside of the wall is carried out and the bentonite/cement mix is cleaned off the outside face. The mix to the back face is allowed to go off and forms a bond with the ground.

If this type of diaphragm wall is encountered then it will be necessary to amend the classification and also to add a further classification for the precast units in Class H.

Coverage Rules

C1 The Specification or Preamble should make it clear where the surplus material is to be disposed of. Removal of services applies to dead and redundant services because live services would be dealt with by the Statutory Authorities in the first instance.

The Contractor would normally price the work on the basis of completing adjacent panels one after the other. Sometimes due to engineering considerations it is specified that the work must be done in a piecemeal fashion, i.e. by going back and forth to different locations at different times. In this case it would be prudent to make the situation clear to the Contractor.

C2 Upholding the sides of excavation would normally be done by the introduction of the bentonite slurry.

C4 It follows that the supporting reinforcement must be detailed by the Engineer (see Measurement Rule M10). Although not specifically mentioned, stiffening and supporting steelwork would also be included because these are specifically included in Rule M10 along with supporting reinforcement.

Additional Description Rules

A1 Driving injection pipes is an operation sometimes used instead of boring separate holes in which to place the grout pipes. It follows that the diameter stated under this rule would not be the diameter of the hole but the diameter of the injection pipe.

A2 Reference should be made to the specification if the item description becomes long or unwieldy.

A3 See Measurement Rule M6.

A4 See also Definition Rule D6 of Class F for clarification of wall thicknesses.

A6 This would normally be in accordance with B.S.5328 (see also Class F). Note that unlike Class F, the item for concrete measured under Class C includes for both the provision and placing of concrete.

Paragraph 5.21

The Commencing Surface must be identified for any work involving boring, driving or excavation for which the Commencing Surface is not the Original Surface. The Excavated Surface must be stated if it is not also the Final Surface.

ITEM MEASUREMENT

C. * * *   Geotechnical and specialist processes generally

Applicable to all items

Divisions   -

Rules   M2 Measure grout holes, holes for ground anchorages and drains along their axes irrespective of the inclination.

Generally State the Commencing Surface unless it is also the Original Surface.
State the Excavated Surface unless it is also the Final Surface.

See also Rules M1, D1, C1.

C. 1-4 * *   Drilling for grout holes and driving injection pipes

Divisions Drilling for grout holes and driving injection pipes are measured in linear metres. State when drilling through rock or artificial hard material; it is not necessary to state nature of artificial hard material. State the angle of inclination in accordance with the Second Division and the depth in accordance with the Third Division.

In addition it is also necessary to measure the number of holes and the number of stages. Measure the number of single and multiple water pressure tests.

Rules   M3 Do not measure drilling through previously grouted holes in stage grouting. For hole extensions measure through the original grouted hole as drilling through rock or artificial hard material.

M4 Measure the number of stages expressly required.

A1 State the diameter of the hole or injection pipe.

Generally It is recommended that staged holes with different stage lengths are kept separate and that the length of the stage is stated if it is otherwise unclear from the drawings or specification. It is also recommended that holes not in stages are kept separate from those which are.

The depth in the Third Division is the maximum depth of each hole and not the maximum depth of the deepest hole in the group (see Definition Rule D1 of Class P). It is the depth from the Commencing Surface to the Final Surface ignoring any intermediate stages in staged holes.

Classifications C. 4 1-4 0 apply equally to driving injection pipes and to drilling grout holes. There is no item for driving injection pipes through hard materials as in practice this would never be done. If hard material was encountered, then it would be removed by other means or the injection point relocated.

It is not clear whether a hole which is not staged should be counted as one number stage for the purposes of C. 4 2 0. One of the purposes of measuring the number of stages is to inform the Contractor how many times he has to set up on the hole to carry out the next stage of the operation. With an unstaged hole this would only be once and would be covered in the item for number of holes. It is not necessary to count it as '1 nr stage' although if the previous recommendations are adopted they will be kept separate from staged holes anyway. In this case a general heading could be given stating that all the following holes are in single stages. Alternatively the stages could be measured in the same number as those for the holes. Whichever procedure is adopted the Preamble should make it clear.

Figure C4 illustrates the various classification of angles of inclination of grout holes.

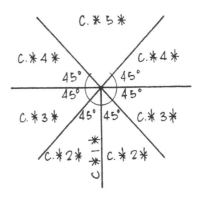

Figure C4

C. 5 1 *   Grout materials

Divisions   Grout materials are measured in tonnes. The materials are the individual components of which the grout is composed.

Rules   M5 Do not include the mass of mixing water.

A2 State the type of materials. The Third Division should be taken as being representative and added to as required.

Generally The actual quantity of injection material measured will by necessity be an approximation and will normally be done in conjunction with the Engineer. Careful post contract administration of the actual grout materials used on site is required in order to ascertain the correct amount of grout injected.

C. 5 2 *   Grout injection

Divisions   The injection of grout is measured in tonnes stating the type of grout. The number of injections must be measured as well as the number of single and double packer settings.

Rules   M6 Measure the number of injections as the total number expressly required.

M7 Do not include the mass of mixing water.

A3 Distinguish between ascending and descending stages in the item descriptions for number of those for injections which are to be carried out in stages.

Generally The Contractor's rates for injection of grout should also include for the mixing of the individual grout components.

C. 6 1-3 * Excavation for diaphragm walls

Divisions   Excavation for diaphragm walls is measured in cubic metres stating the maximum depth in accordance with the Third Division. In accordance with Paragraph 5.21 the depth is measured from the Commencing Surface to the Excavated Surface. Excavation in rock or artificial hard material is measured separately.

Rules   A4 State the thickness of the wall.

A5 State the nature of artificial hard material.

Generally See also Rule C2.

C. 6 4 0 Concrete in diaphragm walls

Divisions   Concrete in diaphragm walls is measured in cubic metres.

Rules   M8 Measure any associated formwork in accordance with Class G.

            M9 Measure from the cut off levels expressly required. Calculate the volume in
            accordance with Measurement Rules M1 and M2 of Class F.

            A4 State the thickness of the wall.

            A6 State the mix specifications or strength (usually in accordance with
            BS5328).

Generally   The joint between adjacent panels of concrete is normally a butt or possibly a
            rebated joint. Occasionally dowels are specified or detailed between the panels
            and these should be measured separately in accordance with Class G.
                The length of individual panels may be left to the Contractor's discretion or
            a maximum length may be specified or the actual length may be given. In the
            case of the latter two these should only be stated in item descriptions if there are
            various walls on the contract with varying panel lengths. In this case it would be
            prudent to group all the relevant items together under a general heading for
            each wall.

            See also Rule C3.

C. 6 5&6 * Reinforcement in diaphragm walls

Divisions   Reinforcement in diaphragm walls is measured in tonnes stating the type and
            diameter. Bars of 32mm diameter or greater are collected together.

Rules   M10 Include the mass of stiffening lifting and supporting steel.

            M11 Take the mass of reinforcement as 7.85t/m3.

Generally   See also Rules D3, C4.

C. 6 7 0 Waterproofed joints

Divsions   Waterproofed joints are measured as a sum.

Rules   -

Generally   Waterproofed joints have always been something of a problem in diaphragm
            walls.
                The normal joint between adjacent panels of walls is a butt joint of some
            description. It may be convoluted to some extent in order to reduce differential
            movement between the panels but generally speaking each panel is quite

separate. Differential movement between the panels together with other factors means that more often than not the joints will seep water. It is necessary in many instances where diaphragm walling is used, to face the wall with something like a skin of blockwork with some method of draining away the seeping water. In cases such as this it would not be necessary to measure waterproofed joints.

It is common practice to specify that ingress of water is acceptable up to a specified limit but that beyond that amount the Contractor must take measures to remedy the situation. Waterproofed joints may have to be used where the ingress of water is excessive.

There are basically two instances when waterproofed joints must be measured. These are:

a) when the Engineer specifically designs them or specifies them in the specification

b) when the ingress of water is allowed to a stated maximum limit but beyond which the Contractor would be responsible for dealing with the excess water.

The unit of measurement for waterproofed joints is the sum and this is acceptable in the second instance where the joints are entirely at the Contractor's risk and also in the first instance where the lengths of the panels and therefore the joints are left to the Contractor's discretion. However, where the joints are expressly required in panels of a specified length it is recommended that they are measured in linear metres.

C. 6 8 0 Guide Walls

Divisions   Guide walls are measured in linear metres.

Rules   M12 Measure each side of the diaphragm wall.

Generally   Give as much additional description as is considered necessary or refer to the drawings or specification. Make it quite clear what is to be included in the item.

C. 7 * * Ground anchors

Divisions   Ground anchors are measured by giving the total length of the anchors and the total number stating the maximum depth. Items must distinguish between those which do and do not contain rock or artificial hard material. Items must also distinguish between temporary and permanent anchors and between those with single or double corrosion protection.

Rules   M13 Measure the lengths between the outer ends of anchorages.

             A7 State the anchor composition, location and working load. Give details of water and grout testing. Give details of pre-grouting and grouting.

Generally   Temporary anchors are those which will normally only be required to withstand the relevant forces for a relatively short period of time, possibly to a maximum of two to three years.

Although not specifically stated the items for ground anchors are fully inclusive and include for their complete installation. Separate items are not required for drilling of holes for ground anchors which could be implied from Measurement Rule M2.

C. 8 * *   Sand, band and wick drains

Divisions   Sand, band and wick drains are measured in linear metres stating the maximum depth in the ranges as classified in the Second Division. The maximum depth is the maximum depth of each hole and not the maximum depth of the deepest hole in a group of holes. Similar holes which fall into the same range would therefore be collected together. The number of drains must also be measured as must the number of pre-drilled holes. The depth of overlying material (if any) must also be given in linear metres. Drains are classified according to their cross-sectional dimension in the Third Division.

Rules   M14 Measure pre-drilling only where expressly required.

M8 Measure sand, band and wick drains separately. State the constituent materials.

Generally   Although not specifically stated, the items for sand, band and wick drains are fully inclusive items and include for their complete installation. Separate items are not required for drilling of holes for drains except for pre-drilling.

All three drains are what are generically known as vertical consolidation drains. These are borings through clay type soils which are filled with a filter media. The drain enables the soil to drain more easily thus accelerating the consolidation i.e. under embankments.

STANDARD DESCRIPTION LIBRARY

**Drilling for 200mm diameter grout holes through
material other than rock or artificial hard material**

Vertically downwards

C. 1 1 2	in holes of depth 5-10m	m
C. 4 1 0	number of holes	nr
C. 4 2 0	number of stages	nr
C. 4 3 0	single water pressure tests	nr

Drilling for 200mm diameter grout holes through rock or artificial hard material

Downwards at an angle of 0-45 degrees to the vertical

C. 2 2 3	in holes of depth 10-20m	m
C. 4 1 0.1	number of holes	nr
C. 4 2 0.1	number of stages	nr
C. 4 3 0.1	single stage pressure tests	nr
C. 4 4 0	multiple stage pressure tests	nr

Driving injection pipes for grout holes

Horizontally or downwards at an angle less than 45 degrees to the horizontal

C. 3 3 3	in holes of depth 10-20m	m
C. 4 1 0.2	number of holes	nr
C. 4 2 0.2	number of stages	nr
C. 4 3 0.2	single water pressure tests	nr
C. 4 4 0.1	multiple water pressure tests	nr

Grout

Materials

C. 5 1 1	cement	t
C. 5 1 3	sand	t

Injection

C. 5 2 1	number of injections	nr
C. 5 2 1.1	number of injections; in descending stages	nr
C. 5 2 3	cement and sand grout	t
C. 5 2 6	single packer settings	nr
C. 5 2 7	double packer settings	nr

Diaphragm walls; thickness 500mm

Excavation in material other than rock or artificial
hard material

| C. 6 1 1 | maximum depth not exceeding 5m | m3 |

Excavation in rock

| C. 6 2 4 | maximum depth 15-20m | m3 |

Excavation in reinforced concrete

| C. 6 3 7 | maximum depth 35m | m3 |

Concrete

| C. 6 4 0 | designed mix to BS5328; grade 20; ordinary portland cement to BS12; 20mm aggregate | m3 |

Plain round bar reinforcement

| C. 6 6 4 | nominal size 12mm | t |

| C. 6 6 8 | nominal size 16mm | t |

Deformed high yield bar reinforcement to BS4449

| C. 6 6 4 | nominal size 12mm | t |

| C. 6 4 7 | nominal size 25mm | t |

Waterproofed joints

| C. 6 7 0 | generally | sum |

Guide walls

| C. 6 8 0 | as drawing number GOSP/DWG/1 | m |

Ground anchors; to wall as located on drawing number GOSP/DWG/2

Temporary with single corrosion protection; composition as specification clause 6.1; working load as indicated; grout and water testing, pregrouting and grouting as specification clause 6.2

C. 7 1 2	number in material other than rock or artificial hard material to 15m maximum depth	nr
C. 7 2 2	total length of tendons in material other than rock or artificial hard material	m

Sand drains

Cross-sectional dimension not exceeding 100mm; filled with granular material as specification clause 6.3

C. 8 1 1	number of drains	nr
C. 8 2 1	number of predrilled holes	nr
C. 8 3 1	depth of overlying material	m
C. 8 6 1	depth of drains of maximum depth 15-20m	m
C. 8 8 1	depth of drains of maximum depth 27m	m

Wick drains

Cross-sectional dimension 200-300mm; preformed liner with granular material as specification clause 6.4

C. 8 1 3	number of drains	nr
C. 8 2 3	number of predrilled holes	nr
C. 8 3 3	depth of overlying material	m
C. 8 4 3	depth of drains of maximum depth not exceeding 10m	m

Chapter 6

CLASS D: DEMOLITION AND SITE CLEARANCE

GENERAL TEXT

Principal changes from CESMM 1

It was no longer a requirement to state the type of land under the item of general clearance (D.1 0 0).

The confusion which previously arose regarding the removal of trees and stumps was removed. (D. 2&3 * 0 and Coverage Rule C2).

Coverage rules were created clarifying the disposal of materials (C1), and the removal of hedge stumps for items of general clearance (C2). It was also necessary to state the nature of the material when backfilling stump holes (Additional Description Rule A3).

Principal changes from CESMM 2

None

Measurement Rules

M1   It is envisaged that this rule would not apply when pipes below 300mm nominal bore were to remain the property of the Employer, in which case they would be measured separately. 'Within buildings' should be read as 'forming part of buildings' because items such as rainwater pipes, which are strictly speaking outside the building, would fall under this rule.

Definition Rules

D1   It follows that the Engineer must state exactly what articles, objects and obstructions, apart from those set out in the class have to be removed. Alternatively he could specify those which have to be retained. This would normally be done in the specification to which reference would be made.

D2   CESMM 3 classifies trees according to their girth and stumps according to their diameter. This rule states that girths of trees are measured 1m above ground level, but does not say where the diameter of stumps shall be taken from, although generally speaking this is not a problem because stumps do not normally exceed 1m in height. However should a scheme have large quantities of stumps for removal of varying heights it is recommended that this rule be extended to state that the stumps shall be measured 1m above ground level if they exceed 1m in height or at their tops if they are less.

D3   Generally buildings are demolished down to the level of the top of the ground floor slab. It is possible for this slab to be part above and part below the Original Surface and there may therefore be a conflict between the volume used for classification and the volume actually demolished. However, as the volume used for classification can only be an approximation, it is felt that this is not a problem.

Coverage Rules

C1   Disposal of the materials arising would normally be off-site to tips supplied by the Contractor. However, it is quite commonplace for suitable materials to be incorporated into adjacent areas of new landform or filling. If this is the case then item descriptions should so state, indicating the areas of fill in which the materials are to be placed. Additional coverage rules should be created if it is necessary to sort the demolition materials beforehand or give them any treatment prior to their incorporation into the works.

C2   The item for general clearance includes for the removal of trees and their stumps below 500mm girth and of stumps below 150mm diameter. It also includes for the removal of all hedges which are required to be uprooted together with their stumps, regardless of their size. However, should one of the hedge plants be set apart from the rest of the hedge then this would then become a separate measurement item if its size exceeded the specified limits.

C3   It follows that if for any reason the stumps are not to be removed then item descriptions must so state.

C4   Supports in this instance could mean either straps or hangers for pipes within buildings or concrete stools for general pipelines. If a pipeline is supported on concrete stools, then care should be exercised in case there is any confliction with Definition Rule D3 with regard to removal of that part of the stool below ground level.

Additional Description Rules

A1   It would also be prudent to state if it is necessary to take them to the Employer's store, the location of which should be stated in the Specification or item description. It is recommended that items which are to remain the property of the Employer should be described as being 'carefully' removed.

A2   This would normally be done by reference to a location plan rather than by detailed description.

A3   It would be necessary to state the exact nature of the materials. For instance it would not be sufficient to say 'backfilled with topsoil' because it is not apparent from this description from where the topsoil is to be obtained. Further description would therefore be required such as 'backfilled with topsoil obtained on site' or 'backfilled with imported topsoil'.

A4　For large sites the simplest method of doing this is to give each building a reference on a location plan and state the reference in the item description.

ITEM MEASUREMENT

D. * * *　Demolition and site clearance generally

Applicable to all items

Divisions　-

Rules　　　A1 Materials which are to remain the property of the Employer shall be so described.

Generally　Items which are to remain the Employer's property as being 'carefully' removed.

　　　　　See also Rule C1.

D. 1 0 0　General clearance

Divisions　General clearance is measured in hectares.

Rules　　　D1 State which articles are to be removed or which are to be retained or refer to the relevant specification clause or drawing.

　　　　　A2 Identify the area included unless it is the total area of the site.

　　　　　A3 State the nature of the filling material if holes left by the removal of trees and stumps are to be backfilled.

Generally　See also Rule C2.

D. 2 * 0　Trees

Divisions　Trees for removal are enumerated stating the girth as classified.　Trees below 500mm girth are deemed to be included with the item of general clearance.

Rules　　　D2 Measure the girth 1m above ground level.
　　　　　A3 State the nature of the filling material if holes left by the removal of trees are to be backfilled.

Generally　See also Rule C3.

D. 3 * 0　Stumps

Divisions　Stumps for removal are enumerated stating the diameter as classified. Stumps below 150mm diameter are deemed to be included with the item of general clearance.

Rules　　　A3 State the nature of the filling material if holes left by the removal of stumps are to be backfilled.

Generally It is recommended that the diameter is measured 1m above ground level or at the top of the stump if it is below 1m high.

D. 4&5 * *　Buildings and other structures

Divisions The demolition of buildings and other structures are given as sums stating the predominant type of material as classified in the Second Division and the volume as classified in the Third Division.

Rules　　　D3 Exclude any volume below the Original Surface in the calculation of the volume.

　　　　　A4 Identify the building or structure or refer to a drawing.

Generally　-

D. 6 * 0　Pipelines

Divisions Demolition of pipelines is measured in linear metres stating the nominal bore as classified.

Rules　　　M1 Only measure the demolition of pipelines in buildings separately if they exceed 300mm nominal bore. However, it is recommended that pipes below 300mm nominal bore which are to remain the property of the Employer are measured separately.

Generally Strictly speaking there is no requirement to state the nominal bore of pipelines over 500mm nominal bore. However, there is obviously a considerable difference in cost between removing pipes of 500mm and 1500mm nominal bore, and it is therefore recommended that either the actual nominal bore is stated or that the categories are extended in continuing increments of 200mm.

　　　　　There is also a significant difference between removing a concrete pipe and a steel pipe of the same nominal bore. The handling techniques could be quite different and the steel pipe would have a greater credit value. It is recommended that either pipelines are identified in the same way as buildings and other structures (Additional Description Rule A4), or that the material of which they consist is stated.

　　　　　See also Rule C4.

STANDARD DESCRIPTION LIBRARY

General clearance

All items except those listed hereinafter; stump holes backfilled with imported topsoil

D. 1 0 0	disposal off site	ha

Trees; stump holes backfilled with subsoil

Girth 1-2m

D. 2 2 0	disposal off site; to remain property of Employer	nr

Stumps

Girth 500mm-1m

D. 3 2 0	disposal off site	nr

Buildings

Brickwork

D. 4 1 2	volume 50-100m3; reference Bldg 1 on drawing number DSC/DWG/1; disposal on site to fill area designated F1	sum

Pipelines

Nominal bore 300-500mm

D. 6 2 0	disposal off site	m

Chapter 7

CLASS E: EARTHWORKS

GENERAL TEXT

Principal changes from CESMM 1

The Earthworks section was changed in several important ways.

It was no longer a requirement to state whether materials were to be for re-use or disposal in item descriptions for excavation (E. 1-4 * *), and this served to reduce the number of Second Division descriptive features. The treatment of the excavated material was the subject of totally separate measurement items and the associated costs were included either in newly created items arising from the disposal classification (E. 5 3 *) or in the filling classification (E. 6 * *).

The uncertainty over where and when to measure trimming and preparation of excavated and filled surfaces was clarified. Topsoil, (E. 5 1-4 1 and E. 7 1&2 1) was also introduced as an additional third level descriptive feature under excavation and filling ancillaries. All trimming and preparation items were also classed according to their inclination to the horizontal.

Pitching was deleted from the Second Division for filling (E. 6 * * in CESMM 1) and became part of the descriptive feature 'to stated depth or thickness' (E. 6 4 *), and covered by Definition Rule D8.

Geotextiles (E. 7 3 0) were introduced as a new item under filling ancillaries.

The classification for hedges with protective fencing was deleted.

It was necessary to state when excavation is within underpinning (Additional Description Rule A3) or borrow pits (Additional Description Rule A5).

It became a specific requirement to state when items measured in accordance with excavation ancillaries (E. 5 * *) were in connection with excavation by dredging (Additional Description Rule A6).

Principal changes from CESSM 2

The term 'ground investigation' replaces 'site investigation' in the 'excludes' list at the beginning of Class E and the word 'underlaying' is replaced by 'underlying' in Measurement Rule M18. Both these corrections were part of the corrigenda.

The term 'existing services' has been changed to 'dead services' in Coverage Rule C1.

An extra Additional Rule, A6 is added to cover hand digging and the subsequent rules are renumbered to accommodate this insertion.

A new M24 Measurement Rule is included dealing with the measurement of laps in geotextiles. The original M24 rule now becomes M25.

Measurement Rules

M1 States that the work is measured strictly net with no allowances for working space etc. (see also Coverage Rule C1).

M2 Clarifies the position where a Contractor executes the work in a different manner from the way in which it is measured. For example, in a bill the road excavation is measured to the Final Surface first and the subsequent drainage work states the Commencing Surface as the Final Surface of the road. However, on site the Contractor may elect to construct the drainage works first commencing from Original Surface and the road excavation thereafter. In such an instance, the remeasurement of the work should be on the basis of the original measurement regardless of how the Contractor carries it out.

M4 If the measurement of excavation by dredging is to be made by any other method other than soundings then item descriptions should so state. An alternative method of measurement for dredging could be to measure the volume of material in the dredger.

M5 This is a very important measurement rule. Separate stages of excavation are only measured were they are expressly required. This rule would not apply to excavating layers of differing materials encountered in the same excavation, which although they are stages in the excavation, they are not expressly required. It would only apply to certain volumes of excavation which are specifically required to be done as a totally separate operation. For example it is sometimes specified that the last 300mm or so of excavation above a road formation is to be excavated separately immediately prior to the laying of the sub-base (in order to protect the formation). Where excavation is measured in accordance with this rule the Commencing and Excavated Surface should be stated in item descriptions in accordance with A4 (see also Measurement Rule D4 and Paragraph 5.21). Where excavation is expressly required to be carried out in stages, it is not envisaged that any trimming item would be measured to the surface of any intermediate level which is then to be excavated further although this should be made perfectly clear in the Preamble.

M6 For a simple square structure, the volume of excavation measured would be simply the

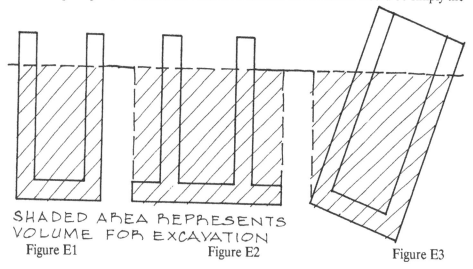

SHADED AREA REPRESENTS
VOLUME FOR EXCAVATION

Figure E1   Figure E2   Figure E3

volume of the structure below the Commencing Surface (Figure E1). However for irregular shapes the volume to be excavated is effectively the volume of the material displaced by the overall plan area multiplied by the net depth (Figures E2 and E3).

M7 The volumes referred to in this rule apply only to excavation work which would be physically submerged at the higher level of fluctuation, and not to work which is only affected by water. For example see Figure E4:-

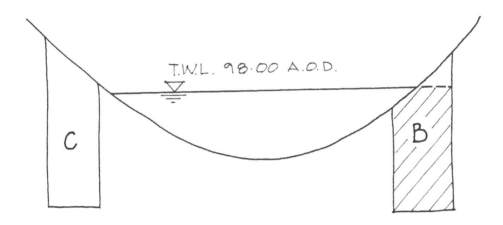

Figure E4

In Figure E4 the Original Surface of excavation A is above the top water level of the river and therefore this would not be indentified separately as being below a body of open water, even though the excavation would be quite 'wet' due to the close proximity of the river. The problems associated with this 'wet' excavation are deemed to be at the Contractor's risk, as are all problems associated with ground water. However, in excavation B the top water level is above most of the Original Surface of the excavation and the excavation represented by the shaded portion is therefore identified as being below a body of open water and measured separately.

M9 States that the volume for excavation in borrow pits shall be the volume of filling materials required. It is therefore obvious that there will not be any actual measurement of borrow pit excavation, but rather a totalling of all the net filling volumes which are to be obtained from the borrow pits. CESMM 3 requires borrow pit excavation to be given in maximum depth bands but this may prove difficult to determine unless the Engineer has specified the maximum depth to which borrow pit excavation can be taken. If the maximum depth is left to the discretion of the Contractor then it is recommended that a preamble is incorporated to this effect which would then allow the maximum depth bands to be omitted from the description of the measured item.

M10 Trimming of excavated surfaces is measured only to surfaces which are left in their excavated state.

M11 Preparation of excavated surfaces is measured to all other surfaces except those:

a) which are to receive any filling measured in accordance with E. 6 * *. It is not clear whether or not preparation is to be measured to surfaces which are to receive sub-bases measured in Class R. Because of the work involved in preparing the excavated surface prior to laying sub-base, it is recommended that it is measured, even though some sub-bases, such as DTp type 1 and 2 granular material are very similar to filling materials measured in accordance with E. 6 * *, and which would not normally warrant measurement. A preamble would have to be incorporated making this clear

b) which are to receive any of the landscaping items measured in accordance with E. 8 * *.

Strictly speaking an area which has been excavated and planted with trees should have trimming measured across its surface with a deduction for the area of the tree pit. This is thought to be too pedantic and it is recommended that the trimming be measured across the whole area. A preamble should make this clear

c) for which formwork has been measured to concrete which is contained within the excavation.

M13 Double handling of excavated material is something which is very rarely required by the Engineer. It should only be measured where the Engineer specifically requests that the material should be temporarily stockpiled in a place other than its final resting place.

An interesting situation can arise in the actual measurement of the volume to be double handled which is defined as the void formed in the temporary stockpile. Take the example shown in Figure E5:

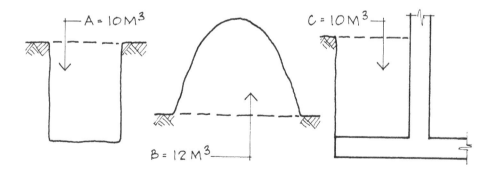

Figure E5

The net volume excavated for A is 10 m3 which is expressly required to be deposited in temporary stockpile B, which after bulking has a volume of 12 m3. The material is required for backfilling to a structure at a later date which has a net volume of 10 m3. For the backfilling two items should be measured. The first is 'filling to structures with excavated material' of net volume of 10 m3, and the second is 'double handling of excavated material'. This volume is the volume of the void formed in the stockpile and this in theory should be 12 m3 because the material should compact to its original volume. Detailed site measurements of stockpiles should be made immmediately prior and subsequent to any excavation in them to overcome discrepancies due to the settlement of the material under self weight.

M16 Basically means that backfilling to working space and overbreak is not measurable.

M17 Accurate levels must be recorded before any temporary roads are placed because it is these levels which the calculation of permanent filling will be based on.

M18 States that only the depth exceeding the first 75mm of filling lost due to settlement or penetration into the material underneath is measured.

In the authors' experience this is a very difficult rule to administer. Unless accurate instrumentation is used then it is almost impossible to check whether the 75mm has been exceeded. It can be done by comparing the volume of the fill on site with the theoretical quantity placed. However, even if the quantities placed on site have been accurately surveyed the use of notional factors to convert tonnes to cubic metres to calculate the theoretical quantity makes the overall calculation unsatisfactory and it suggested that unless there is a strong reason for retaining it, this rule is written out.

M19 Only refers to Classes E and T. However filling materials could be obtained from other classes particularly D, I, J, K and P and these should not automatically be discounted if, due to the nature of the work, these classes produced large quantities of suitable filling materials. This measurement rule should be amended accordingly in the Preamble.

M20 This would normally be done by taking the weight of the rock and applying a
& conversion factor to calculate the volume (it is recommended that the conversion
M21 factor is stated in the description). Conversely the unit of measurement used could be changed to tonnes if it is known at tender stage that this is the method of remeasurement to be adopted. Although M20 specifically states rock, it could also equally apply to any filling, particularly hard materials.

M22 Trimming of filled surfaces is measured only to surfaces which are left in their filled state.

M23 Preparation of filled surfaces is measured to all other surfaces except those:

a) which are to receive any filling measured in accordance with E. 6 * *. It is not clear whether or not preparation is to be measured to surfaces which are to receive sub-bases measured in Class R. Because of the work involved in preparing the filled surface prior to laying sub-base, it is recommended that it is measured, even though some sub-bases, such as DTp type 1 and 2 granular material are very similar to filling materials measured in accordance with E. 6 * * and which would not therefore normally warrant measurement. A preamble would have to be incorporated making this clear

b) which are to receive any of the landscaping items measured in accordance with E. 8 * *.

Strictly speaking an area which has been filled and planted with trees should have trimming measured across its surface with a deduction for the area of the tree pit. This is thought to be too pedantic and it is recommended that the trimming be measured across the whole area. A preamble should make this clear

c) for which formwork has been measured to concrete which is contained within the filling.

M24 This rule states that in accordance with general measurement practice, the additional laps of geotextile materials shall not be measured.

Definition Rules

D1 A definition of rock has to be given in the Preamble by the Engineer. Artificial hard materials not exposed at the Commencing Surface would usually comprise mass concrete, reinforced concrete and masonry. Artificial hard materials exposed at the Commencing Surface would normally comprise mass concrete, reinforced concrete, masonry and hard pavings. Bases to hard paving and loose hard fill materials such as hardcore would not normally come under the definition of artificial hard materials and should be included in the Contractor's rate for digging out 'normal' materials.

D2 Examples of cutting excavation are given in Figures E6 and E7:-

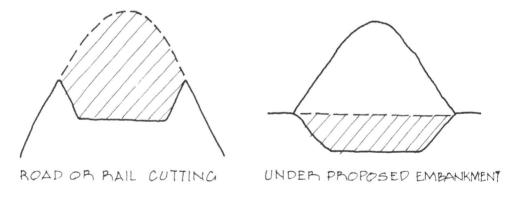

ROAD OR RAIL CUTTING   UNDER PROPOSED EMBANKMENT

Figure E6   Figure E7

Paragraph 5.21

Although this paragraph appears in Section 5, the last sentence is particularly relevant to excavation work and is discussed fully here.

This paragraph should be read very closely in association with Additional Description Rule A4, Measurement Rule M5 and Paragraphs 1.12 and 1.13 in Section 1 and can be best explained by the following example (see Figure E8):

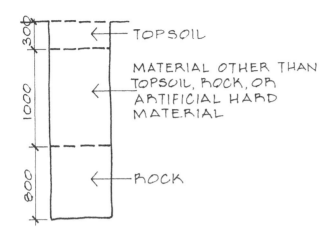

In this particular instance there would be three distinct measurement items. These are:-

1. Excavation of topsoil.

2. Excavation of material other than topsoil, rock, or artificial hard material.

3. Excavation of rock.

Figure E8

The depth classification for all these three items would be 2-5m. This is because the second paragraphs of Definition 1.12 and 1.13 make it clear that although the excavation is carried out in three separate stages of materials, the Commencing Surface and the Excavated Surface for all three items are the same i.e. the top of the void and the bottom of the void.

The key factor is not the depth of the individual layers, but the overall depth of the void itself. However, if the work was expressly required to be done in stages in accordance with M5 and the material was the same throughout, then this would create two measured items for a similar size void each with its own depth classification, because each would have its own Commencing Surface and Excavated Surface in accordance with Additional Description Rule A4 (see Figure E9).

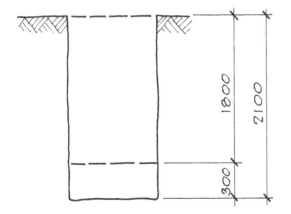

1. Excavation 1-2m deep, Excavated Surface 300mm above above Final Surface.

2. Excavation 0.25-0.5m deep; Commencing Surface 300mm above Final Surface or 1800mm below Original Surface.

Figure E9

Confusion could arise if, by coincidence, the lower half of the excavation in the example was a different material such as rock. Should the depth classification be from the top of the void to the bottom for both measured items in accordance with the second paragraph of Definition Paragraph 1.13 or would this be superseded by Measurement Rule M4 and separate depth classifications be applicable because each item has its own Commencing Surface and Final Surface?

It is apparent that for a similar sized excavation which would require similar plant that entirely different measurement items could be created depending on whether Measurement Rule M5 is applicable. However, because M5 would not commonly be used, the main factor is generally the overall void depth and if Measurement Rule M5 does apply then a preamble should be incorporated stating exactly which measurement philosophy has been adopted.

Notwithstanding the above, where excavation of individual layers of differing materials occurs over very large areas, as opposed to in 'holes' it may be more relevant to state the depth of the individual layers as opposed to the overall depth of the excavation. If this is the case then a suitable preamble should be incorporated.

D4 Normally to tips supplied by the Contractor unless otherwise stated in item descriptions.

D7 The constraints governing the use of excavated rock as filling material would have to be clearly stated by the Engineer in the Specification. Filling materials should be described as rock only where expressly required by the Engineer. This obviates potential problems on site where a Contractor may choose as filling material excavated rock which is suitable, but not expressly required by the Engineer.

D8 The second part of this rule states that bulk filling should always be measured under items E. 6 1-3 * and not E. 6 4 * even though it is deposited and compacted in layers of constant depth.

Coverage Rules

C1 Services in this context include redundant drains and sewers. Where excavation has to be carried out around existing live services, it is suggested that a further coverage rule is included to cover this. When it is anticipated that the incidence of live services in an excavation will be of major cost significance, the item descriptions should state that the excavation is carried out around live services. The other main item which is not stated here but which would normally be included in the rates for excavation is the disposal of water.

C2 This states that the removal and replacement of overburden and unsuitable material is deemed to be included in the borrow pit excavation. This is acceptable provided that the Contractor is supplied with sufficient detailed ground investigation information over the borrow pit area. If this is not possible it is felt that items for removal and replacement of overburden should be measured separately and a suitable preamble incorporated to override this rule.

C4 It may also be necessary to extend this rule to include other items such as root dip, insecticide, weed killer, ameliorates and similar items.

Additional Description Rules

A1 Under normal circumstances reference would be given in the item description to the drawings which would define precisely the location and limits of excavation by dredging.

A2 This would normally form part of a heading under which would appear all the relevant excavation items.

A4 See Paragraphs 1.12, 1.13 and 5.21 and Measurement Rule M5.

A6 This important new rule states that where hand digging is 'expressly required', it must be identified separately.

A8 This means that all trimming and preparation is between 0 and 10 degrees to the horizontal unless otherwise stated.

A9 This could be achieved by reference to the drawings or by providing a detailed description. Where the precise location is not known the item description should state haul distances.

A10 It is necessary to state the precise nature of imported filling material or refer to the Specification.

A11 There should be two separate measured items for filling with identical material which has two different compaction requirements and not one item with two compaction requirements specified. This adheres to the general philosophy of CESMM 3 that different requirements should be measured separately.

A13 This rule mainly refers to items such as pitching or beaching which uses large stone particles of specified size. It is common practice to refer to the Specification where this rule has to be complied with.

ITEM MEASUREMENT

E. 1-4 * *   Excavation Work

Applicable to all items

Divisions   -

Rules   M1   Measure the work net.

          M5   Measure separate items for separate stages of excavation but *only* where they are expressly required.

          A2   Where excavation is below a body of open water identify it in the item descriptions.

A4   State the Commencing Surface and Excavated Surface if applicable.

Generally   See also Rules M2, M6, M7, M8, D1, C1.

E.1 * 0   Excavation by dredging

Divisions   Dredging is measured in cubic metres stating the type of material to be excavated. State the type of any artificial hard material. It is not necessary to state the maximum depth in bands.

E.1 * 0   Excavation by dredging continued

Rules   M4   State the alternative method of measurement if not done by soundings.

A1   State the location and limits if not clear or refer to a Drawing.

Generally   See also Rule M3.

E. 2 * 0   Excavation for cuttings

Divisions   Excavation for cuttings is measured in cubic metres stating the type of material to be excavated. State the type of any artificial hard material. It is not necessary to state the maximum depth in bands.

Rules   -

Generally   See also Rule D2.

E. 3 * *   Excavation for foundations

Divisions   Excavation for foundations is measured in cubic metres stating the type of material to be excavated and the maximum depth of the excavation as classified. State the type of any artificial hard material. The actual maximum depth of the excavation must be stated where this exceeds 15 metres.

Rules   A3 State the location and limits of the excavation where this would otherwise be uncertain or refer to a drawing.
State when excavating around pile shafts.
State when excavating for underpinning.

Generally   See also Paragraph 5.21.

E. 4 * *   General excavation

Divisions   General excavation is measured in cubic metres stating the type of material to be excavated and the maximum depth of the excavation as classified. State the type of any artificial hard material. The actual maximum depth of the excavation must be stated where this exceeds 15 metres.

Rules   A5 State when general excavation is in borrow pits.

             A6 State separately any hand digging that is expressly required.

             M9 Measure the volume of excavation in borrow pits by totalling the net
                 volume required for filling.

Generally   See also Paragraph 5.21 and Rules D3, C2.

E. 5 * *   Excavation ancillaries

Applicable to all items

Divisions   -

Rules   M1 Measure the work net.

             A7 State when items are in connection with excavation by dredging.

Generally   See also Rule D5.

E. 5 1&2 *   Trimming and preparation of excavated surfaces

Divisions   Trimming and preparation of excavated surfaces are measured in square
             metres stating the type of material as classified. State the type of any artificial
             hard material.

Rules   A8 State the angle of inclination as classified.

Generally   See also Rules M10, M11.

E. 5 3 *   Disposal of excavated material

Divisions   Disposal of excavated material is measured in cubic metres stating the type of
             material as classified. State the type of any artificial hard material.

Rules   A9 State the location of the disposal area for on-site disposal of materials

Generally   See also Rules M12, D4.

E. 5 4 *   Double handling of excavated material

Divisions   Double handling of excavated material is measured in cubic metres stating the
             type of material as classified. State the type of any artificial hard material.

Rules   M13 The volume measured is the void formed in the temporary stockpile.

Generally   -

E. 5 5 0 Dredging to remove silt

Divisions   Dredging to remove silt is measured in cubic metres.

Rules   -

Generally   See also Rule M14.

E. 5 6 0 Excavation of material below the Final Surface and backfilling

Divisions   Excavation of material below the Final Surface and backfilling is measured in cubic metres. State the type of backfilling material.
                 This item is generally used for excavation of 'soft spots' below the formation level of a structure. The rate inserted by the Contractor for soft spot excavation covers excavation to any reasonable depth. Should the Contractor encounter soft material which was of an unreasonable depth, then he may have recourse through the Contract for additional payment. It is not clear how the excavated material should be dealt with. It could be treated under E. 5 3 * (disposal of excavated material) or the disposal could be included in the item description and covered by a preamble.

Rules   -

Generally   -

E. 5 7&8 0 Timber and metal supports left in

Divisions   Timber and metal supports left in are measured in square metres.

Rules   -

Generally   See also Rule M15.

E. 6 * * Filling

Applicable to all items

Divisions   -

Rules   M1   Measure the work net.

                 M17   Do not deduct the volume of any temporary roads approved by the Engineer which are subsequently to be incorporated into the permanent filling.

                 M18   Measure additional filling necessitated by settlement or penetration into underlying material only were it exceeds 75mm.

E.6** Filling (cont'd)

A10   State the type of material when it is imported.

A11   State the compaction requirements where more than one type of compaction is required to any one filling material.

A12   State the rate of deposition of material where this is limited.

Generally   Particle or piece sizes of filling material should also be stated where this would otherwise not be clear from the Drawings or Specification (e.g. for pitching etc.).

See also Rules M19, M20, M21, D6, D7, C3.

E. 6 1* Filling to structures

Divisions   Filling to structures is measured in cubic metres stating the type of material. As well as backfilling around structures this item would also cover filling items inside structures, e.g. stone filling to transformer pits or filter material inside filter buildings in water treatment works. Where material is contained inside buildings, it would be prudent to state this in item descriptions.

Rules   M16   Only measure filling around structures for which excavation has been measured in accordance with M6.

Generally   -

E. 6 2&3 *   Filling to embankments and general filling

Divisions   Filling to embankments and general filling is measured in cubic metres stating the type of material as classified.

Rules   -

Generally   -

E. 6 4 *   Filling to stated depth or thickness

Divisions   Filling to stated depth or thickness is measured in square metres stating the type of material and the depth or thickness.

Rules   A13   State the type of material particularly the particle size for items such as pitching, beaching, etc., where it is unclear.

A14   State the inclination of the filling as classified.

Generally   See also Rule D8.

E. 7 * *   Filling ancillaries

Applicable to all items

Divisions   -

Rules   M1 Measure the work net.

Generally   -

E. 7 1&2 * Trimming and preparation of filled surfaces

Divisions   Trimming and preparation of filled surfaces is measured in square metres stating the type of material as classified. State the type of any artificial hard material.

Rules   A15 State the angle of inclination as classified.

Generally   See also Rules D9, M22, M23.

E. 7 3 0 Geotextiles

Divisions   Geotextiles are measured in square metres.

Rules   A15 State the angle of inclination as classified.

               A16 State the type and grade of material.

               M24 Only measure net area of geotextiles excluding laps.

Generally   -

E. 8 * * Landscaping

Applicable to all items

Divisions   -

Rules   M1 Measure the work net.

Generally   See also Rule C4.

E. 8 1-3 * Turfing, hydraulic mulch grass seeding, other grass seeding

Divisions   Turfing, hydraulic mulch grass seeding and other grass seeding are measured in square metres.

Rules   A17 State where turfing is pegged or wired.

E. 8 1-3 * Turfing, hydraulic mulch grass seeding, other grass seeding (cont'd)

A18 State when applied to surfaces exceeding 10 degrees to the horizontal.

Generally   State the type of other grass seeding.

E. 8 4-6 0 Plants, shrubs and trees

Divisions   Plants, shrubs and trees are enumerated stating their species (usually a Latin name) and size (usually in accordance with B.S.3936).
For larger specimens it may also be prudent to specify whether it is bare root, containerised or root balled, because this could have a significant bearing on cost.

Rules   -

Generally   -

E. 8 7 * Hedges

Divisions   Hedges are measured in linear metres stating their species (usually Latin name), size (usually in accordance with B.S.3936) and spacing. State whether single or double row.
It may also be prudent to specify whether the hedge plants are bare root, containerised or root balled because this could have a significant bearing on cost.

Rules   M25 Measure hedges along their developed lengths.

Generally   -

STANDARD DESCRIPTION LIBRARY

Excavation by dredging; in Tilhouse dock; River Flow

Material other than topsoil, rock or artificial hard material

E. 1 2 0   generally   m3

Excavation for cuttings; located as drawing number EWKS/DWG/1

Topsoil

E. 2 1 0   generally; Excavated Surface underside of topsoil   m3

Excavation for cuttings; located as drawing number EWKS/DWG/1 (cont'd)

Material other than topsoil, rock or artificial hard material

| E. 2 2 0 | generally; Commencing Surface underside of topsoil | m3 |

Rock

| E. 2 3 0 | generally | m3 |

Excavation for foundations

Topsoil

| E. 3 1 2 | maximum depth 0.25-0.5m; Excavated Surface underside of topsoil | m3 |

Material other than topsoil, rock or artificial hard material

E. 3 2 3	maximum depth 0.5-1m	m3
E. 3 2 4	maximum depth 1-2m	m3
E. 3 2 5	maximum depth 2-5m; Commencing Surface underside of topsoil; Excavated Surface 2m above final surface	m3
E. 3 2 8	maximum depth 17.5m; Commencing Surface underside of topsoil	m3

Excavation for foundations; around pile shafts

Material other than topsoil, rock or artificial hard material

| E. 3 2 3.1 | maximum depth 0.5-1m | m3 |

General excavation

Topsoil

| E. 4 1 1 | maximum depth not exceeding 0.25m | m3 |
| E. 4 1 2 | maximum depth 0.25-0.5m; Excavated Surface underside of topsoil | m3 |

Rock

| E. 4 3 5 | maximum depth 2-5m | m3 |

CLASS E: EARTHWORKS

General excavation; within borrow pits

Materials other than topsoil, rock or artificial hard material

| E. 4 2 5 | maximum depth 2-5m | m3 |

Excavation ancillaries

Trimming of excavated surfaces

E. 5 1 2	material other than topsoil, rock or artificial hard material	m2
E. 5 1 2.1	material other than topsoil, rock or artificial hard material; vertical	m2
E. 5 1 3	rock; surfaces inclined at an angle of 10-45 degrees to the horizontal	m2

Preparation of excavated surfaces

E. 5 2 2	material other than topsoil, rock or artificial hard material	m2
E. 5 2 2.1	material other than topsoil, rock or artificial hard material; vertical	m2
E. 5 2 3	rock; surfaces inclined at an angle of 10-45 degrees to the horizontal	m2

Disposal of excavated material

E. 5 3 1	topsoil; to spoil heaps 100m from excavation	m3
E. 5 3 2	material other than topsoil, rock or artificial hard material	m3
E. 5 3 3	rock	m3

Dredging to remove silt

| E. 5 5 0 | generally | m3 |

Excavation of material below the Final Surface and replacement with

| E. 5 6 0 | hardcore | m3 |

Excavation ancillaries (cont'd)

Supports left in

E. 5 7 0	timber	m2
E. 5 8 0	metal	m2

Excavation ancillaries; in connection with excavation by dredging

Disposal of excavated material

E. 5 3 2.1	material other than topsoil, rock or artificial hard material	m3

Filling

To structures

E. 6 1 4	selected excavated material other than topsoil or rock	m3

To structures; compacted by 10 tonne vibrating roller

E. 6 1 4.1	selected excavated material other than topsoil or rock	m3

Embankments

E. 6 2 6	excavated rock	m3

General

E. 6 3 3	non-selected excavated material other than topsoil or rock; compacted in accordance with specification clause 8.1	m3

To 250mm depth or thickness

E. 6 4 1	excavated topsoil	m2
E. 6 4 2	imported topsoil; surfaces inclined at an angle of 10-45 degrees to the horizontal	m2
E. 6 4 7	imported granular material type A40; compacted by 10 tonne vibrating roller	m2

Filling ancillaries

Trimming of filled surfaces

E. 7 1 2	material other than topsoil, rock or artificial hard material	m2
E. 7 1 2.1	material other than topsoil, rock or artificial hard material; surfaces inclined at an angle of 45-90 degrees to the horizontal	m2
E. 7 1 3	rock	m2
E. 7 1 4	hardcore	m2

Preparation of filled surfaces

E. 7 2 2	material other than topsoil, rock or artificial hard material	m2
E. 7 2 2.1	material other than topsoil, rock or artificial hard material; surfaces inclined at an angle of 45-90 degrees to the horizontal	m2
E. 7 2 3	rock	m2
E. 7 2 4	hardcore	m2

Geotextiles; Terram; laid upon a surface at an angle to the horizontal

E. 7 3 0	not exceeding 10 degrees	m2
E. 7 3 0.1	10-45 degrees	m2
E. 7 3 0.2	45-90 degrees	m2
E. 7 3 0.3	vertical	m2

Landscaping

Turfing

E. 8 1 0	not exceeding 10 degrees	m2
E. 8 1 0.1	exceeding 10 degrees; pegged	m2

Plants; daffodil

E. 8 4 0	bulb	nr

Landscaping (cont'd)

Shrubs; Prunus spinosa 'Purpurea'

E. 8 5 0	35-45cm high	nr
E. 8 5 0.1	45-60cm high	nr

Trees; Acer platanoides; containerised

E. 8 6 0	feather	nr
E. 8 6 0.1	standard	nr
E. 8 6 0.2	heavy standard	nr

Hedges; privet 60 - 90cm high; 1 metre centres

E. 8 7 1	single row	m
E. 8 7 2	double row	m

Chapter 8

CLASS F: IN SITU CONCRETE

GENERAL TEXT

Principal changes from CESMM 1

This class was changed to comply with the requirements of BS5328 - Specifying Concrete, which supercedes CP110. Apart from this, which affects the provision of concrete items, the only change to the divisions was that the item for placing of concrete in bases, footings and ground slabs (F. 6-8 2 * in CESMM 1) was expanded to incorporate concrete in pile caps (F. 4-6 2 *).

Notes F5 and F8 in CESMM 1 were qualified to say that columns and piers integral with walls, and beams integral with slabs were measured with the walls and slabs respectively except where they were expressly required to be cast separately.

Note F10 of CESMM 1 was expanded to include the volume of joints (Measurement Rule M1).

Notes F10(d) and F11 of CESMM 1 were amended so that the maximum cross-sectional area for inclusion in the volume of concrete measured was changed from 0.005m2 to 0.01m2 (Measurement Rules M1(d) and M2).

It was a requirement that the placing of concrete against excavated surfaces which was expressly required should so be stated (Additional Description Rule A2). This was previously only implied from Note 1 of Class G.

Note F2 of CESMM 1 did not appear as a rule of CESMM 2, but as a note at the bottom of the page.

Principal changes from CESMM 2

The descriptions of types of concrete in the 1st Division together with the references to cement and aggregate in the 3rd Division have been re-arranged and altered to comply with the revisions to BS5328.

Definition Rule D2 now refers to a standard mix instead of a designed mix. D3 refers to a designed mix instead of a special prescribed mix and a new D4 rule defines a prescribed mix. The old definition Rules D4 to D8 are re-numbered to D5 to D9.

Measurement Rules

M1 States that the volume of concrete measured shall include the space occupied by the items listed including:

CLASS F: IN SITU CONCRETE

a) steel sections measured under Classes M and N as well as bar and fabric reinforcement

b) anchorages, ducts and tendons

c) bolts, precast or timber members, pipework and fittings, gullies, ducts, pipe supports and similar items

d) refers to chamfers or internal splays, an example of which is shown in Figure F1, and to fillets. This is somewhat confusing because in building terms a fillet is something which physically occupies a volume as opposed to being absent from it as a chamfer or external splay (see Figures F1 and F2) and there appears to be a conflict of terms. It is therefore recommended that the reference to fillets in M1(d) is ignored

e) pockets for bolts and holes for pipes which are expressly required to be cast into preformed openings

f) joint fillers, waterstops, sealing materials and dowels.

CHAMFER OR INTERNAL SPLAY (ABSENT FROM VOLUME).

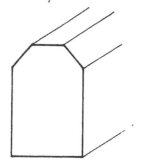

FILLET (OCCUPIES A VOLUME).

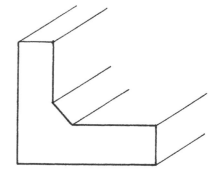

Figure F1   Figure F2

M2 States that the volume of concrete measured shall exclude the space occupied by nibs and external splays not exceeding 0.01m2. Examples of these are shown in Figure F3.

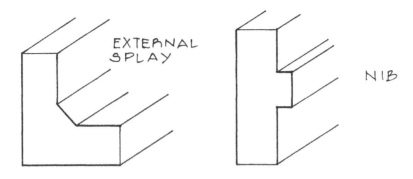

Figure F3

However, there are times when it is recommended that the volume of nibs below 0.01m2 in cross-sectional area should be measured. An example of this would be where they form a surface feature as shown in Figure F4.

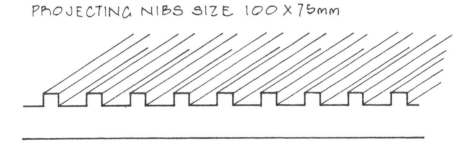

Figure F4

In this particular instance, it would be patently unfair to the Contractor to omit the measurement of all the nibs as shown and it is recommended that in cases such as these a suitable preamble is incorporated in accordance with paragraph 5.4.

M3 States that columns and piers integral with walls, and beams integral with slabs shall
and be measured and described as walls and slabs respectively, except where they are
M4 expressly required to be cast separately. It is difficult to envisage an example where columns or beams would be cast separately in this instance, as by definition they are integrated with the walls or columns. If they were cast separately they would be columns or beams within their own right and it would be more appropriate to measure them under F. 4-6 5&6 *. Although not specifically stated, it is recommended that these rules also apply to casings to metal sections because of their similarity to beams and columns. Due to their potentially complex nature it is felt that special beam sections should be measured separately in every case.

Definition Rules

D2 A standard mix is defined as on which is selected by the contractor from the list of constituent materials in BS5328 Part 2 Section 4.

D3 Designed mixes are those in which the required performance is specified by the Engineer in terms of grade and the Contractor is responsible for selecting the mix proportions to produce the required performance.

D4 Prescribed mixes are those in which the mix proportions are specified by the Engineer who is then responsible for ensuring that the proportions prescribed will produce a concrete which has the performance he requires.

D6 The thickness used for the classification of blinding is its minimum thickness, irrespective of whether its maximum thickness takes it into another classification. The classification applies to individual areas of blinding and areas of blinding which are separate from each other should be classified separately.

D7 States that where ground slabs, suspended slabs and walls have integral beams, columns, piers and other projections, then the thickness used for the classification shall be the thickness of the ground slabs, suspended slabs and walls excluding the thickness of the projections. As an integral beam or column becomes progressively wider it is not clear when a separate classification should be used (Figure F5).

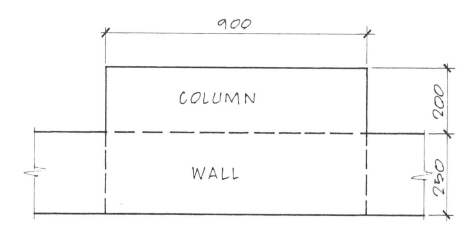

Figure F5

In Figure 5 it is not clear whether all the concrete shown should be measured as walls 150-300mm thick or whether there should be two classifications. The first would be 150-300mm thick for that length of wall beyond the sides of the column, and the second would be 300-500mm thick for the column itself plus that length of wall contained within the sides of the column. Although not strictly applicable, it is felt that the answer lies in Definition Rule D8.

This states that slabs less than 1 metre wide are classed as beams, and walls less than 1 metre wide as columns. There shall only one classification for integral beams or columns less than 1 metre wide which shall be the thickness of the slab or wall itself and for integral beams or columns 1 metre wide or over there shall be the additional classification which shall be the thickness of the slab or wall plus the thickness of the beam or column (Figure F6 A and B).

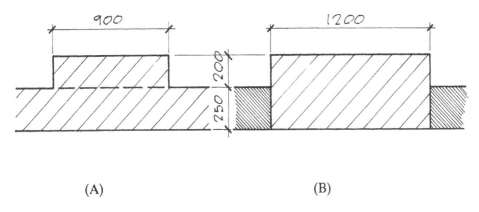

(A)   (B)

Both the column and the wall
measured as one item and
classified as walls 150-300mm
thick

Dark portion measured as
one item classified as walls
150-300mm thick

Light portion measured as
second item classified as
walls 300-500mm thick

Figure F6

D8 Suspended slabs less than 1 metre wide are classed and described as beams irrespective of their depth. Walls less than 1 metre wide are classed and described as columns irrespective of their width.

Although not specifically mentioned, it should be taken that this rule also applies to ground slabs less than 1 metre wide.

D9 Explained in a simpler manner this rule says that beams are classed as special beam sections only where the length of the special beam section exceeds one fifth of the total beam length. Examples of special beam sections would be I, T, and U section beams, and D8 also implies that non-rectangular beams are also classed as special beam sections.

Additional Description Rules

A1 States that the specification of concrete shall be given in item descriptions in accordance with BS5328 unless a mix reference is stated elsewhere in the Contract.
Unfortunately CESMM 3 only goes so far in the three divisions towards giving the necessary descriptive features and detailed reference must be made to the British Standard Specification.

A2 This rule enables the Contractor to allow for the additional wastage of concrete associated with casting it against uneven surfaces but situations can arise where large volumes of concrete cast within excavations only have one face cast against excavated surfaces (Figure F7).

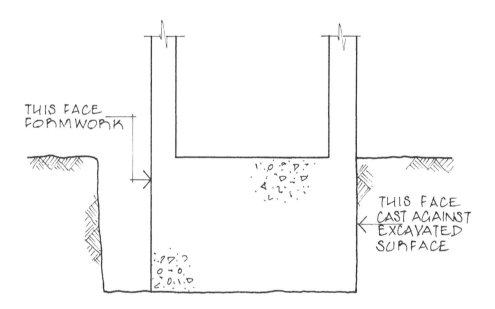

Figure F7

In this instance it would be inapplicable to describe the whole of the base slab as being 'cast against excavated surfaces'. For cases such as this it is recommended that a new measured item is created in square metres for the actual area of concrete in contact with the ground and a suitable preamble incorporated explaining the deviation from CESMM 3.

A3 It would be simpler to refer to a drawing or section for the cross-section sizes of special beam sections because stating anything more than the overall size would lead to excessive and over complicated descriptions.

A4 Other concrete forms include all those items not specified in the divisions. Examples of these would be stairs, lintols, machine bases and upstand kerbs. Alternatively the descriptive features contained within the divisions could be extended to cover similar items or a further coverage rule could be included to extend the item coverage of any one of the descriptive features.

ITEM MEASUREMENT

F. 1-3 * * Provision of Concrete

Applicable to all items

Divisions   -

Rules   M1 The volume of concrete measured shall include the space occupied by:

a) reinforcement and steelwork

b) prestressing components

c) items which are cast into concrete and which do not exceed 0.1m3 each

d) rebates, grooves, chamfers, throats, internal splays and similar items but only where their cross-sectional area does not exceed 0.01m2

e) any pockets, holes and similar items for which formwork has been or will be measured in accordance with G. 1-4 7 *

f) joint or joint components measured in accordance with G. 6 * *

M2 The volume of concrete measured shall exclude the space occupied by nibs, external splays or other projections of a similar nature protruding beyond the main body of the concrete providing they do not exceed 0.01m2 in cross-sectional area.

D1 Provision of concrete shall be classified in accordance with BS5328.

A1 The specification of concrete shall be in accordance with BS5328 unless otherwise stated in the Contract and shall be stated in item descriptions.

Generally   Although Additional Description Rule A1 states that the specification should be given in accordance with BS5328 it is not practical to give every single detail such as water/cement ratios, air content, etc. in item descriptions, and these should be included in the specification.

The Second Division of the class for design mixes is somewhat confusing in that it gives some of the compressive strength grades in BS5328 but not all of them. BS5328 was amended on 30 November 1990 and as far as CESMM is concerned the measurement of the provision of concrete has been rationalized and simplified. The taker-off will need to arm himself with a copy of the BS in cases where the design engineer does not specify precisely the concrete mix, strength, water/cement ratios etc.

In practice, the physical measurement of provision of concrete does not occur. At the end of the measurement of placing of concrete all the volumes calculated should be abstracted and totalled according to their various concrete grades or types. It follows therefore that when measuring the placement of concrete, two identical concrete components with different concrete grades or types must be measured separately.

See also Rules D2, D3, D4

F. 1 * *   Provision of standard mix concrete

Divisions   Provision of standard mix concrete is measured in cubic metres stating the compressive strength in accordance with Table 3 of BS5328, Section 4. State the type of cement to be used and the type and nominal maximum size of aggregate in accordance with the Table 4.

Rules   -

Generally   -

F. 2-3 * *   Provision of designed mix concrete

Divisions   Provision of designed mix concrete is measured in cubic metres stating the compressive or flexural strength grade in accordance with Tables 1 and 2 of BS5328 as appropriate. State the type of cement to be used and the type and required nominal maximum size of aggregate.

Rules   -

Generally   -

F. 4 * *   Provision of prescribed mix concrete

Divisions   Provision of prescribed mix concrete is measured in cubic metres stating the mix proportions in kilograms of each constituent. State the type of cement to be used and type and required nominal maximum size of aggregate.

Rules   -

Generally   -

F.5-7 * *   Placing of concrete

Applicable to all items

Divisions   State whether the concrete is mass, reinforced or prestressed.

Rules   M1 The volume of concrete measured shall include the space
occupied by:

  a) reinforcement and steelwork

  b) prestressing components

  c) items which are cast into concrete and which do not exceed
  0.1m3 each

  d) rebates, grooves, chamfers, throats, internal splays and
  similar items but only where their cross-sectional area does
  not exceed 0.01m2

  e) any pockets, holes and similar items for which formwork has
  been or will be measured in accordance with G. 1-4 7 *

  f) joint or joint components measured in accordance with
  G. 6 * *.

M2 The volume of concrete measured shall exclude the space
occupied by nibs, external splays or other projections of a
similar nature protruding beyond the main body of the concrete
providing they do not exceed 0.012m2 in cross-sectional area.

A2 State when concrete is expressly required to be cast against an
excavated surface (except for blinding).

Generally   When measuring placing of concrete the different concrete grades and types as
well as components should be kept separate. If for example, there are various
slabs with different concrete grades they should be measured separately so that
they can be readily identified for the calculation of the provision of concrete
items. When abstracted and billed for the placing of concrete items, they will
appear as one item because there is no requirement to differentiate the placing
of concrete by grades.

See also Rule D5.

F. 5 1 * Blinding

Divisions   Blinding is measured in cubic metres stating the thickness as classified.

Rules   -

Generally   Due to its nature blinding is often measured in square metres stating the actual thickness. If this is adopted, then a suitable preamble stating the deviation from CESMM 3 should be incorporated. Note that although CESMM offers the options, blinding would only be used as mass concrete not reinforced or prestressed.

         See also Rule D6.

F. 5-7 2 * Bases, footings, pile caps and ground slabs

Divisions   Bases, footings, pile caps and ground slabs are measured in cubic metres stating the thickness as classified.

Rules   M4 Include any integral beams or other similar projections in the measurement of ground slabs.

Generally   See also Rules D7, D8.

F.5-7 3 * Suspended slabs

Divisions   Suspended slabs are measured in cubic metres stating the thickness as classified.

Rules   M4 Include in the measurement of suspended slabs any integral beams or other similar projections.

Generally   See also Rules D7, D8.

F. 5-7 4 * Walls

Divisions   Walls are measured in cubic metres stating the thickness as classified.

Rules   M3 Include in the measurement of walls any integral columns and piers or other similar projections.

Generally   See also Rules D7, D8.

F. 5-7 5 1-5 Columns and piers

Divisions   Columns and piers are measured in cubic metres stating the cross-sectional area as classified. Tapering components which have a cross-sectional area which extends into more than one of the area classifications should be divided into the separate classifications.

Rules   M3 Any columns or piers which are integral with concrete walls should be classed and measured as walls.

Generally   See also Rules D7, D8.

F. 5-7 6 * Beams

Divisions   Beams are measured in cubic metres stating the cross-sectional area as classified. Tapering components which have a cross-sectional area which extends into more than one of the area classifications should be divided into the separate classifications.
    Special beam sections are measured in cubic metres.

Rules   M4 Any beams which are integral with slabs or suspended slabs should be classed and measured as slabs or suspended slabs. This rule should not apply to special beam sections.

    A3 State the cross-sectional dimensions of special beam sections or give a type or mark number or refer to the drawings.

Generally   See also Rules D7, D8, D9.

F. 5-7 7 1-5 Casings to metal sections

Divisions   Casings to metal sections are measured in cubic metres stating the cross-sectional area as classified. Tapering components which have a cross-sectional area which extends into more than one of the area classifications should be divided into the separate classifications.

Rules   M3 Any casings to metal sections which are integral with walls or slabs should be classed and measured as walls or slabs.

Generally   Casings to metal sections are similar items to columns and beams and it may be appropriate to expand the description to distinguish between casings which are akin to columns and those which are similar to beams.

    See also Rules D7, D8.

F. 5-7 8 0 Other concrete forms

Divisions   Other concrete forms are measured in cubic metres.

Rules   A4 Identify the component or form and either state the principal dimensions, give the type or mark number of the component for which dimensions are given on a drawing or refer to the drawing on which the component is detailed.

Generally   -

STANDARD DESCRIPTION LIBRARY

Provision of concrete

Ordinary prescribed mix to BS5328

F. 1 3 3	ST3; ordinary portland cement to BS12; 20mm aggregate to BS882	m3

Designed mix to BS5328

F. 2 4 7	grade C20; sulphate resisting cement to BS4027; 20mm aggregate to BS882	m3
F. 2 9 3	grade F3; rapid hardening cement to BS12; 20mm aggregate to BS882	m3
F.3 2 2	grade F4; ordinary portland cement to BS12; 14mm aggregate to BS882	m3

Prescribed mix to BS5328

F. 4 9 3	mix 1:3:6; ordinary portland cement to BS12; 20mm aggregate to BS882	m3

Placing of mass concrete

Blinding

F. 5 1 1	thickness not exceeding 150mm	m3

Bases, footings, pile caps and ground slabs

F. 5 2 2	thickness 150-300mm	m3
F. 5 2 4	thickness exceeding 500mm	m3

Placing of mass concrete (cont'd)

F. 5 2 4.1	thickness exceeding 500mm; placed against excavated surfaces	m3

Placing of reinforced concrete

Suspended slabs

F. 6 3 2	thickness 150-300mm	m3
F. 6 3 3	thickness 300-500mm	m3

Walls

F. 6 4 3	thickness 300-500mm	m3
F. 6 4 4	thickness exceeding 500mm	m3

Columns and piers

F. 6 5 4	cross-sectional area 0.25-1m2	m3
F. 6 5 5	cross-sectional area exceeding 1m2	m3

Casing to metal sections

F. 6 7 5	cross-sectional area exceeding 1m2	m3

Placing of prestressed concrete

Beams

F. 7 6 3	cross-sectional area 0.1-0.25m2	m3
F. 7 6 6	I Section; type B12	m3

Other concrete forms

F. 7 8 0	bridge units; mark number 6 on drawing number 1C/DWG/1	m3

Chapter 9

CLASS G: CONCRETE ANCILLARIES

GENERAL TEXT

Principal Changes from CESMM 1

Class G of the CESMM 1 contained several anomalies which were corrected.

It was clearly defined in the CESMM 2 when formwork should be measured (Measurement Rule M2). It was previously stated that formwork was not to be measured for blinding concrete (Note G2 of the CESMM 1) but this was deleted and it was necessary to state when formwork was required to blinding concrete (Additional Description Rule A3).

The measurement of rebates, grooves and fillets was clarified and these are measured lineally (G. 1-4 8 5&6).

It was an additional requirement to state when formwork is to be left in (Additional Description Rule A1).

Formwork to components of constant cross-section which are curved were to be described as such stating the radius (Additional Description Rule A6).

Note G11 in CESMM 1 referring to non-circular reinforcing bars was deleted.

A new item was created for the measurement of special joints in reinforcement (G. 5 5 0).

Additional coverage items were included to cover intermediate surface treatments and formwork to joints (Coverage Rules C2 and C3).

It was an additional requirement to state the position of inserts in relation to the concrete (Additional Description Rule A15). It must also be stated when inserts are expressly required to be grouted into preformed openings (Additional Description Rule A16).

An additional item was created for grouting under plates (G. 8 4 *).

Principal changes from CESMM 2

The references to the BS numbers for reinforcing bars have been amended: BS 4461 is deleted and those to BS 4449 updated. This affects item descriptions G. 5 1 *, G. 5 2 *, Definition Rule D6 and Additional Description Rule A9.

Open surface and formed surface joints are now stated in G.6, 1-4, 1-3 as 'average width' not 'width or depth'. This also affects the wording of Definition Rule D9.

Measurement Rule M2 is slightly reworded so that the blinding concrete is referred to as having width not depth.

The wording of Measurement Rule M6 has been altered to define more clearly the measurement of formwork over voids, inserts and the like.

The old Definition Rule D9 is omitted and the subsequent rules are re-numbered.

A new Measurement Rule M13 is included which states that treatments to surfaces formed by the Contractor at his own discretion shall not be measured. The old rules M13, M14 and M15 are renumbered.

Measurement Rules

M1 States that formwork shall be measured to the surfaces of concrete requiring temporary support during casting except where otherwise stated. It is not necessary to measure formwork to the surface of a concrete wall which is cast against an old brick surface because it is permanently supported by the brickwork and does not require temporary support.

M2 Formwork is not measured to the following:

a) edges of blinding concrete not exceeding 0.2m wide. This could pose a problem to the Contractor when pricing if there were two different thicknesses of blinding within the same contract at say 175mm and 225mm thick. When billed these would appear as a single item classed as 150-300mm thick. For the thinner layer the Contractor would either have to allow for formwork to the edges within his rates for the concrete itself, or for increased waste of concrete. For the thicker layer, formwork will have been measured and the Contractor's rates would not require the aforesaid allowances. This gives two different rates for the same billed item. One method of obviating this problem would be to bill them separately stating the actual thickness

b) joints or associated rebates or grooves. This means that the Contractor must allow for the cost of the formwork in his rates for the joints

c) surfaces formed at the discretion of the Contractor, e.g. day or construction joints

d) surfaces which are expressly required to be cast against an excavated surface. This applies to surfaces of any angle which the Engineer specifically requires the concrete to be cast against the ground

e) surfaces of concrete which are cast against excavated surfaces inclined at an angle less than 45 degrees to the horizontal.
   Upon first reading this rule would seem to be stating the obvious because one would not normally measure formwork to any concrete face which is cast against surfaces regardless of its angle of inclination.
   When read in a slightly different way this rule could have far reaching implications. It implies that formwork *is* measured to surfaces of concrete which are cast against excavated surfaces inclined at an angle *exceeding* 45 degrees to the horizontal. As stated previously, formwork is not measured to concrete faces which are expressly required to be cast against excavated surfaces because it is not possible, but this rule may be stating the contrary and could be the cause of a dispute. It is usually quite clear from the drawings where formwork should be measured or when concrete should be described as 'cast against excavated surfaces' and if not, then clarification should be sought from the Engineer. However, under no circumstances should both items be measured as this rule implies, and the

Preamble should make this clear.

M3 Formwork is always measured to upper surfaces of concrete which exceed 15 degrees to the horizontal. For surfaces up to and including 15 degrees to the horizontal, formwork may still be measured, but only if expressly required by the Engineer (see also Definition Rule D1 and Additional Description Rule A2).

M4 See Definition Rule D3.

M5 There could be instances when it may not be desirable to measure projections and intrusions in accordance with M5 and Definition Rule D5. An example of this could be where the concrete is to have a ribbed effect as shown in Figure G1.

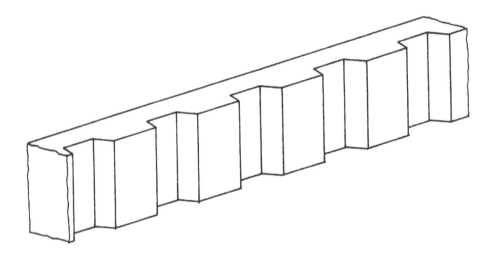

Figure G1

Should the concrete be measured to the external thickness and the ribs be measured as intrusions, or should it be measured to the minimum thickness and the ribs measured as projections? Regardless of this and whether the projections/intrusions exceed 0.01m2 it is felt that this would be better measured as a surface feature stating the size and spacing of the ribs.

M6 Formwork is measured across projections i.e. nibs and external splays not exceeding 0.01m2 in cross-sectional area, and intrusions i.e. rebates, grooves, internal splays, throats and chamfers not exceeding 0.01m2 in cross-sectional area.

Formwork is also measured across all inserts although there is no indication what the maximum area for inclusion should be. For instance some types of pipework may have cross-sectional areas of several square metres and it may not be desirable to include these in the measurement. It is also not clear whether or not the area of formwork should be deducted at the intersections of beams with the vertical faces of columns and walls. In both these cases guidance is taken from Definition Rule D3 and it is recommended that the deduction is made only when the area does not exceed 0.5m2.

M8 This means that the supports to top reinforcement must be measured. The Engineer must always ensure that the supports to top reinforcement are detailed and scheduled. If this is not done or if the Engineer wishes to leave these supports to the Contractor's discretion, then this rule should be omitted by way of a suitable preamble and Coverage Rule C1 should be amended accordingly.

M9 States that the additional fabric reinforcement in laps shall not be measured. This rule is quite clear although somewhat superfluous because all quantities are measured net from the drawings with no allowance for bulking, shrinkage or waste unless otherwise stated (Paragraph 5.18). Because this rule is specific about fabric reinforcement only, could it be argued that laps in bar reinforcement should also be measured? The answer of course is no, and this should be considered as being a 'belt and braces' rule only.

M10 This rule is to prevent the Contractor claiming for payment of day and construction joints which are done with the Engineer's permission but at the Contractor's own volition.

M11 No guidance is given on how the lengths of sealed rebates are to be measured. Because they are generally of a small cross-section it would be simpler if they were measured at their exposed faces.

M12 External jacking operations are not often used these days for prestressing operations. One of the few examples where they are still used is in post-tensioning runway slabs. Banks of jacks are placed in lines of pockets across the slab and the force applied. External jacking should be measured only once. If subsequent operations are required due to the Contractor's inefficiencies or incorrect work these should not be measured.

M13 This rule falls in line with the philosophy of rule M10 in that the Contractor will not be paid for work to surface treatment to concrete surfaces which are temporary and/or formed at his own discretion.

M14 Finishing of top surfaces is only measured where a specific separate operation is required in addition to the general concrete laying and levelling.

M15 It is recommended that this also applies to areas occupied by the bases of columns, pipe support plinths and the like.

M16 The rule means that it must be clearly defined by the Engineer when inserts, particularly items such as holding down bolts, are to be grouted into openings already formed or cast into the main body of concrete as it is poured (see also Definition Rule D3 and Additional Description Rule A16).

Definition Rules

D1   This can be more easily explained by way of a diagram as shown in Figure G2.

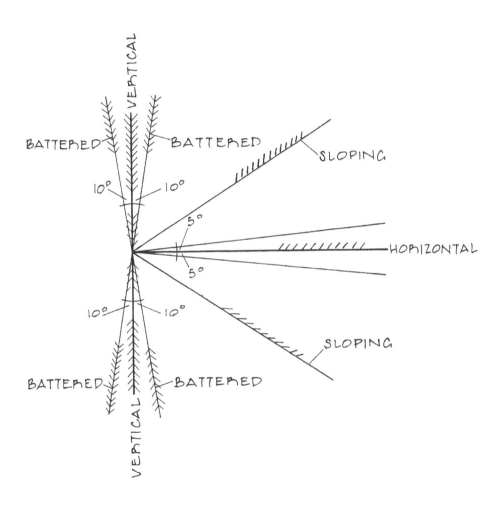

Figure G2

Figure G2 shows the various classifications of formwork. The shaded areas on the diagram indicate the position of the concrete. It can be seen that for vertical or battered classifications the concrete can be on either side of the formwork. However, for sloping and horizontal classifications, the concrete must be in the position shown otherwise it would have to be described as to upper surfaces (see also Additional Description Rule A2).

D3   Defines the criteria for measuring large and small voids. Voids are items such as sumps, permanent openings, temporary pockets and openings for the building in and grouting of inserts at a later date, but in this context the classifications are too broad. For example, the smallest classification for circular voids is up to 350mm diameter x 500mm deep. For building in a foundation bolt which may only be 100mm long x 12mm diameter this is an excessive size. Additional Description Rule A16 requires the actual sizes of openings for the grouting of inserts to be stated so it may also be prudent to state the actual size of the void as it is formed. If this approach is adopted a suitable preamble would need to be incorporated stating the deviation from CESMM 3.

Problems may also arise when trying to ascertain the depth of the void because D3 states that the depth is measured perpendicularly to the adjacent surface. Difficulties are encountered when sloping or tapering components are involved as shown in Figures G3 and G4.

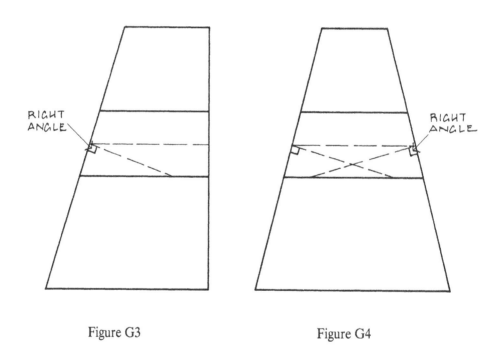

Figure G3   Figure G4

Both figures have two perpendiculars to the adjacent concrete surface as indicated by dotted lines. In Figure G3 the answer to the problem is straightforward because the correct depth is the one which is perpendicular to the vertical side and in this case the description could be amended to read 'average' depth. In Figure G4 neither perpendicular gives the correct overall depth of the void which is the length measured along the centre line running horizontally through the void. It is recommended that the words 'perpendicular to the adjacent surface of concrete' are not taken too literally and that situations such as these should be treated on their own merit.

D7 Although not actually stated, special joints in reinforcing bars would only be measured where expressly required by the Engineer and not when the Contractor fixes them of his own volition or where they are given only as an optional method of jointing to conventional lapping and tying.

D8   In its strictest sense the term 'formed joints' could mean any joints which are not horizontal. Measurement Rule M3 implies that concrete will generally only need support when its upper surface exceeds 15 degrees to the horizontal and it may be desirable to expand Rule D8 to incorporate the wording contained within Measurement Rule M3 with regard to the angle. Formed joints would then be any joints requiring temporary support over 15 degrees to the horizontal.

D10 'The outside face of anchorage' is the outside face of the anchorage itself and not the end of the bars or strands which may project beyond the anchorage.

D11 Inserts would include ties, bolts, pipes, fittings, conduits and the like. When building in substantial items such as large diameter pipework it is essential to ensure that the reinforcement is designed around it or that the item coverage is extended to include for cutting and lacing the reinforcement around the insert.
    One possible area for confusion here is whether a steel member encased in concrete measured and described as 'casing to metal sections' under F. 4-6 7 * should also be measured as an insert and measured under G. 8 3 *. One of the purposes of the measurement of inserts is to enable the Contractor to price for the additional cost of placing concrete around obstructions which he has not been able to include in his normal concrete rates. The other intention is to allow for the supply and installation of the insert. In the case of encased steelwork, the member will already have been included in the supply and fix item measured under Class M. The additional cost of placing the concrete around the obstruction will also have been measured under F. 4-6 7 * and there should be no necessity to also include an item under the inserts classification.

Coverage Rules

C1   All supports and tying wire except supports to top reinforcement are deemed to be included in the Contractor's rates (see Measurement Rule M8 also).

C2   Intermediate surface treatments would only include labours on the concrete such as trowelling. Any surface applied treatments such as bitumen paint or similar items would be described separately.

C3   This would include formwork in forming the groove for the sealant.

C6   It follows from this that the nature of the surface finish must be stated by the Engineer and any joints in the applied finished must be detailed in full.

C7   An example of those items for which the supply would also be included would be ties, bolts and the like. Items whose supply would be included under other classes would be pipes, ducts and similar items.

Additional Description Rules

A1 Although not specifically stated, formwork would be described as left in only when expressly required by the Engineer.

A2 Figure G5 shows when formwork is to upper surfaces. Between 0 and 15 degrees to the horizontal it is only measured where it is expressly required by the Engineer. Between 15 and 80 degrees to the horizontal it is always measured. Rule A2 states that it is not necessary to describe formwork to upper surfaces when it does not exceed 10 degrees to the vertical. This is because at this angle there would be very little difference in the supporting falsework whether it was to either upper surfaces or side surfaces. It is not possible to have descriptions 'battered upper surface formwork' (see also Measurement Rule M3 and Definition Rule D1).

A4 These are the simpler forms of curved formwork. Where curved formwork is more complicated it is recommended that item descriptions make reference to the drawings.

A5 The sizes of components of constant cross-section must be stated except for projections and intrusions which by definition are only a maximum of 0.01m2 in cross-sectional area.

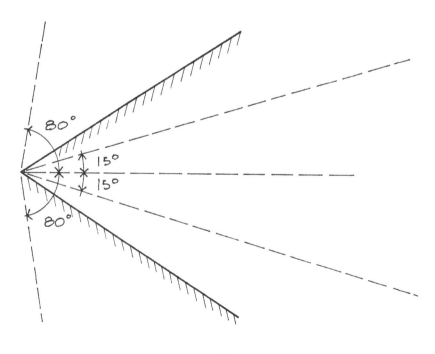

Figure G5

A7 These are complete lengths of steel without joints because the steel would normally be bent before any joints are made.

A10 It is further recommended that the spacing of the bars in both directions is given.

A12 The various British Standards covering the components of post-tensioned prestressing are 5896, 4757, 4486 and 4447. Most post-tensioned prestressing these days is usually done by one of the many proprietary systems available such as BBRV, CCL, PSC Freyssinet, and Macalloy.

A15 Inserts are classed according to their position in the concrete as shown in Figure G6.

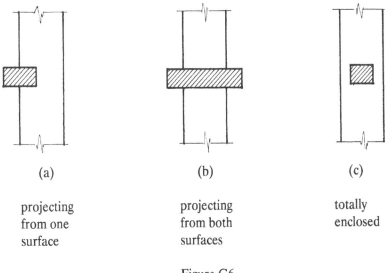

(a)	(b)	(c)
projecting from one surface	projecting from both surfaces	totally enclosed

Figure G6

It is feasible to have an insert projecting from more than two surfaces and item descriptions should so state.

A16 Inserts would only be described as 'grouted into preformed openings' when the opening has also been measured under G.1-4 7 * (see also Measurement Rule M16 and Definition Rule D3).

A17 It is not necessary to state the thickness of materials for grouting under plates because this will be dependent upon the accuracy with which the Contractor has laid the foundation base and erected the steelwork.

ITEM MEASUREMENT

G. 1-4 * * Formwork

Applicable to all items

Divisions State whether formwork is rough, fair, of other stated finish or of stated surface feature. The Specification will often give specific references for formwork types, i.e. Type 1, Type 2, etc., and these should be used in preference to CESMM 3 classifications.

Stated surface features are those in which the concrete is given a particular shape or feature by the formwork itself and not those for which further treatment is necessary afterwards, (these would be measured under G. 8 2 *). An example of a stated surface feature would be as shown in Figure G7.

This ribbed effect could be measured as projections or intrusions (G. 1-3 8 5 and 6) if the ribs did not exceed 0.01m2 cross-sectional area, or as plane surfaces (G. 1-3 1-4 *) (measured as the developed area) if the ribs exceeded 0.01m2, or as a stated surface feature regardless of the cross-sectional area. If measured as a stated surface feature it must be made clear whether it is the developed area that has been measured or the undeveloped area. It would also be necessary to refer to a drawing to enable the Contractor to ascertain the work involved precisely.

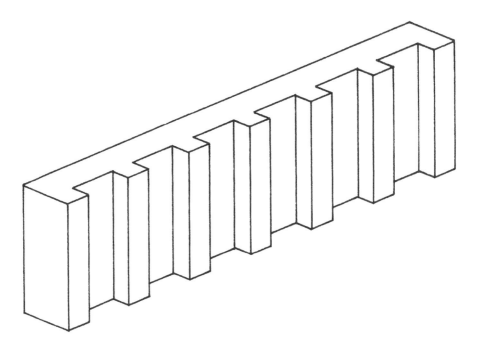

Figure G7

It is also quite common for a high grade concrete finish to be obtained by specifying a lower grade formwork, with further treatment after the formwork has been struck to obtain the higher grade. In instances such as this, the recommended measurement technique would be to specify the higher grade finish in the actual item description and to incorporate a coverage rule stating that the rate for the item should include for the lower grade formwork plus the additional work in achieving the final finish.

Rules   M3   Measure and describe formwork to upper surfaces of concrete inclined
          and   between 15 and 80 degrees to the horizontal. Measure formwork to other
          A2   upper surfaces only where it is expressly required by the Engineer.

A1 State when formwork is to be left in.

A3 State when formwork is to blinding concrete.

M6 Include the area of formwork obscured by forms for large and small voids measured under G. 1-4 7 *, for projections and intrusions measured under G. 1-4 8 5&6 and for all inserts regardless of size.

Generally   It is not clear whether curved formwork which is also to upper surfaces should be so described. It would be difficult to apply Rule A2 to curved formwork because of the constantly changing angle and it is recommended that A2 should not apply to curved formwork. A suitable preamble would have to be incorporated to this effect.

See also Rules M1, M2, D1, D2.

G. 1-4 1-5 * Horizontal, sloping, battered, vertical and formwork curved to one radius in one plane

Divisions   Horizontal, sloping, battered, vertical and formwork curved to one radius in one plane is measured in linear metres up to 0.2m wide and square metres over 0.2m wide, stating the width as classified. The width is taken as the narrowest width.

Rules   A4 State the radius of curved formwork.

Generally   There are two instances which can occur which should be explained in the classification of formwork in accordance with the Third Division. Firstly where narrow widths of formwork are caused by voids, projections and intrusions and secondly by joints. Measurement Rule M6 states that formwork is measured across voids which are measured as large and small voids and across projections and intrusions not exceeding 0.01m2 cross-sectional area. Narrow widths of formwork caused by these items are not measured separately but where narrow widths are caused by voids which exceed the sizes stated in Definition Rule D3 or by projections and intrusions which exceed 0.01m2 cross-sectional area they should be measured separately.

All narrow widths of formwork between joints in concrete should be measured separately.

G. 1-4 6 0 Other curved formwork

Divisions   Other curved formwork is measured in square metres. It is not necessary to state the width according to any classification.

Rules   A4 State when formwork is curved to one radius in two planes (spherical) and give the radius.

State when formwork is curved to varying radii (conical) and give the minimum and maximum radii.

Generally   -

G. 1-4 7 * Formwork to voids

Divisions   Formwork to voids is enumerated stating whether they are large or small voids calculated in accordance with Definition Rule D3. State the depth as classified.

Rules   -

Generally   See also Rules M4, D3.

G. 1-4 8 * Formwork to components of constant cross-section

Divisions   Formwork to components of constant cross-section is measured in linear metres stating the type of component.

Rules   A5 Except for projections and intrusions, state the principal cross-sectional dimensions and mark number, location or other unique identifying feature.

A6 State when components are curved and give the radius.

Generally   Items should be measured under this category whenever possible because it provides the Contractor the best information with regard to the number of formwork uses. Components measured here must be of exactly the same cross-sectional shape throughout their length. Problems may arise at the ends of components of constant cross-section. For instance formwork would have to be measured separately to the end of a free-standing wall that had been measured as a component of constant cross-section.

It is not necessary to give the location or dimensions of projections or intrusions as the maximum cross-sectional size of these is 0.01m2 and has small cost significance.

See also Rules M5, D4, D5.

G. 5 1-4 * Bar Reinforcement

Divisions   Bar reinforcement is measured in tonnes stating the type of material and the diameter. Materials for plain round and high yield steel bars are defined in BS4449. Stainless steel bars and bars of other material should state the quality.

Rules   M8 Measure the mass of steel supports to top reinforcement.

               A7 State the lengths of bars to the next highest multiple of 3m where they
                  exceed 12m before bending.

Generally   Where total quantities are small it may be prudent to take the measurement
               to one place of decimals.

               See also Rules M7, D6, C1.

G. 5 5 0   Special joints in reinforcement

Divisions   Special joints in reinforcement are enumerated.

Rules   D7 Special joints are defined as being welded, swaged or screwed sleeve
                  joints.

               A8 State the type of joint and type and size of reinforcing bar.

Generally   Measure special joints only where they are expressly required.

G. 5 6&7 *   Fabric reinforcement

Divisions   Fabric reinforcement is measured in square metres stating the type of
               material.

Rules   M9 Do not measure fabric in laps.

               A9 Fabric to BS4483 shall state the reference number in accordance with
                  table 1 or 2 of the Standard. Note that BS4483 does not have any
                  reference in the 7-8kg/m2 classification.

               A10 State the size and nominal mass per square metre for other types of fabric
                  reinforcement.

Generally   Although not a specific requirement it is recommended that the spacing of the
               bars is also given.

G.6 * *   Joints

Applicable to all items

Divisions   -

Rules   M10 Measure only when expressly required.

             M11 Measure over widths occupied by rebates, grooves, fillets and waterstops.

             A11 State the dimensions and nature or type of the components and materials involved. This rule applies to the components of the joints only.

Generally   See also Rules D8, D9, C2, C3.

G. 6 1-4 * Open surface and formed surface joints

Divisions   Open surface and formed surface joints are measured in square metres stating the width as classified. State the actual width where it exceeds 1m. State the nature of the material in filled joints.

Rules   -

Generally   -

G. 6 5&6 * Waterstops

Divisions   Waterstops are measured in linear metres stating the type and the actual width. More commonly a manufacturer's type and reference will be given.

Rules   M11 Measure along centre line with no deductions for intersections or other fittings.

Generally   See also Rule C4.

G. 6 7 0 Sealed rebates or grooves

Divisions   Sealed rebates or grooves are measured in linear metres.

Rules   -

Generally   It also may be prudent to state whether the groove is horizontal or vertical. In other words whether it is possible to pour sealant into the groove or if it has to be placed in by trowel or gun.

G. 6 8 * Dowels

Divisions   Dowels are enumerated stating whether they are plain, greased, sleeved or capped.

Rules   -

Generally   -

G. 7 * *   Post-tensioned prestressing

Applicable to all items

Divisions   -

Rules   A12 Identify the component to be stressed.

Generally   See also Rules M12, C5.

G. 7 1-4 *   Prestressing with tendons

Divisions   Prestressing with tendons is measured by enumeration of the tendons. State whether horizontal or inclined and whether in in situ or precast concrete. State the length of the tendon as classified.

                Strictly speaking there is no such thing as post-tensioned prestressed in situ concrete. Post-tensioning can only be applied to a unit which has previously been cast and would always be a precast unit. The classifications are left as they are to distinguish between tendons to units which have been measured under Class F: In Situ Concrete and Class H: Precast Concrete.

Rules   A12 State the composition of the tendon and type of anchorage.

Generally   See also Rules D10, D11.

G. 7 5 0   Prestressing by external jacking only

Division   External jacking operations are enumerated.

Rules   -

Generally   -

E. 8 1&2 *   Finishing of top and formed surfaces

Applicable to all items

Divisions   -

Rules   M15 Do not deduct openings less than 0.5m2 each.

Generally   -

G. 8 1 * Finishing of top surfaces

Divisions   Finishing of top surfaces is measured in square metres stating the type of finish required.

Rules   A13 State the materials, thicknesses and surface treatments of granolithic and other stated applied finishes.

Generally   See also Rules M13, M14, C6.

G. 8 2 * Finishing of formed surfaces

Divisions   Finishing of formed surfaces is measured in square metres stating the type of treatment required.

Rules   -

Generally   See also G. 1-4 * * Formwork, Rule M13

G. 8 3 * Inserts

Divisions   Linear inserts are measured in linear metres. All other inserts are enumerated.

Rules   M16   Measure formwork to voids separately for inserts which are expressly required to be grouted into preformed openings.

          C7   State when the supply of inserts is included elsewhere.

          A14   Identify the insert and state the principal dimensions.

          A15   Distinguish between inserts which are at the surface, project from one surface, project from both surfaces or which are totally enclosed.

          A16   Where inserts are to be grouted into preformed openings state the size of the opening and the composition of the grout.

Generally   See also Rule D11.

G. 8 4 * Grouting under plates

Divisions   Grouting under plates is enumerated stating the area as classified. The actual area must be stated where this exceeds 1m2.

Rules   A17   State the materials used as grout.

Generally   -

STANDARD DESCRIPTION LIBRARY

Formwork rough finish

Plane horizontal

G. 1 1 1	width not exceeding 0.1m	m
G. 1 1 1.1	width not exceeding 0.1m; left in	m

Plane vertical

G. 1 4 2	width 0.1-0.2m; to blinding concrete	m
G. 1 4 5	width exceeding 1.22m	m2

Formwork type 1 finish

For voids

G. 3 7 1	small void depth not exceeding 0.5m	nr
G. 3 7 7	large void depth 1-2m	nr

For components of constant cross-section

G. 3 8 1	beams 200 x 600mm; reference B2	m
G. 3 8 1.1	beams 200 x 600mm; reference B3; curved to mean radius 10m	m
G. 3 8 5	projections	m
G. 3 8 6	intrusions	m

Formwork troughed surface feature; as drawing number CA/DWG/1

Plane sloping

G. 4 2 3	width 0.2-0.4m	m2
G. 4 2 3.1	width 0.2-0.4m; to upper surfaces	m2

Plane curved to 2m radius in one plane

G. 4 5 3	width 0.2-0.4m	m2

Formwork troughed surface feature; as drawing CA/DWG/1 (cont'd)

Curved to more than one radius

G. 4 6 0.1	to 2m radius in two planes	m2
G. 4 6 0.2	maximum radius 2m, minimum radius 1m	m2

Reinforcement

Plain round bars to BS4449

G. 5 1 1	nominal size 6mm	t
G. 5 1 1.1	nominal size 6mm; 15-18m long	t

Deformed high yield steel bars to BS4449

G. 5 2 3	nominal size 10mm	t
G. 5 2 3.1	nominal size 10mm; 18-21m long	t

Special joints

G. 5 5 0.1	welded joints; 6mm diameter mild steel bar	nr
G. 5 5 0.2	swaged joints; 12mm diameter high yield steel bar	nr

Steel fabric to BS4483

G. 5 6 2	nominal mass 2-3kg/m2; reference A142	m2

Fabric of stainless steel

G. 5 7 2	nominal mass 2.5kg/m2; 6mm bars at 100mm centres one way; 8mm bars at 150mm centres other way	m2

Joints

Open surface plain

G. 6 1 1	average width not exceeding 0.5m	m2

Formed surface with 25mm thick Korkpak

G. 6 4 2	average width 0.5-1m	m2

Joints (cont'd)

Plastic waterstop; Expandite

| G. 6 5 1 | average width 150mm; centre bulb type | m |

Sealed rebates or grooves

| G. 6 7 0 | 25 x 25mm; filled with 2 part polysulphide sealant | m |

Dowels

| G. 6 8 1 | plain; 25mm diameter x 250mm long mild steel bar | nr |

| G. 6 8 2 | capped; 12mm diameter x 150mm long mild steel bar; 75mm cardboard cap | nr |

Post-tensioned prestressing

Horizontal internal tendons in in situ concrete; seven wire standard strand to BS5896; relax class 1; nominal diameter 11mm; concrete cone anchors to BS4447

| G. 7 1 3 | length 7-10m; beam reference B2 | nr |

Concrete accessories

Finishing of top surfaces

| G. 8 1 1 | wood float finish | m2 |

| G. 8 1 4 | granolithic finish; 1:1:2; 50mm thick; steel trowelled | m2 |

Finishing of formed surfaces

| G. 8 2 1 | aggregate exposure using retarder | m2 |

| G. 8 2 3 | point tooling | m2 |

Inserts

| G. 8 3 1 | plastic conduit; 100mm diameter; totally enclosed | m |

| G. 8 3 1.1 | plastic conduit; 100mm diameter; totally enclosed; supply included elsewhere | m |

Concrete accessories (cont'd)

G. 8 3 2	mild steel pipe; 150mm diameter x 150mm long; projecting from two surfaces	nr
G. 8 3 2.1	galvanised mild steel foundation bolt 12mm diameter x 250mm long; projecting from one surface; grouting into preformed openings 100 x 100 x 150mm deep with cement mortar (1:3)	nr
	Grouting under plates	
G. 8 4 1	area not exceeding 0.1m2; cement mortar (1:3)	nr

Chapter 10

CLASS H: PRECAST CONCRETE

GENERAL TEXT

Principal changes from CESMM 1

This class was not changed significantly.

Note H7 of CESMM 1 was omitted and it was not necessary to state when weir blocks are laid to precise levels.

Item descriptions for units for subways, culverts, ducts, copings, sills and weir blocks must state the mass per metre (Additional Description Rule A6.)

Principal changes from CESMM 2

None

Measurement Rules

M1 The measurement of units for subways, culverts and ducts, copings, sills and weir blocks is linear metres for the length of identical units. It is thought that this unit of measurement should only apply to units of constant cross-section throughout. It may be necessary to enumerate certain items such as end units or units which do not have a constant cross-section.

Definition Rules

D1 The mass stated shall be the mass of each unit. Additional Description Rule A6 says that the mass per metre shall be stated in the item descriptions for units for subways, culverts, ducts, copings, sills and weir blocks. It would be superfluous to state both and because the Third Division classifications only apply to H. 1-6 * * and also because subways, culverts, ducts, copings and weir blocks are measured in linear metres, it would seem appropriate to state the mass per metre. However, if these items were measured by number (see Measurement Rule M1) then the mass per unit should be given.

D3 This rule covers the situation where to a large extent the unit may be considered as cast in-situ ie. subject to all the limitations of weather, setting out, access, placing etc, but is not cast in its final position. Units should only be measured in accordance with this rule when the work is expressly required by the Engineer to be precast on site and not when the Contractor chooses to do so. The question arises under what section of Class A the Temporary Works should be included, either specified requirements or Method Related Charges. Because it would be an express requirement for the site precasting to be done, it is recommended that it is included in the specified requirements, so it could then be adjusted if necessary.

Coverage Rules

C1 It follows that the Drawings and/or Specification must give clear details of reinforcement, joints and finishes unless the units are to be Contractor designed. Any cast-in inserts would also normally be included.

Joints in this respect means joints within the unit itself and not joints between individual units. These should be measured separately in accordance with the relevant classifications of Class G. It would be prudent to state in item descriptions that the joints are between precast concrete units.

Additional Description Rules

A1 It would only be necessary to give the precise location of individual units if it had a particularly important bearing on cost. Otherwise a general position would suffice.

A2 It follows that the Drawings and/or Specification must have detailed references for each type of unit.

A3 One purpose of A3 is to distinguish between pre-tensioned and post-tensioned work. Although not specifically mentioned pre-tensioned work measured under this class would include for the tendons and jacking operations etc., and details of the tendons should be given in item descriptions. Post-tensioned items would not include for any of the post tensioning operations or materials and these would be measured separately under G. 7 3-4 *.

A4 Examples of cross-section types are square, rectangular, circular, triangular and the like. When stating the principal dimensions of units for subways, culverts, ducts, copings, sills and weir blocks, the length of each individual unit should be stated even though the items are measured by linear metre. This is to enable the Contractor to ascertain the weight of the individual unit by applying the weight per metre as given in Additional Description Rule A6.

ITEM MEASUREMENT

H. * * * Precast concrete generally

Applicable to all items

Divisions   -

Rules   A1 State the specification of the concrete and the position in the works
              for each type of precast member.

A2 State the mark or type number for each precast unit. Units with different dimensions to have different marks or numbers.

A3 State the particulars of tendons and prestressing for pre-tensioned prestressed units (this rule applies to all pre-tensioned units and not just to beams).

Generally   One problem with Class H is that it does not cover the whole spectrum of precast concrete units and does not have an all embracing descriptive feature such as other concrete forms in Class F. There are other numerous items which are commonly precast such as steps, bridge members, floor and roof planks, cladding panels, lintels etc. and many of these, as well as those units listed in the First Division can either be reinforced or prestressed. The First Division should therefore be taken as being indicative only and further items created as required.

This class generally relies heavily on the Drawings and Specification to enable the Contractor to ascertain and price the work precisely. It follows that the Engineer must ensure that all the components of precast units are adequately detailed.

See also Rules D2, D3, C1.

H. 1-4 * *   Beams, prestressed pre-tensioned and post-tensioned beams and columns

Divisions   Beams, prestressed pre-tensioned and post-tensioned beams, and columns are enumerated stating the length and the mass as classified. Because it is necessary to state the principal dimensions including the actual length of the unit, it would not be necessary to also give the length in ranges as classified in the Second Division and these should only be used for coding purposes (Paragraph 3.10).

Rules   A4 State the cross-section type and principal dimensions.

Generally   See also Rule D1.

H. 5 * *   Slabs

Divisions   Slabs are enumerated stating the plan area as classified and the mass as classified.

Rules   A5 State the average thickness.

Generally   It is recommended that the actual area is stated where this exceeds 50m2.

See also Rule D1.

H. 6 0 *   Segmental units

Divisions   Segmental units are enumerated stating the mass as classified.

Rules   A4 State the cross-section type and principal dimensions.

Generally   It should be made clear whether the item includes all the segments which form the complete item or each individual segment only.

See also Rule D1.

H. 7 0 0   Units for subways, culverts and ducts

Divisions   Units for subways, culverts and ducts are measured in linear metres.

Rules   A4 State the cross-section type and principal dimensions.

A6 State the mass per linear metre.

Generally   See also Rule M1.

H. 8 * 0   Copings, sills and weir blocks

Divisions   Copings, sills and weir blocks are measured in linear metres stating the cross-sectional size. Although the Second Division classifies the cross-sectional area in ranges it is not necessary to give the area in these bands because the actual size of the unit has to be stated (Paragraph 3.10). The classification given should be used for coding purposes only.

Rules   A4 State the cross-section type and principal dimensions. Include the length of each unit to enable the mass to be calculated.

A6 State the mass per linear metre.

Generally   See also Rule M1.

STANDARD DESCRIPTION LIBRARY

Designed mix to BS5328 grade 30; ordinary portland cement to BS12; 10mm aggregate to BS882

Beams; rectangular section

H. 1 1 3   200 x 600 x 3000mm long; mass 500kg-1t; mark 10;
roof   nr

**Designed mix to BS5328 grade 30; ordinary portland cement to BS12;
10mm aggregate to BS882 (cont'd)**

Prestressed pre-tensioned beams; rectangular section; 3nr seven wire
standard strands to BS5896; relax class 1; nominal diameter 11mm

H. 2 2 7	900 x 1500 x 6000mm long; mass 10-20t type B2; first floor	nr

Columns; square section

H. 4 1 4	400 x 400 x 3500mm long; mass 1-2t; all columns third floor generally	nr

Slabs

H. 5 2 4	area 1-4m2; average thickness 200mm; mass 1-2t; reference S1; valve chamber roof	nr

Units for subways, culverts and ducts; square section

H. 7 0 0	2000 x 2000 x 5000mm long; mass 4t/m; mark 4; chainage 1750m	m

Copings, sills and weir blocks; weathered section

H. 8 2 0	500 x 500 x 2000mm long; mass 1.25t/m; type C1; top of weir wall	m

Chapter 11

CLASS I: PIPEWORK - PIPES

GENERAL TEXT

Principal changes from CESMM 1

The only change to the divisions in CESMM 2 was that the classification of trench depths was altered (I.**2-8).

Measurement Rule M1 stated that the Commencing Surface adopted in the Bill of Quantities shall also be the Commencing Surface for remeasurement.

Pipes in trenches were measured through fittings and valves. Pipes not in trenches exclude lengths occupied by fittings and valves (Measurement Rule M3). Previously all pipes were measured through fittings and valves.

Item coverage was extended to include for removal of existing services (Coverage Rule C2).

Trench depths in excess of 4m are stated in multiples of 0.5m (Additional Description Rule A6). The previous multiple was 2m.

Paragraph 5.20 of CESMM 1 required that items of work affected by bodies of water (other than groundwater) should be described. The classes which this paragraph commonly affected were E, I, J and P. Paragraph 5.20 in CESMM 2 does not have the same requirement but states that the body of water shall be identified in the Preamble. Additional Description Rule A2 in Class E requires that excavation below the body of water to be so described but Class I does not have any equivalent rule and it would seem that it was no longer necessary to state when work is below water in this class. Where work under this class or any of the associated Classes J, K and L is below a body of open water it is recommended that item descriptions shall so state in accordance with Paragraphs 5.10 and Additional Description Rule A2 of Class E.

Principal changes from CESMM 2

The first division of this class has been reformed with asbestos cement and pitch fibre pipes being deleted and other pipes made of more commonly used materials introduced.

A new Additional Description Rule A7 is introduced which confirms the amendment in A6 of Class E stating that where hand digging is expressly required it must be identified separately.

Measurement Rules

M1 This overcomes the problem where there are two possible Commencing Surfaces for the excavation of trenches. For example in road drainage it may be possible for the

Contractor to commence the trench excavation from either the original ground level or from the formation level of the road after the general excavation. In this instance the remeasurement of the completed work would always follow the philosophy of the original measurement irrespective of the Contractor's modus operandi.

M2 States that the backfilling of trenches with excavated and selected excavated material is not measured (see also Coverage Rule C2).

M3 Effectively means that fittings and valves on pipes in trenches are 'extra over' and fittings and valves not in trenches are full value. Lengths of pipes in trenches also exclude the length of any pipes or components comprising backdrops to manholes because these are included with the items for the manholes themselves. In circumstances such as at tapers where the pipe bore changes it is customary to measure the larger bore pipe through the fitting.

M4 The items for pipes not in trenches include for the provision, laying and jointing of the pipes only. The remainder of the items in connection with these pipes are measured in Classes K and L. These items include manholes, chambers, headings, thrust boring pipe jacking, concrete stools and thrust blocks and other pipe supports. Although not specifically stated, work classified under Classes E and T would also be measured separately.

M5 Where a straight pipe in a trench enters a manhole or chamber the length of pipe measured in trenches includes the built in length up to the inside face of the manhole or chamber. This does not apply to pipes or components comprising backdrops to manholes which are included with the items for the manholes themselves.

   If a pipe projected a short distance past the inside face of a manhole or chamber then it may not be appropriate to use the inside face of the chamber as the cut off point for the measurement of pipes in trenches but include the short length projecting into the measurement as well. If the pipeline was several miles long with only a few intermittent small air valve chambers for which no additional excavation beyond the nominal pipe trench width was necessary it may also be appropriate to measure the pipeline through the chambers and fittings. If this method was adopted then suitable preambles would have to be incorporated stating the deviation from the CESMM 3.

   Problems with this rule can also arise on major pipelines such as ductile iron water pipes because the pipe which passes through the chamber wall is usually a fitting i.e. a puddle flanged pipe. This usually projects someway into the chamber as well as outside it and can cause confusion in its measurement and the measurement of the pipe leading up to it (Figure I1).

   To measure the puddle flanged pipe strictly in accordance with CESMM 2 would entail the following items having to be measured.

1. 'Pipe in trenches' up to the inside face of the chamber.

2. 'Extra over' item 1 for *part* of the puddle flanged fitting 'in trenches'.

3. The other part of the fitting measured 'not in trenches' for the length of the fitting actually inside the chamber.

Clearly the method of measurement for this situation is not satisfactory and the most appropriate way of dealing with this situation would be to measure the puddle flanged pipe full value and described as 'not in trenches' and to measure the incoming pipe 'in trenches' up to the flexible coupling. In practice there would not be any trench excavation for the length of the fitting projecting beyond the chamber because this would be accommodated in the overbreak or working space of the excavation for the chamber.

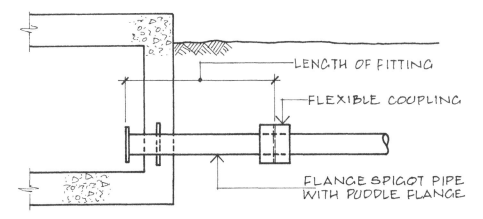

Figure I1

It is shown that the strict application of this rule can lead to problems and it is recommended that certain situations are treated on their own merits and preambles are included as required.

Definition Rules

D1 Pipes suspended or supported above the ground could include pipes inside or outside structures or buildings. Whilst it is not a specific requirement to do so it is recommended that item descriptions distinguish between both cases unless it is otherwise clear from the drawings or the way in which the items are billed i.e. under a group heading.

   The requirement to measure pipes laid within volumes measured separately for excavation as not in trenches can produce odd situations (Figure I2).

  In the above example, the excavation for the structure is completed first. The measurement for the pipe would be as follows:-

Length A	-	Pipe in trench.
Length B	-	Pipe laid within volumes measured separately for excavation.
Length C	-	Pipe in trench (Rule M5) or pipe supported above the ground or other surface in chambers (Rule D1)
Length D	-	Pipe supported above the ground or other surface in chambers.

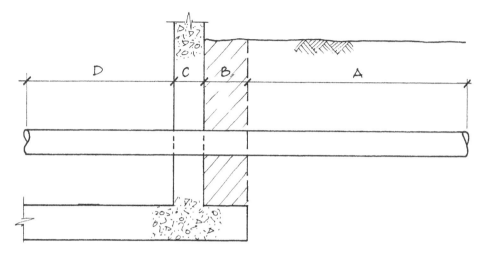

Figure I2

To measure the pipework in the above manner would not be logical although it would be in accordance with CESMM 3. The most appropriate method of measurement in this instance would be to ignore the separately excavated volume completely and simply have two items of measurement. These would be:

1.   Pipe in trench up to the inside face of the chamber (A + B + C)

2.   Pipe supported above the ground or other surface in chambers (D)

Alternative methods of measurement may be appropriate in certain instances (Measurement Rule M5).

D3 The depth stated is the maximum depth and should be measured perpendicularly from the pipe and not as the vertical distance from the Commencing Surface although the difference between these two is marginal unless the ground slopes dramatically (Figure I3).

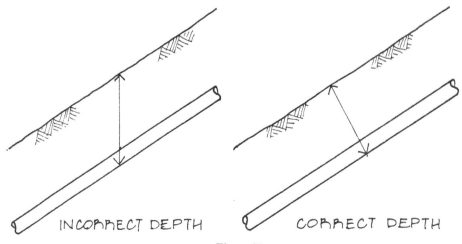

Figure I3

CLASS I: PIPEWORK - PIPES

Where a pipe runs through several Third Division depth classifications then the length of each part of the pipe contained within each of the classifications should be measured separately (Figure I4).

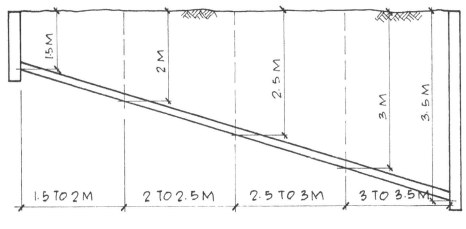

Figure I4

When the Commencing Surface for excavation constantly undulates at or around the boundary between two different depth classifications it would not be practical to try and measure each small length of pipe as it alternately 'jumps' from one classification to another. In this instance it would be more appropriate to measure the depth of the pipe at regular intervals and calculate the average. It would be wise to state in item descriptions that the depth range is an 'average' and also include a preamble stating how it has been calculated.

Coverage Rules

C1 On contracts which involve the installation of iron or steel pipes which may have long delivery periods it is common practice for the Client to purchase the pipes under a separate contract with installation only by the Contractor. In these cases great care must be taken in the item descriptions for the laying of the pipes concerning the Contractor's obligations with regard to the supply of materials, particularly jointing materials, and the provision of plant and labour for unloading the pipes at the Works together with any specialist jointing equipment.

It may also be necessary to expand the coverage rules to include items such as making good, wrapping and lining when cutting through wrapped and lined pipes and for cutting back wrapping when building through walls.

C2 Disposal of excavated material would normally be off the site unless the contract specifically stated otherwise. Care should be exercised in certain circumstances where for instance a scheme contains large amounts of pipework and also large quantities of imported filling in landform or similar filled areas. It may be advisable to measure the amount of surplus material arising from the pipework separately which is then deducted from the amount of imported filling, otherwise the Contractor may eventually be paid

145

for removing the surplus material off site when he is not doing so and also for importing filling materials for landform.

The Contractor can only allow for backfilling with selected excavated material providing sufficient suitable material can be obtained from the excavations. It would also be reasonable for the Contractor to assume that the suitable material must come from and be placed back into the same trench or part of trench. If neither of these two conditions are met then the Contractor may have just cause to claim for additional costs if he could prove he was not aware of the ground conditions previously and there would appear to be no fair method of overcoming this problem.

The item refers to dead services because live services would have to be dealt with by the Statutory Authorities.

Additional Description Rules

A1 Pipework locations could be given for individual lengths between termination points or for a complete run of lengths over several termination points depending on the nature of the work. Type of pipework refers to the function which the pipework performs i.e. drain, rising main, etc.

A2 Details of coatings or wrappings should also be given. There are instances where it would be prudent to state the wall thickness.

A3 See Definition Rule D1.

A4 See Measurement Rule M1.

A5 For pipes in shared trenches it is necessary to identify the precise length of pipe which is in the combined trench and not just the run. The reason is that not all of the run may be in shared trenches but only part of it.

French or rubble drains which contain pipes are also measured in this class and they include the trench and associated items. The excavation for unpiped French or rubble drains is included in Class K and this would seem to be inconsistent. It is often the case that French and rubble drains on an individual scheme are of both the piped and unpiped variety but both have the same cross-sectional shaped trenches and other characteristics. French and rubble drains often have unique cross-sections and it is recommended that trenches for them are excluded from Class I by way of a suitable preamble and measured under Class K. It would become necessary to add to the item description of pipework in French or rubble drains 'trench measured elsewhere'.

A6 The additional depth ranges would commence 4-4.5m, 4.5-5m, 5-5.5m, etc.

ITEM MEASUREMENT

I. * * * Pipework - pipes

Divisions   State the material of which the pipes consist.

State the actual nominal bore (Additional Description Rule A2). As the actual nominal bore is given it is not also necessary to state the range as classified in the Second Division and these should be used for coding purposes only (Paragraph 3.10).

State whether not in trenches or in trenches. If in trenches state the depth as classified.

Rules   M3   Measure pipes in trenches through fittings and valves but exclude backdrops. Measure pipes not in trenches excluding fittings and valves.

    M5   Measure lengths of pipes contained within the inside faces of chambers and manholes as not in trenches. Measure lengths of pipes entering manholes and chambers up to the inside face as in trenches except for backdrops which are included in Class K.

    A1   State the location or type of pipework for each item or group of items.

    A2   State the materials, joint types, linings and wrapping or coating requirements.

    A3   State when pipes 'not in trenches' are in headings, tunnels, shafts, installed by thrust boring and pipe jacking or within volumes measured separately for excavation.
        State also when they are suspended or supported above the ground or other surface and identify when contained within a structure or building.

    A4   State the Commencing Surface if it is not also the Original Surface.

    A5   State when pipes are in shared trenches. Identify the precise length of trench.

    A6   State trench depths exceeding 4m deep to the next highest multiple of 0.5m.

    A7   State separately any hand digging that is expressly required.

Generally The Rules governing the measurement of pipework and associated work in Classes J, K and L should not be strictly adhered to without consideration of the individual circumstances of a pipework contract. It has been shown that deviations from CESMM 3 may be desirable in certain instances and it is recommended that the measurement of pipework generally should be judged separately for different contracts.
    State when work is below a body of open water in accordance with Paragraph 5.10 and Additional Description Rule A2 of Class E.

See also Rules M1, M2, M4, D1, D2, D3, C1, C2.

STANDARD DESCRIPTION LIBRARY

Clay pipes; BS65 surface water type; extra strength; spigot and socket flexible mechanical joints; glazed internally

Nominal bore 150mm

I. 1 1 2	in trenches depth not exceeding 1.5m; run 21	m
I. 1 1 3	in trenches depth 1.5-2m; between MH21 and MH31	m

Concrete pipes; prestressed; BS5178 Class M; sulphate resisting core and cover coat; rebated flexible joints; bituminous coating internally and externally; SWMH1 - SWMH 20

Nominal bore 450mm

I. 2 3 4	in trenches depth 2-2.5m	m
I. 2 3 6	in trenches depth 3-3.5m	m

Nominal bore 750mm

I. 2 4 8	in trenches depth 4-4.5m	m
I. 2 4 8.1	in trenches depth 4.5-5m	m

Nominal bore 1000mm

I. 2 5 8	in trenches depth 5.5-6m	m
I. 2 5 8.1	in trenches depth 7-7.5m	m

Concrete pipes; precast; ARC Slimline Class H; reinforced; spigot and socket flexible joints

Nominal bore 150mm

I. 2 1 3	in trenches depth 1.5-2m; run A-B	m
I. 2 1 4	in trenches depth 2-2.5m; run B-C	m

Iron pipes; BS4772; spun ductile; spigot and socket flexible mechanical joints; 5mm nominal thickness cement mortar lining internally; hot coal tar coating externally; run MH1 - draw off chamber; below Lake Waterside

Nominal bore 150mm

I. 3 1 3	in trenches depth 1.5-2m	m
I. 3 1 4	in trenches depth 2-2.5m	m

Steel pipes; GKN Steelstock grade ERW 320; flanged joints; hot black bitumen coating internally and externally; between buildings A and B

Nominal bore 250mm; wall thickness 5mm

I. 4 2 1	not in trenches; supported above ground	m
I. 4 2 1.1	not in trenches; suspended above ground	m

PVC pipes; Terrain Buried Drain System; double socket coupler solvent cement joints; generally all contract pipework

Nominal bore 100mm

I. 5 1 2	in trenches; depth not exceeding 1.5m	m
I. 5 1 2.1	in trenches; depth not exceeding 1.5m; Commencing Surface underside of topsoil	m
I. 5 1 2.3	in trenches; depth not exceeding 1.5m; in French or rubble drains (trench measured elsewhere)	m

Chapter 12

CLASS J: PIPEWORK - FITTINGS AND VALVES

GENERAL TEXT

Principal changes from CESMM 1

The only change to the divisions was the addition of straight specials as a further type of fitting (J.1-78*). Straight specials were further defined in Definition Rule D2.

Only vertical bends in pipework which exceed 300mm nominal bore were identified separately (Additional Description Rule A3).

Due to the fact that fittings and valves in trenches were 'extra over' and those not in trenches are 'full value', it was necessary to distinguish between the two and fittings and valves in trenches must be so described (Additional Description Rule A4).

Principal changes from CESMM 2

The first division of this class has also been reformed and some of the materials which are not used so often have been replaced with those in more common use.

Measurement Rule M2 dealing with the measurement of straight special pipes has been reworded whilst Definition Rule D2 has been replaced with a clearer definition of a special.

A new Additional Description Rule A5 has been inserted to deal with fittings to relined water mains in the expanded Class 7. The previous A5 has been renumbered to A6 and a new A7 is included and refers to valves and fittings also in Class Y.

Measurement Rules

M2 The measurement of straight specials needs careful consideration of where and when they should be taken. They should not be measured where the length of the pipe is not a critical factor. For instance the Engineer would never expressly require the last pipe entering a surface water manhole to be a specific length, although invariably the last pipe in this case would be a short length . On the other hand ductile iron pipework in large pumping stations is more often than not designed in detail and all the pipes and fittings are designed to specific lengths. In this case any short lengths of pipe would be *expressly* required and therefore measurable. The main criterion is that straight specials are those which are *designed* to be a specific length and are shown as such (see also Definition Rule D2).

Definition Rules

D1 This does not mean that only the largest bore of the fitting is stated in item descriptions but that the largest bore is used for classification/coding. For fittings with multiple bores, each bore should be stated.

D2 Normally these would be short lengths but it may be possible to have longer than standard lengths made to order and these would also be classed as straight specials.

Problems may arise in defining exactly what constitutes a non-standard length of pipe. Take for instance ductile iron pipes with welded flanges. BS4772 states the standard length of these pipes is four metres. The company which is probably the country's leading supplier of ductile iron pipes, stocks standard lengths of these pipes at five metres and six metres and states that the standard lengths of four metres in BS4772 are only supplied to order. If an item description stated, 'Ductile iron pipes to BS4772 it may become necessary to order the pipes specially at four metres long which may cost more per metre when the five or six metre lengths would have probably sufficed and cost less. Shorter lengths of pipe can also be purchased 'off the shelf' and these are non-standard lengths and are measured as straight specials. It should be stated in the item description for pipework or in the Preamble the standard length of pipe on which the measurement is based and *all* pipes which are expressly required to be longer or shorter than this length would be classed as specials. This may seem to be a rather cautious approach especially when a certain type of pipe may be available from different manufacturers at various standard lengths. The alternative would be for the taker-off to ascertain beforehand exactly how many standard lengths of pipe there are available but this is usually an impracticable solution.

Coverage Rules

C1 Some fittings and valves have long delivery periods and it is common for these to have been pre-purchased under a separate contract. In these cases, careful consideration must be given to item descriptions for fixing the fittings and valves with particular regard to the supply of jointing materials and the provision of plant and labour for unloading at the Works.

C2 This rule only applies to fittings in trenches (see also Coverage Rule C2 in Class I).

C3 On wrapped and lined pipes this coverage rule should be extended to include making good the wrapping and lining.

Additional Description Rules

A1 Details of coatings or wrappings should also be given. Details of the type of puddle flange i.e. single or double, integral or split should be stated. All the bores in multiple bore fittings should be stated.

A2 CESMM 3 distinguishes between iron fittings above and below 300mm and for those above 300mm the principal dimensions of the fittings have to be stated. This implies that iron fittings are all of a similar price range where they are of one particular type and bore below 300mm. There is a large difference in price, however, between long radius and short radius bends whatever the diameter and it would be prudent to make the distinction. It would also be desirable to give the dimensions for *all* fittings measured 'not in trenches' as these are full value items and not 'extra over'.

A4 Additional Description Rule D3 in Class I requires that item descriptions for pipes 'not in trenches' distinguish between the various categories listed in Definition Rule D1 of that class. Class J has no such requirement but it is recommended that the same procedure be adopted for fittings and valves because these are also measured 'full value'.

A6 Lengths of spindles should also be given.

ITEM MEASUREMENT

J. 1-8 * * Fittings
Divisions Fittings are enumerated stating the type of material, the type of fitting and the nominal bore. The Second Division only gives a representative number of the types of fitting commonly encountered and this should be expanded upon to include further items as required.

Rules   M2 Measure straight specials only where they are designed to be a specific length.

A1 State the method of jointing, lining and coating requirements and when puddle flanges are attached.

A2 State the principal dimensions for iron pipes over 300mm nominal bore and for all steel pipes.

A3 State when bends are vertical in pipes over 300mm diameter.

A4 State when fittings are not in trenches.

A5 Fittings to relined water mains to be stated separately.

A7 Valves and perstocks to be stated separately.

Generally State when work is below a body of open water in accordance with Paragraph 5.10 and Additional Description Rule A2 of Class E.

It is felt that as much further information as possible is given about fittings, particularly for iron and steel fittings 'not in trenches' and therefore measured full value. This would include effective lengths, angles of bends and branches and additional description of the type of fitting e.g. bends would be short radius, long radius, duckfoot, etc.

Although CESMM 3 classifies fittings and valves separately from pipes, it may be desirable to group them together with the main body of the pipes, particularly for fittings and valves in trenches because these are 'extra over' the pipe itself. This will reduce unnecessary repetition of item descriptions for the specification of materials and joints etc.

There are instances when CESMM 3 does not adequately deal with situations which commonly arise on major pipeline contracts. One instance of this is where the end joint on a run of pipes is different from those in the running length (Figure J1).

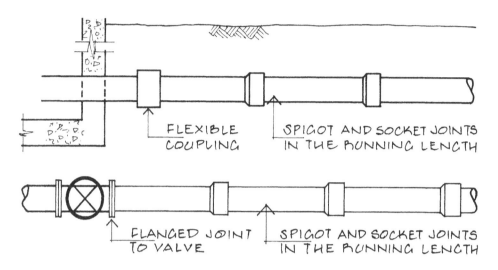

Figure J1

In the above examples the main pipework would be described as having spigot and socket joints but strictly speaking this would not be correct for the last pipe in the line. This can be overcome by measuring an enumerated item for the flexible coupling in the first diagram and a flange in the second. This is not a specific requirement and a suitable preamble should be incorporated stating the deviation from CESMM 3.

See also Rules M1, M2, D1, D2, C1, C2, C3.

G. 8 * *   Valves and Penstocks

Divisions Valves and penstocks are enumerated stating the type of valve or penstock and the nominal bore.

Rules   A5 State the material type, joint requirements, and details of any further items such as draincocks extension spindles and brackets.

Generally The Second Division should be taken as covering only a small proportion of the many different types of valve available. The provision of the power supply would not normally be included in the items for power operated valves but it should be made perfectly clear whether the connection to the power is to be included or not.

See also Rules C1, C2.

STANDARD DESCRIPTION LIBRARY

Clay pipe fittings; BS65 surface water type; extra strength; spigot and socket flexible mechanical joints; glazed internally

Bends

J. 1 1 1	nominal bore 150mm	nr

Bends; vertical

J. 1 1 3	nominal bore 350mm	nr

Junctions and branches

J. 1 2 2	nominal bore 225 x 225 x 150mm	nr
J. 1 2 2.1	nominal bore 225 x 225 x 225mm	nr

Concrete pipe fittings; prestressed; BS5178 Class M; sulphate resisting core and cover coat; rebated flexible joints; bituminous coating internally and externally

Tapers

J. 2 3 4	nominal bore 750-600mm	nr

Straight specials

J. 2 8 2	nominal bore 300mm	nr
J. 2 8 2.1	nominal bore 300mm; not in trenches	nr

Concrete pipe fittings; precast; ARC Slimline Class H; reinforced; spigot and socket flexible joints

Bends

J. 2 1 1	nominal bore 150mm	nr
J. 2 1 1.1	nominal bore 150mm; not in trenches	nr

Iron pipe fittings BS4772; spun ductile; spigot and socket flexible mechanical joints; 5mm nominal thickness cement mortar lining internally; hot coal tar coating externally

Bends

J. 3 1 1	nominal bore 150mm	nr
J. 3 1 1.1	nominal bore 150mm; not in trenches	nr

Tapers

J. 3 3 2	nominal bore 225-150mm	nr
J. 3 3 2.1	nominal bore 300-225mm; not in trenches	nr

Bellmouths

J. 3 7 2	nominal bore 300mm	nr
J. 3 7 2.1	nominal bore 300mm; not in trenches	nr

Straight specials

J. 3 8 1	nominal bore 150mm; single integral puddle flange	nr
J. 3 8 1.1	nominal bore 150mm; double integral puddle flange	nr

Steel pipe fittings; GKN Steelstock grade ERW 320; welded flanged joints; hot black bitumen coating internally and externally

Blank flanges

J. 3 9 3	nominal bore 500mm	nr
J. 3 9 3.1	nominal bore 500mm; not in trenches	nr

PVC pipe fittings; Terrain Buried Drain System; double socket coupler solvent cement joints

Bends

J. 4 1 1	nominal bore 100mm	nr
J. 4 1 1.1	nominal bore 150mm	nr

Valves and penstocks

Gate valves hand operated; BS5163; cast iron; double
flanged with draincock; as specification clause 13.2

J. 8 1 1   nominal bore 150mm   nr

Butterfly valves hand operated; BS5155; gunmetal;
single flange wafer type; as specification clause 13.3

J. 8 4 3   nominal bore 350mm   nr

Chapter 13

CLASS K: PIPEWORK - MANHOLES AND PIPEWORK ANCILLARIES

GENERAL TEXT

Principal changes from CESMM 1

There were several changes to the divisions in this class. The depth classifications for manholes, chambers and ducts and metal culverts were changed (K. 1&2 * *, K. 5 * *). An additional item was created for trenches for pipes or cables not laid by the Contractor (K. 4 8 *). New items for the crossing of existing sewers or services were created (K. 6 7&8 *), and the classification of the pipe bores was changed for crossings (K. 6 * *), and reinstatement (K. 7 1-5 *). A new item was created for the stripping and reinstatement of topsoil from easements (K. 7 6 0). Reinstatement of field drains are measured in linear metres instead of enumerated (K. 8 1 0).

There were no further major changes to the class except for expansion and clarification of the Notes of CESMM 1.

Principal changes from CESMM 2

Coverage Rule C3 has been re-worded to state that all items of metal work and pipework within the manhole, chamber or gulley are deemed to be included.

Additional Description Rule A3 has been included to repeat one of the earlier changes, viz where hand digging is expressly required, it must be stated. Existing rules A3, A4, A5, A6, A7, A8, A9 A10 and A11 are renumbered.

Measurement Rules

M1 See Additional Description Rule A5 in Class I.

M3 Generally speaking, it would have been better if the measurement of ducts and metal culverts had been included in Class I because all the rules of Class I are applicable. It would be prudent to refer to a cross-section drawing for multi-way cable ducts because a six-way duct could be any of the configurations given in Figure K1, each of which gives rise to different cost considerations.

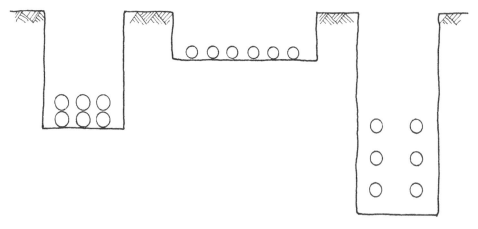

Figure K1

Ducts and metal culverts are measured through all fittings regardless of whether they are in trenches or not. Fittings are not measurable and the Contractor's rates for ducts and metal culverts are deemed to include these items.

M4 This rule implies that crossings are not measurable for any of the items measured under the French drains classification (K. 4 * *). These items have a similar work content to pipes, ducts and metal culverts (apart from the actual pipes, ducts and culverts themselves) and the cost of crossing any of the obstructions listed under K. 6 * by any of the items in K. 4 * would be similar to any other crossing. It is recommended that crossings are measured for items under K. 4 * * and a suitable preamble amending this rule is incorporated.

M5 Width in this context means maximum width when the water is at the higher level of fluctuation in accordance with Paragraph 5.20.
   It is not clear whether water crossings are measured in addition to or instead of stating that the work is done under a body of open water in accordance with Paragraph 5.10 (see the general section of Item Measurement under Class I) although it would seem inappropriate to accommodate both. If it was done in accordance with Paragraph 5.20 it would be necessary for the taker-off to state exactly how he arrived at the measured length, whether he had measured it as the net crossing width or whether an allowance had been made for projecting beyond the crossing. It is felt that the identification and measurement of the crossing itself gives sufficient information to the Contractor for his pricing and that it would be superfluous to also measure the pipework in Class I as being 'under water'. If thought to be helpful however, the item description for the crossing could identify the body of water itself.

M6 Comments on Measurement Rule M4 to also apply here.

M7 Lengths of pipes occupied by manholes and other chambers requires some explanation. Manholes and chambers can be constructed in various sizes, from small precast units to large in situ concrete types which are tens of metres long. In the latter case it would be inappropriate to measure the reinstatement through the manhole or chamber and as a

160

general rule of thumb this rule can be taken as meaning that reinstatement shall be measured through manholes and other chambers only where the manholes and other chambers are measured by number in accordance with K. 1&2 * *. Manholes and chambers measured in detail would automatically have the reinstatement covered in other classes and it would not be necessary to include their reinstatement with the pipeline. Where pipes of different bores enter and leave the same manhole, the bore used for the classification of the pipe should be the largest bore.

M8 Usually the stripping and reinstatement of topsoil over the normal trench width would be covered by items for reinstatement of land measured in accordance with K. 7 5 *. On large pipeline contracts in particular, in order to prevent damage to adjacent areas of topsoil to the pipeline by items of plant, the Engineer may specifically require that the topsoil is stripped off over a greater width than the normal trench width and put to one side. It is normal practice for the topsoil to be removed from one side of the pipeline only and for it to be stored on the opposite side. Contractor's plant would then carry out pipelaying operations on the stripped side only, leaving the stored topsoil untouched. It follows that the Engineer must give the Contractor sufficient width to enable his pipelaying operations to be carried out and also to enable the topsoil to be stored after allowing for bulking.

M9 Comments on Measurement Rule M4 to also apply here.

M10 A potential problem arises under this rule where field drains occur in the actual topsoil thickness. If the stripping of the topsoil is measured under K. 7 6 0 (stripping of topsoil from easement and reinstatement) then the length of the land drain to be reinstated would be the width of the easement itself not the nominal trench width (Figure K2). If the measurement was in accordance with this rule there would be a large difference between the amount of reinstatement measured and the actual amount to be done on site. In situations such as this it is recommended that the length of land drain reinstatement is measured to the full easement width and a preamble incorporated to that effect.

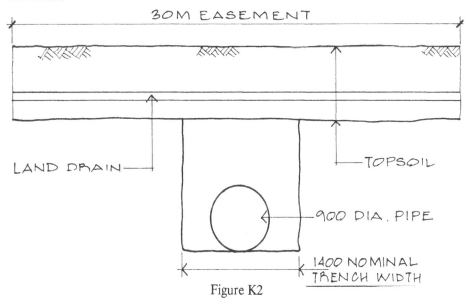

Figure K2

It is also not clear whether the length of field drain reinstatement should be always measured at right angles to the line of the pipe, which would not be the true length if the field drain were at an oblique angle to the line of the pipe (Figure K3).

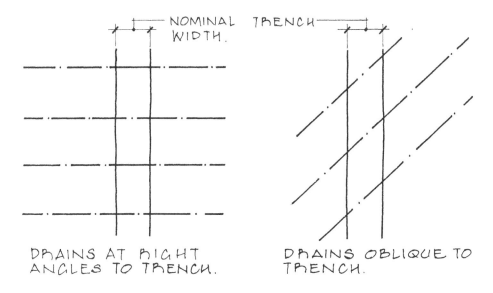

Figure K3

It would be unfair to expect the Contractor to stand the cost of the additional length and it is felt that the length measured should be the actual length of the land drain calculated on the basis of the nominal trench width as defined in Definition Rule D1 in Class L. This rule should be amended accordingly if this procedure is adopted.

Definition Rules

D1 This rule applies to multi-way cable ducts. Examples of centre lines are given in Figure K4.

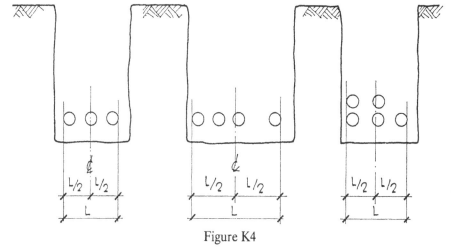

Figure K4

It follows that if in any one length of trench the number of ducts should change or if their relationship to each other in the trench should change, then the centre line may also vary.

D2 It is not thought possible to have a channel invert below the bottom of the base slab and therefore generally speaking the invert will always be to the top of the base slabs, (Figure K5).

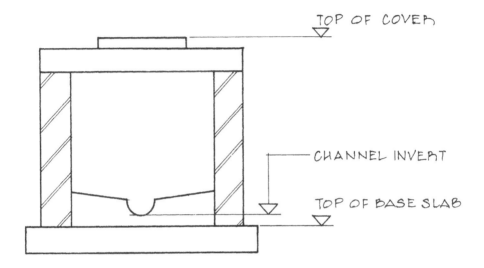

Figure K5

D4 When measuring non-circular metal culverts, it would be necessary to alter the wording in the Second Division from 'nominal internal diameter' to 'maximum nominal internal cross-sectional dimension'.

D5 See Measurement Rule M3.

D6 Water crossings are measured in the ranges 1-3m (grouped together) 3-10m (grouped together) and for crossings in excess of 10m the actual width has to be stated. The width is taken as being the maximum width as defined in Paragraph 5.20. In summer months, smaller streams often dry up completely, but the Contractor would still be entitled to the inclusion of the item in the remeasurement.

D7 Examples of various classifications are indicated by the letter C shown in Figure K6.

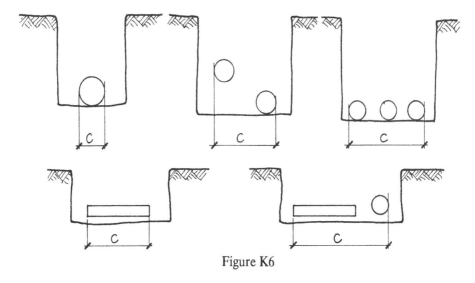

Figure K6

D8 It is not necessary to measure an item for crossing of roads and paths when the reinstatement of these is measured under K. 7 1-4 *.

Coverage Rules

C1 Disposal of excavated material would normally be off the site unless otherwise stated. Certain instances can arise for which further consideration of the disposal of excavated material should be given. (See second paragraph of Coverage Rule C2 in Class I).

C2 It follows that drawings for manholes, chambers and other such items have to be fully detailed by the Engineer.

C3 This is a very general rule and should not be regarded as being definitive. Depending upon the level of detail and how the manholes have been referenced, it may be desirable or even necessary to measure the manholes in detail in accordance with the note at the bottom of the page.

Metalwork would include covers and frames, ladders and step irons. In general terms, items for enumerated manholes should include for everything shown on the manhole details. As well as the items already listed in the coverage rules, further items would be channels, branch bends, benchings, finishing to benchings, ventilation pipes, cover slabs, stoppers and chains etc.

C5 See Measurement Rule M3.

C6 To enable the Contractor to price items of reinstatement in accordance with this rule, it would be necessary to define precisely what the item of reinstatement was, i.e. it would not be sufficient to state 'wall' but further description what the wall actually comprised of would be necessary e.g. 'random rubble wall 300m thick x 2 metres high etc...' It is possible that only partial reinstatement of the item is required e.g. the wall may not be re-built to its full former height. It may also be that complete reinstatement is physically not possible e.g. it may not be possible to purchase hedge plants 2 metres high to replace the ones which were removed. It follows that the precise extent of the reinstatement must be stated in item descriptions for crossings where the extent would otherwise be unclear.

The question of reinstatement for crossings is further complicated because a Contractor may elect not to take down the obstruction (for instance a wall), but rather support it temporarily and go under it. In this case it is recommended that no adjustment to the rate for the measured item be made, even if it does specify 'reinstatement' providing always of course that the Engineer is satisfied with the modus operandi of the work.

C7   This rule would not normally apply to reinstatement of topsoil measured under K. 7 6 0 because the width of the topsoil stripped and reinstated would normally be greater than the area of manholes and chambers.

C8   Kerbs and channels would also include their beds and backings.

C9   The item for reinstatement would include for all items necessary to restore the land to its original condition including seeding. Trees and single plants etc. would be dealt with under D. 2 * 0 and E. 8 4-6 0.

Additional Description Rules

A1   Manholes and other chambers can only be enumerated where they are fully designed and detailed in the tender documentation. If there are any areas of doubt or confusion then they should be measured in detail in accordance with the note at the bottom of the page.

A8   Reinstatement of hard surfaces other than roads and footpaths are also measurable where they are required to be reinstated. The road and footpath material and thickness are stated, the thickness including the depth of bases and sub bases.

A9   The width of the easement can often vary and it is recommended that the minimum and maximum width is stated.

A10 Where any doubt or confusion arises on what type of land is involved, then it may be prudent to identify the precise location.

A11 A coverage rule for marker posts should be included defining precisely what is to be included in the item e.g. foundations etc.

A12 This is a fairly general rule and leaves the scope for describing the work quite open. The item description should not go into the minutest detail but should be kept fairly broad. State the nature of the existing service to be connected into e.g. surface water manhole, foul water drain etc. Connections to existing pipes should give details of adaptors, connectors, saddles and the like which should be included here in preference to Class J.

ITEM MEASUREMENT

K. 1&2 * * Manholes and other stated chambers

Divisions	Manholes and other stated chambers are enumerated stating the type and the depth as classified.
Rules	D2 Measure depths from top of cover to top of base slab.
	A1 State the type of mark number.
Rules	A2 State the type and loading duties of covers and frames.
	A3 State if hand digging is expressly required.
Generally	Manholes can be measured in detail in accordance with other classes in CESMM 3 if required.
	See also Rules D3, C1, C2, C3, C4.

K. 3 * 0 Gullies

Divisions	Gullies are enumerated stating the type.
Rules	A1 State the type or mark number.
	A2 State the type and loading duties of covers and frames.
	A3 State if hand digging is expressly required.
Generally	See also Rules C1, C2, C3.

K. 4 1&2 0 Filling of French and rubble drains

Divisions	Filling of French and rubble drains is measured in cubic metres.
Rules	A4 State the nature of the filling materials.
Generally	See also Rules M1, C1, C2.

K. 4 3-8 * Trenches and ditches for French and rubble drains, pipes and cables

Divisions	Trenches and ditches for French and rubble drains, pipes and cables are measured in linear metres stating the type of the trench or ditch as classified in the Second and the cross-sectional area as classified in the Third Division. State the actual cross- sectional area when it exceeds 3 m2.

Rules   M2 Measure cross-sectional areas to the Excavated Surface.

A7 State the materials and thicknesses of the lining material in lined ditches.

Generally   When the cross-sectional area of the trench or ditch falls into several of the classifications in any one length or run, the length of each individual length falling within each of the ranges must be measured separately.

         The items for trenches for French and rubble drains and for ditches will include for disposal of all the excavated material as they would not normally be backfilled. Trenches for pipes or cables not to be laid by the Contractor may or may not be backfilled pursuant to the laying of the pipes and the item description or preamble should make it clear whether or not the trenches are to be backfilled by the Contractor after the laying of the pipes or cables by others, or whether the Contractor has to dispose of all or part of the excavated material.

See also Rules M1, C1, C2.

K. 5 * *   Ducts and metal culverts

Divisions   Ducts and metal culverts are measured in linear metres. For cable ducts state the number of ways, 1 way, 2 way, etc. State the nominal internal diameter of metal culverts. State whether in trenches or not in trenches. If in trenches state the depth as classified.

Rules   M3 Measure through all fittings.

D4 For non-circular metal culverts, state the maximum nominal internal cross-sectional dimension instead of the nominal internal diameter.

M3 All the rules of Class I: Pipework - Pipes apply to ducts and D6. metal D5 culverts except that all fittings are deemed to be included. Where any of & the rules in Class I conflict with the rules in this class then the rules in this A6 class should take precedence.

         The most important rule which applies from Class I is Additional Descriptions Rule A2 which requires the materials, joint types, nominal bores etc. to be stated in item descriptions. Multi-way cable ducts which have ducts of different types and nominal bores must give details of each duct in the item description.

         Additional Description Rule A6 of Class I would not apply to multi-way cable ducts because by definition the trench will contain more than one duct.

Generally See also Rules D1, C1, C2, C5.

See also Item Measurement for Class I: Pipework - Pipes.

K. 6 * *   Crossings

Divisions   Crossings are enumerated stating the type of crossing as classified in the Second Division and the nominal bore as classified in the Third Division. State the actual bore when it exceeds 1800mm.

Rules   M5 Measure the crossing of streams only when their higher level of fluctuation exceeds 1 metre.

            A7 State the type of lining material which is to be broken through and reinstated where rivers, streams or canals are lined.

Generally   Although not specifically stated, crossings are in effect 'Extra over' the pipes in which they occur and therefore pipes and other items are measured through the item being crossed.

            The Second Division classifications should be taken as being representative only and any other surface obstructions which have to be crossed should also be measured.

            It is not necessary to state the actual bore of the pipe because the ranges given are sufficient to enable the Contractor to price the additional costs involved. For non-circular metal culverts the words 'pipe nominal bore' in the Third Division should be taken as meaning 'metal culvert maximum nominal internal cross-sectional dimension'. In instances where items measured under K. 4 3-8 * (French drains etc.) cross surface obstructions, these should also be measured and in the absence of any specific classifications for these items (because they are unpiped), the Third Division classification would be the same as the classification for the continued trenches themselves.

            There are two main cost aspects to the crossing of surface obstructions. The first is the excavation of the trench itself and the second is the stringing and laying of the pipe. The relative importance of these to each other is dependent on the type of crossing being undertaken. For a small stream or canal crossing the important factor would be the excavation work because the more expensive aspect would be keeping the trench dry by either pumping operations or temporary cofferdams etc. The laying of the pipes would only be marginally affected because they would probably be strung out by the same plant as that laying the normal run but working from the bank. If the crossing were a wide river however, then the laying of the pipe may become a more important factor because it may be necessary to bring in additional plant or equipment to string them, such as longboom or floating cranes.

            Clearly as much additional information should be given as possible regarding crossings to enable the Contractor to ascertain the precise nature of the work. It would therefore be appropriate to state the number of pipes as well as their bores.

See also Rules M4, D6, D7, C1, C2, C6.

K. 7 1-4 * Reinstatement of roads and footpaths

Divisions   Reinstatement of roads and footpaths is measured in linear metres stating the type of reinstatement and the pipe nominal bore as classified. State the actual bore when it exceeds 1800mm.

Rules   M7 Measure through manholes and other chambers.

C8 Measure through kerbs, channels and edgings.

A8 State the type and depth of surfacing including base and base courses.

Generally   The depths of the pipes measured in Class I to which the items of reinstatement relates would be measured from the top of the hard surface and the items of reinstatement are therefore 'extra over' and trench excavation.
Reinstatement should also be measured for items measured under K. 4 3-8 * (trenches for unpiped French drains, rubble drains and ditches). Temporary reinstatement should only be measured where it is expressly required that the permanent reinstatement of the surfacing is not to be carried out under the Contract. Other surfaces which are similar to roads and footpaths would also be measured here e.g. runways.

See also Rules M6, D7, D8, C1, C2, C7.

K. 7 5 * Reinstatement of land

Divisions   Reinstatement of land is measured in linear metres stating the nominal bore as classified. State the actual bore when it exceeds 1800mm.

Rules   M7 Measure through manholes and other chambers.

A10 State whether grassland, gardens, sports fields or cultivated land.

Generally   The depths of pipes measured in Class I to which the items of reinstatement of land relates would be measured from the top of the land and the items of reinstatement are therefore 'extra over' the trench excavation. Reinstatement should also be measured for items measured under K. 4 3-8 * (trenches for unpiped French drains, rubble drains and ditches).
On most pipeline contracts the majority of the material excavated will be returned to the trench as backfilling. It is not necessary to state the depth of the topsoil to be reinstated because the cost of the work will not differ greatly whether the material is topsoil or other excavated material. The main additional cost will be in the surface preparation and seeding which is not dependent upon the topsoil thickness.

See also Rules M6, D7, D8, C1, C2, C7.

K. 7 6 0   Strip topsoil from easement and reinstate

Divisions   Strip topsoil from easement and reinstate is measured in linear metres.

Rules   M7   Measure through manholes and other chambers.

           M8   Only measure where a greater width than the nominal trench width is expressly required.

           A9   State any limitations on the width.

           A10   State whether grassland, gardens, sports fields or cultivated land.

Generally   Strip topsoil from easement and reinstate should also be measured for items measured under K. 4 3-8 * (trenches for unpiped French drains, rubble drains and ditches).

           The work measured under this classification is full value work and the calculation of the depth of pipes installed after the initial strip will be from the underside of the topsoil. There is no requirement to state the depth of topsoil and this is acceptable providing that the depth is not excessive and that it is known by all parties (from trial holes or bore hole logs etc.). In certain parts of the country the topsoil thickness can greatly exceed the 'normal' thickness of 150mm. In situations where the topsoil depth is not known it would be impractical and unfair to expect the Contractor to allow in his rates for stripping to any depth. Where the depth is unknown or uncertain then it is recommended that an actual depth is stated in item descriptions which would then form the basis of new rates should the actual thickness vary greatly. Any variation on the actual thickness from that stated in the description would also affect the depth of the subsequent trench excavation.

           The unit of measurement is linear metres and this is acceptable providing the actual width is stated in accordance with Additional Description Rule A9. The easement can often be of a non-constant width, varying because the physical constraints encountered on the site may affect the easement boundary. In cases such as these it would be more appropriate to measure the item in square metres stating the minimum and maximum widths. A suitable preamble would have to be incorporated stating the deviation from CESMM 3.

See also Rules D7, C1, C2, C9, C11.

K. 8 1 0   Reinstatement of field drains

Divisions   Reinstatement of field drains is measured in linear metres.

Rules   M10   Length should be based on the nominal trench width defined in Rule D1 of Class L.

Generally   There is no specific requirement to state the type or size of field drains requiring reinstatement but clearly an item appearing in the Bill of Quantities which simply said 'reinstatement of field drains' would be impossible to price. It is recommended that types and sizes of drains be stated in item descriptions and several common types could be given even if the actual type was unknown. Even if these where subsequently proved to be wrong they could still form the basis of a negotiated re-rate. Reinstatement of field drains should also be measured for items measured under K. 4 3-8 * (trenches for unpiped French drains, rubble drains and ditches).

See also Rules M10, C10.

K. 8 2 0 Marker Posts

Divisions   Marker posts are enumerated.

Rules   A11   State size and type.

Generally   It would be appropriate to state details of foundations etc., or refer to a drawing. Marker posts should also be measured for items measured under K. 4 3-8 * (trenches for unpiped French drains, rubble drains and ditches).

See also Rules M9, C1, C2.

K. 8 3&4 0 Timber and metal supports left in excavations

Divisions   Timber and metal supports left in excavations are measured in square metres.

Rules   M11   Measure only where expressly required to be left in. The area measured shall be the undeveloped area.

Generally   Timber and metal supports left in should also be measured for items measured under K. 4 3-8 * (trenches for unpiped French drains, rubble drains and ditches).

See also Rules M9, C1, C2.

K. 8 5&6 * Connection of pipes to existing manholes, chambers and pipes

Divisions   Connections of pipes to existing manholes, chambers and pipes are enumerated stating the nominal bore of the pipe as classified. State the actual nominal bore where it exceeds 1800mm.

Rules   A11   State the nature of the existing service and the work to be included.

Generally   It is not necessary to state the actual bore of the pipe except where it exceeds 1800mm. Any fittings supplied and fitted such as saddles, etc. should have the exact diameters stated.
See also Rules M9, C1, C2, D7.

STANDARD DESCRIPTION LIBRARY

Manholes

Brick; cover and frame to BS497 reference MB1-55

| K. 1 1 1 | depth not exceeding 1.5m; type SW1 | nr |
| K. 1 1 4 | depth 2.5-3m; type SW3 | nr |

Brick with backdrop; cover and frame to BS497
reference MB1-55

| K. 1 2 7 | depth 5.6m; type SW7 | nr |

Other stated chambers

Brick silt pits; cover and frame to BS497 reference MB1-55

| K. 2 1 2 | depth 1.5-2m; type SP1 | nr |
| K. 2 1 3 | depth 2-2.5m; type SP2 | nr |

Gullies

Clay; cover and frame to BS497 reference GB-325

| K. 3 1 0 | internal size 300mm diameter x 600mm deep; 150mm outlet; type G1 | nr |

French drains, rubble drains and ditches

Filling French and rubble drains with graded material

| K. 4 1 0 | type A filter material | m3 |

Trenches for unpiped rubble drains

| K. 4 3 2 | cross-sectional area 0.25-0.5m2 | m |

Rectangular section ditches unlined

| K. 4 4 3 | cross-sectional area 0.5-0.75m2 | m |

Rectangular section ditches lined with Filtram

| K. 4 5 6 | cross-sectional area 1.5-2m2 | m |

French drains, rubble drains and ditches (cont'd)

Trenches for pipes or cables not to be laid by the Contractor

K. 4 8 6   cross-sectional area 1.5-2m2   m

Trenches for pipes or cables not to be laid by the Contractor; backfilling

K. 4 8 6.1   cross-sectional area 1.5-2m2   m

Ducts; Hepduct; polythene sleeve joint with compression flange

Cable ducts 2 way

K. 5 2 3   in trenches; depth 1.5-2m   m

Metal culverts; as specification clause 15.1

Sectional corrugated metal culverts nominal internal diameter 750mm

K. 5 6 1   not in trenches; suspended above ground; building A-B   m

Crossings

River, stream or canal width 1-3m

K. 6 1 1   pipe nominal bore not exceeding 300mm   nr

River, stream or canal width 3-10m

K. 6 2 5   two pipes each nominal bore 300mm; in shared trench   nr

Wall

K. 6 5 2   pipe nominal bore 300-900mm; excluding reinstatement   nr

300mm diameter undergound gas main

K. 6 8 2   pipe nominal bore 300-900mm   nr

Reinstatement

Breaking up and temporary reinstatement of roads; 75mm macadam surfacing on 350mm hardcore base

K. 7 1 1   pipe nominal bore not exceeding 300mm   m

Reinstatement (cont'd)

Breaking up, temporary and permanent reinstatement of roads;
200mm reinforced concrete paving on 250mm hardcore base

K. 7 3 2	pipe nominal bore 300-900mm	m

Reinstatement of land; grassland

K. 7 5 1	pipe nominal bore not exceeding 300mm	m

Reinstatement of land; sports fields

K. 7 5 1.1	pipe nominal bore not exceeding 300mm	m

Strip topsoil from easement and reinstate; cultivated land

K. 7 6 0	10m wide	m
K. 7 6 0.1	minimum 10m, maximum 30m wide; average depth 600mm	m2

Other pipework ancillaries

Reinstatement of field drains; pitch fibre

K. 8 1 0	nominal bore 75mm	m

Marker posts; including foundations

K. 8 2 0	precast concrete; size 150 x 150 x 1000mm; as drawing number MPA/DWG/1	nr
K. 8 2 0.1	H.J. Baldwin; reference M3	nr

Timber supports left in excavations

K. 8 3 0	generally	m2

Connection of pipes to existing surface water manhole;
dealing with flow; breaking out and adapting benching

K. 8 5 2	pipe nominal bore 200-300mm	nr

Connection of pipes to existing foul water pipes;
dealing with flow; 150mm nominal bore saddle to 600mm
nominal bore pipe

K. 8 6 1	pipe nominal bore not exceeding 200mm	nr

CLASS L: PIPEWORK - SUPPORTS AND PROTECTION, ANCILLARIES TO LAYING AND EXCAVATION

GENERAL TEXT

Principal changes from CESMM 1

The main change to the divisions in this class was that the extras to excavation and backfilling had to be classified according to whether they were in trenches, manholes and chambers, headings, thrust boring or pipe jacking (K. 1 1-5 *). Other changes to the divisions included the addition of pipe jacking as a new Second Division classification (K. 2 3 *) and the changing of 'selected granular material' to 'selected excavated granular material' (K. 3-5 2 *).

Rules were given for the calculation of the volumes for items measured as extras to excavation and backfilling in headings, thrust boring and pipe jacking (Measurement Rule M6) and the criteria for the excavation of isolated volumes of rock was established (Measurement Rule M8).

Measurement Rule M10 and Coverage Rule C3 clarified the position regarding crossings and the provision of access pits, shafts and jacking blocks for pipes in headings, thrust boring and jacking which was not clear previously.

Coverage Rules C1 and C2 in CESMM 2 defined precisely the items deemed to be included with regard to Earthworks, Concrete and Concrete Ancillaries.

The volume of concrete stools and thrust blocks was clearly defined as excluding the pipe volume (Definition Rule D4).

Additional Description Rule A3 required beds, haunches and surrounds to multiple pipes to be so described.

Principal changes from CESMM 2

Definition Rule D2 has been reworded to ensure that the term 'surround' includes haunches and beds of the same material.

A new Additional Description A1 is introduced which confirms the amendment in rule A6 of Class E stating that where hand digging is expressly required it must be identified separately. Previous rules A1 to A6 are renumbered to accommodate this insertion.

Measurement Rules

M1 Rule M1 states the items in Classes I, K and Y for which further items shall be measured in accordance with this class. Strictly speaking, the reference to manholes and other chambers should also include gullies. As with manholes, items for gullies

(K. 3 * 0) also include for excavation and any hard materials encountered in the excavation would need to be measured as well. Beds and surrounds to gullies would not be measured separately but included with the the the gully.

M2 This rule defines how to calculate the nominal width of the trench for metal culverts and multi-way ducts when read in conjunction with Definition Rule D1, (Figure L1).

It also defines the dimensions to be used for the classification in the Third Division for items measured under L. 3-8 * * and in Figure L1 this is represented by the dimension D.

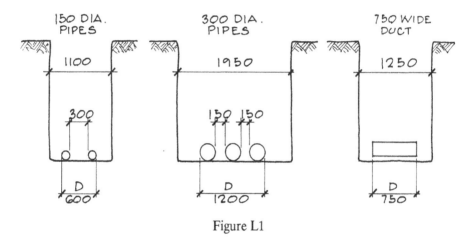

Figure L1

M3 Removal of hard surface materials would only be measured under Class K where there is temporary or permanent reinstatement involved. Where there is no reinstatement the items should be measured under Class L.

M4 It is common practice to use the lengths of the pipes in trenches measured in Class I for the calculation of extras to excavation and backfilling but this is not strictly correct. Pipes in trenches are measured to the inside face of manholes and chambers and the length which is built in should not be included in the calculations for volumes of extras.

M5 Gullies measured under K. 3 * 0 should also be included in this rule. The maximum plan area of manholes should include the area of backdrops because they are included with the items of manholes.

M6 It follows that if a specific size of heading is stated which is greater than the size of the pipe then the Contractor would have to allow in his rates for the additional volume of 'extras' between the pipe and the heading wall. Alternatively this rule could be amended so that the volume of material is based on the actual size of the heading concerned. A suitable preamble stating the deviation from CESMM 3 would have to be included accordingly.

M7 It follows that the volume of material which the Contractor has to allow for disposal is the same volume as that measured for backfilling. Excavation below the Final Surface will generally take the form of excavation of unsuitable material below the trench bottom (generally termed 'soft spot excavation').

Generally speaking soft spot excavation will always be 'expressly required' because leaving unsuitable material in place may cause uneven settlement of the pipe.

M9 Should the Contractor decide to lay a pipe by special pipe laying methods instead of by normal 'cut and cover' operations as measured in the Bill of Quantities the remeasurement should still be on the basis of cut and cover irrespective of whether his chosen modus operandi costs him more or less.

M10 This rule sets out the criteria for the measurement of access pits, shafts and jacking blocks. It basically states that these items are measured when they are designed or specified by the Engineer in the Contract. Where the Engineer has no specific requirement regarding these items, they are not measured separately but are deemed to be included with the items of special pipe laying methods themselves (see also Coverage Rule C3).

M11 The reference to manholes and chambers in this rule should also include gullies. Rule M11 does not define how to measure beds, haunches and surrounds to multiple pipes but it is reasonable to assume that Definition Rule D1 of Class K will apply.

M11 As in Measurement Rule M6 it is common to use the lengths of the pipes
and themselves measured under Class I for the lengths of the beds, haunches and surrounds
M12 but again this would be a slight overmeasurement due to the lengths built in to
manholes and chambers.

Definition Rules

D1 Examples of nominal trench widths are given in Figure L2.

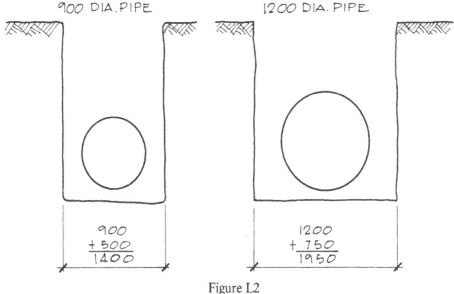

Figure L2

For nominal trench widths of multiple pipe trenches see Measurement Rule M2.

D2 Figure L3 gives examples of how the measurement of beds, haunches, and surrounds operates.

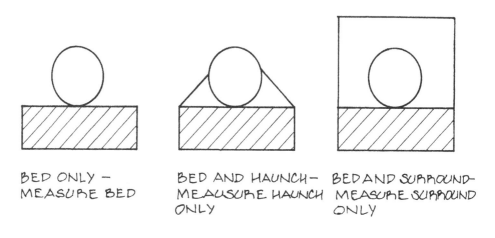

BED ONLY —
MEASURE BED

BED AND HAUNCH—
MEAUSURE HAUNCH
ONLY

BED AND SURROUND-
MEASURE SURROUND
ONLY

Figure L3

Rule D2 is required because it is unusual to have either a haunch or surround to a pipe without a corresponding bed and it would be superfluous to state that the bed exists in all descriptions. This would only apply if the material were the same throughout but if the bed is composed of material different to either the material in the haunch or surround then it would be necessary to state the bed separately in the item description or to measure it separately altogether in which case a suitable preamble would have to be incorporated.

D4 The volume is the net volume of concrete after deduction of the volume of the pipe. See Figure L4 for examples.

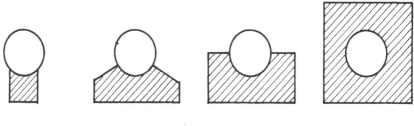

SHADED AREA REPRESENTS VOLUME MEASURED

Figure L4

D5   See Figure L5 for examples of the pipe supports classification.

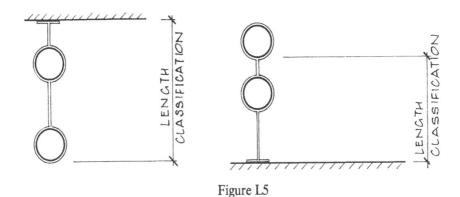

Figure L5

It can be seen from the above example that if this rule is strictly adhered to it can produce different lengths for similar situations depending upon whether the pipes are supported from above or below and it is necessary to state this.

D6   This will involve adding together the nominal bores of all the individual pipes to which the support relates. Where the nominal bore is aggregated it is recommended that item descriptions state this to inform the Contractor that more than one pipe is supported.

Coverage Rules

C1   See Coverage Rule C1 in Class K.

C2   It follows that drawings for components containing these items have to be fully detailed by the Engineer.

C3   This rule only applies if the items for access pits, shafts and jacking blocks have not been detailed and itemised separately in accordance with Measurement Rule M10.
It is not clear from either Class I or Class L where the Contractor should allow for the excavation of 'normal' materials in connection with special pipe laying methods. Class I clearly includes for the cost of the pipe itself and Class L includes the ancillaries to the pipe laying operations such as the access pits etc. Coverage Rule C2 of Class I specifically mentions excavation to pipes in trenches and does not include excavation for special pipe laying methods. This rule has a global 'catch all' phrase which states that the items measured under this class include other work associated with special pipe laying methods so it is presumed that the excavation is included here.

C4   It follows that the Contractor must allow in his rates for the extra labour and material cost involved in wrapping fittings, valves and joints.

C5   See Coverage Rule C2.

CLASS L: PIPEWORK - SUPPORTS AND PROTECTION, ANCILLARIES TO LAYING AND EXCAVATION

Additional Description Rules

A1 State separately any hand digging that is expressly required.

A2 Although this rule only requires that the type of packing is to be stated for pipes in headings, it is sometimes a specific requirement that the annulus of pipes installed by thrust boring should be packed. In such instances the type of packing should also be described and this rule amended accordingly in the Preamble.

A4 In the case of multiple pipes it is often better to state the overall size of the bed, haunch or surround stating the number and sizes of pipes or to refer to a drawing cross-section.

A6 Where concrete stools and thrust blocks form a major part of the cost of the works it may be more appropriate to measure them separately in accordance with other classes or reference should be made to drawings or details. This is also true for larger, more complicated types. In any case stools and thrust blocks which have the same volume but different reinforcement or other unique features should be billed separately.

A7 It is usually difficult for the Contractor to price pipe supports from the brief descriptions given and it is essential that reference is given to drawings or details.

ITEM MEASUREMENT

L. 1 * * Extras to excavation and backfilling generally

Applicable to all items

Divisions State the nature of the excavation or backfilling materials in accordance with the Third Division.
    Item 4 of the Third Division simply states 'other artificial hard material'. It is felt that this is not intended to include for all types of artificial hard materials and that the nature of the material should be stated in item descriptions.
    Although not specifically stated, the specification of the concrete in items 5 and 7 of the Third Division should be stated where it would otherwise be unclear.

Rules M8 Only measure isolated volumes of hard materials when their volume exceeds 0.25m3.

Generally See also Rules M1, M2, M3, C1, C2.

L. 1 1&2 * Extras to excavation and backfilling in pipe trenches, manholes and other chambers

Divisions Extras to excavation and backfilling in pipe trenches, manholes and other chambers are measured in cubic metres.

M4 For trenches calculate the volume based on the nominal trench width and
and the average depth and length.
D1

M5 For manholes, other chambers and gullies, calculate the volume based on the
maximum plan area and the average depth.

M7 Measure backfilling and excavation above and below the final surface only
where it is expressly required.

Generally -

L. 1 3-5 * Extras to excavation and backfilling in special pipe laying methods

Divisions Extras to excavation and backfilling in special pipe laying methods are measured
in cubic metres stating the nature of the method of laying in accordance with the
Second Division.

Rules M6 Calculate the volume based on the internal cross-sectional area and the
average length.

Generally -

L. 2 * * Special pipe laying methods

Divisions Special pipe laying methods are measured in linear metres stating the nature of
the laying method and the nominal bore of the pipe as classified in the Third
Division.

Rules M9 Only measure when expressly required.

M10 Give items in specified requirements in Class A for access pits, shafts and
jacking blocks but only when they are designed and specified by the
Engineer, otherwise they are deemed to be included.

A1 Identify the pipe run. State the type of packing for pipes laid in headings.

Generally See also Rules M1, M2, M3, C1, C2, C3.

L. 3-5 * * Beds, haunches and surrounds

Divisions Beds, haunches and surrounds are measured in linear metres stating the nature
of the material (Additional Description Rule A2) and the nominal bore of the
pipe as classified in the Third Division.

Rules   M11 Measure along centre lines through fittings and valves but exclude
manholes and chambers.

           D2   Do not measure beds separately when measuring haunches and surrounds.

           D3   Classify multiple pipes according to the maximum nominal distance
between the inside faces of the outer pipe walls.

           A4   State when items are to multiple pipes and state the maximum
nominal distance between the inside faces of the outer pipe walls.

Generally   Refer to drawings or details where this would be appropriate.

           Item 2 in the Second Division refers to 'selected excavated granular material'.
Generally speaking this item would refer to materials such as sand and gravels
for which minimum grading would be required. If the Contract required that
excavated materials such as rock or granite had to be broken down to form the
material then it would be prudent to extend the coverage rules accordingly.

           See also Rules M1, M2, M3, C1, C2.

L. 6 0 *   Wrapping and lagging

Divisions   Wrapping and lagging are measured in linear metres stating the bore of the pipe
as classifed in the Third Division.

Rules   M12 Measure along centre lines through pipes and fittings but exclude manholes
and chambers when the wrapping or lagging does not continue through them.

           A5   State the type of materials.

Generally   See also Rules M1, M2, M3, C1, C2, C4.

L. 7 * *   Concrete stools and thrust blocks

Divisions   Concrete stools and thrust blocks are enumerated stating the volume as classified
in the Second Division and the nominal bore of the pipe as classified in the Third
Division. State the actual volume where it exceeds 6m3. State the actual
nominal bore where it exceeds 1800mm.

Rules   D4   Exclude the volume of the pipes when calculating the volume.

           A6   State the specification of the concrete. State whether it is
reinforced.

CLASS L: PIPEWORK - SUPPORTS AND PROTECTION, ANCILLARIES
TO LAYING AND EXCAVATION

Generally Unlike 'Other isolated pipe supports' (L. 8 * *) there is no rule defining how to
classify concrete stool and thrust blocks where they carry more than one pipe. It
is recommended that Definition Rule D6 is adopted here also and the
classification should be the aggregated bore of the pipes and that item
descriptions should so state.

See also Rules M1, M2, M3, C1, C2, C5.

L. 8 * * Other isolated pipe supports

Divisions Other isolated pipe supports are enumerated stating the height as classified in
the Second Division and the nominal bore as classified in the Third Division.
State the actual length where it exceeds 6m. State the actual nominal bore where
it exceeds 1800mm.

Rules D6 Classify supports carrying more than one pipe by aggregating the bores
together. State so in item description.

A7 State principal dimensions and materials. Although this rule requires
that the principal dimensions are to be stated, paragraph 3.10 would not
apply here because the principal dimension given may not necessarily be
the height of the support.

Generally Refer to drawing or details.

See also Rules M1, M2, M3, D5, C1, C2.

STANDARD DESCRIPTION LIBRARY

Extras to excavation and backfilling

In pipe trenches

L. 1 1 1	excavation of rock	m3
L. 1 1 4	excavation of mass masonry	m3
L. 1 1 5	backfilling above the Final Surface with special prescribed mix concrete to BS5328 mix 1:3:6; 20mm aggregate	m3

In manholes and other chambers

L. 1 2 2	excavation of mass concrete	m3
L. 1 2 3	excavation of reinforced concrete	m3

Extras to excavation and backfilling (cont'd)

L. 1 2 8	excavation of natural material below the Final Surface and backfilling with hardcore	m3

In thrust boring

L. 1 3 1	excavation of rock	m3
L. 1 3 2	excavation of mass concrete	m3
L. 1 3 3	excavation of reinforced concrete	m3

Special pipe laying methods

In headings; MH1 to MH2

L. 2 1 3	nominal bore 300-600mm; packed with grout	m

Thrust boring; run TB2

L. 2 2 4	nominal bore 600-900mm	m

Pipe jacking; all pipework on drawing number
SPALE/DWG/1

L. 2 3 5	nominal bore 900-1200mm	m

Beds

Sand

L. 3 1 1	nominal bore not exceeding 200mm; bed depth 150mm	m
L. 3 1 5	nominal bore 900-1200mm; bed depth 150mm; to 3nr pipes; 1200mm maximum nominal distance between the inside faces of outer pipe walls	m

Type A granular material

L. 3 3 3	nominal bore 300-600mm; bed depth 300mm	m
L. 3 3 6	nominal bore 1200-1500mm; bed depth 300mm; to 3nr pipes; 1300mm maximum nominal distance between the inside faces of outer pipe walls	m

Haunches

Mass concrete, ordinary prescribed mix concrete to BS5328
grade C15P; 20mm aggregate

L. 4 4 2	nominal bore 200-300mm; bed depth 150mm	m
L. 4 4 4	nominal bore 600-900mm; bed depth 150mm; to 2nr pipes; 900mm maximum nominal distance between the inside faces of outer pipe walls	m

Surrounds

Imported granular material type A10

L. 5 3 2	nominal bore 200-300mm; bed depth 200mm	m
L. 5 3 7	nominal bore 1500-1800mm; bed depth 300mm; to 3nr pipes; 2100mm maximum nominal distance between the inside faces of outer pipe walls; as drawing number SPALE/DWG/2	m

Wrapping and lagging

Denso tape

L. 6 0 1	nominal bore not exceeding 200mm	m
L. 6 0 2	nominal bore 200-300mm	m

Concrete stools and thrust blocks; designed mix concrete to BS5328 grade 20; 20mm aggregate

Volume not exceeding 0.1m3

L. 7 1 1	nominal bore not exceeding 200mm	nr
L. 7 1 2	nominal bore 200-300mm	nr

Volume 6.75m3

L. 7 8 7	nominal bore 1500-1800mm	nr
L. 7 8 8	nominal bore 2100mm	nr

**Concrete stools and thrust blocks; designed mix concrete
to BS5328 grade 20; 20mm aggregate; reinforced**

Volume 1-2m3

L. 7 5 2	nominal bore 200-300mm	nr
L. 7 5 2.1	nominal bore 200-300mm; aggregated	nr

**Other isolated pipe supports; hangers; galvanised mild
steel; 20 x 5 x 500mm girth as drawing number
SPALE/DWG/3**

Height not exceeding 1m; supported from above

L. 8 1 1	pipe nominal bore not exceeding 200mm	nr

**Other isolated pipe supports; hangers; aluminium; 100 x
50 x 2000mm long; as drawing number SPALE/DWG/4**

Height 1.5-2m; supported from above

L. 8 3 3	pipe nominal bore 300-600mm	nr

Chapter 15

CLASS M: STRUCTURAL METALWORK

GENERAL TEXT

Principal changes from CESMM 1

There were no changes to the Divisions apart from a minor reshuffling of the items.

Note M2 of CESMM 1 required temporary structural metalwork to be identified. This was deleted.

Coverage Rules were created clarifying the items of fabrication, erection and site bolts (C1, C2 and C3).

Additional Description Rule A3 required that item descriptions for fabrication of cranked members other than portal frames should be stated.

Item descriptions for off-site surface treatments must also state the materials and number of applications (Additional Description Rule A8).

Note M4 of CESMM 1 was qualified to state that the mass shall be calculated from the overall length of members except for flats and plates (Measurement Rule M2).

Principal changes from CESMM2

The wording of Measurement Rule M6 dealing with the weights of steel has been changed to conform to BS4360.

Measurement Rules

M2 This means that for flats and plates the mass of the material is calculated on the net size of the member whereas other members are calculated on their gross size (Figure M1).

M4 Rule M4 is relatively clear in meaning except for the treatment of bolts. There are basically two types of bolts and these are shop bolts i.e. those which are fixed in the factory, and site bolts. Site bolts are enumerated separately and covered in the Divisions under M. 5-7 3-7 * but there is no separate provision for the measurement of shop bolts which would seem to be inconsistent. It is possible that the intention is to include the cost of the shop bolts in the rates for the overall mass of the steelwork. This would involve the Engineer in having to specify whether every bolt on the drawings was to be fixed in the shop or on the site and this rule specifically excludes the mass of all bolts from the calculation of the mass of steel. In order to save confusion it is recommended that all bolts shown on the drawings are measured in accordance with M. 5-7 3-7 *, and that a suitable preamble is incorporated stating the deviation from CESMM 3.

The cost of all welds, fillets and rivets have to be included by the rates for the steelwork.

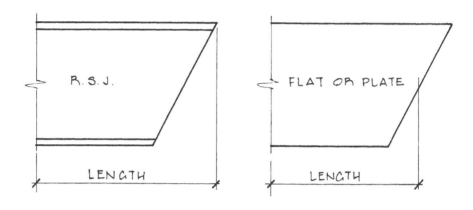

Figure M1

M7   An item would have to have some substance before it would be measured under this classification. For instance a series of holding down bolts cast into a concrete base would not merit measurement here but under Class G (Concrete Ancillaries). If the bolts were supported by a framework of steel sections it would be appropriate to include it here.

Coverage Rules

C1   Delivery would also include unloading. Fabrication of metalwork generally includes for the cutting of steel sections to the required shapes and sizes together with pre-drilling of bolt holes etc. Individual pieces are generally given references on the shop drawings for ease of identification and the sections are marked as such in the fabrication shop.

C2   Erection of members includes for all work subsequent to the delivery of the fabricated members to the site. This would include for haulage around the site, hoisting into position, site bolting, welding, riveting, and cutting etc.

Additional Description Rules

A1   The principal types of steel used in structural metalwork are those to BS4360, 2989 and 1449.

A2   Tapered and castellated members are obviously more expensive to produce as they are non-standard sections and are manufactured by cutting and welding. Other odd shaped pieces of steelwork may also warrant additional description and this rule should be construed as applying to these as well. Where additional description would be lengthy or unwieldy reference should be made to the relevant drawing instead.

A3   The implication of this rule is that portal frames are cranked which is not necessarily the case. Generally speaking portal frames consist of straight members which are joined together by various types of connection to form the portal itself. There would not generally be any cranked members. It is recommended that cranked members to portal frames are also described.

A4   This rule makes it necessary for the taker-off to give sizes and weights of the individual components of these built up assemblies.

A5   The taker-off should not get involved in lengthy or detailed descriptions but refer to the relevant drawing.

A7   It follows that the erection of crane rails with fixing clips and resilient pads should be identified separately although there is no specific requirement to do so.

ITEM MEASUREMENT

M. 1-4 * * Fabrication of structural metalwork generally

Applicable to all items

Divisions   Fabrication of structural metalwork is measured in tonnes except for anchorages and holding down bolt assemblies which are enumerated.

Rules   M2 Calculate the weight from the overall length of members except for plates or flats which are measured net.

          M3 The weight shall be the total weight including the weight of fittings but excluding the weight of welds, bolts, nuts, washers, rivets and protective coatings.

          M5 Do not deduct the weight of holes and notches unless they exceed 0.1m2.

          M6 Take the weight of steel to BS4360 as 7.85t/m3.

          A1 State the materials and grades.

          A2 Identify tapered or castellated members.

          A3 Identify cranked members.

Generally   Class M treats the measurement of structural metalwork in a simple manner. For instance M. 1 * * does not require the sizes of the sections to be stated or even to state what the sections are. It would be difficult for an estimator to price with any degree of accuracy item descriptions prepared strictly in accordance with CESMM 3. It is recommended that as much further additional description be given as is thought necessary with regard to types, sizes and weights of members, and/or reference be made to drawings or details.

See also Rules M7, C1.

M. 1 1-3 * Fabrication of main members for bridges

Divisions   State the type of member in accordance with the Second Division and the shape in accordance with the Third Division.

Rules   -

Generally   -

M. 2 1 * and M. 2 2&3 0 Fabrication of subsidiary members for bridges

Divisions   State the type of member in accordance with the Second Division. If the members are deck panels then also state the shape in accordance with the Third Division.

Rules   -

Generally   -

M. 3&4 1-6 * and M. 3&4 7&8 0 Fabrication of members for frames and other members

Divisions   State the nature of the component or member in accordance with the Second Division. Except for grillages and anchorages state the shape of the component in accordance with the Third Division.

Rules   A4 State the sizes and types of sections for trestles, towers, built up columns and trusses and built up girders.

A5 State the particulars of the type of assembly for anchorages etc., or refer to the relevant drawing.

Generally   It is recommended that the type of member is stated for fabrication of other members (M. 4 * *). (See also M. 5-7 1&2 0 - Generally).

CLASS M: STRUCTURAL METALWORK

M. 5-7 1&2 0   Erection - members

Divisions   Erection of members is measured in tonnes. Items are given for trial erection and permanent erection.

Rules   A6   Identify and locate separate bridges and frames or parts thereof.

A7   State when fixing clips and resilient pads are used to fix crane rails.

Generally   Items for permanent erection would include for the normal tasks of lining and levelling the steelwork prior to tightening up and final fixing. Lining and levelling should not be taken as being trial erection and a separate item for trial erection should not be measured in this case. Trial erection is sometimes a specific requirement of the Engineer where the accuracy of the steelwork fabrication is of prime importance i.e. on power stations. Although not specifically stated, trial erection would also normally include dismantling and should only be measured where it is expressly required by the Engineer that the steelwork, or some part of it is erected prior to final erection. Trial erection is more often than not carried out in the fabrication shop although in certain circumstances it may be done on the site or even in its final location. If the trial erection is carried out in its final location and it is subsequently found that it is not necessary to take it down again then some credit may be due to the Employer. If at tender stage it is thought that there is a possibility that this may occur, then it would be prudent to divide the trial erection into erection and dismantling, so that the latter may be deducted at a later stage. Under normal circumstances, trial erection, if required, should only be measured once without any further measurement for additional trial erection necessitated because of inaccuracies in the steelwork found in the first trial erection. It should be made perfectly clear in the documents that the Contractor should allow for trial erections in his programme and that any delays caused due to the failure of the first trial erection are the liability of the Contractor.

Strictly speaking anchorages and holding down bolt assemblies should also be weighted and included as an erection item although these items are the only ones enumerated under the fabrication classifications. The situation is also unclear with regard to the bolts to these items and whether these should be included in the weight measured for erection or whether they should be measured under M. 5-7 3-7 *. The entire position can be clarified and made simpler by adding a preamble stating that the item for anchorages and holding down bolt assemblies measured under M. 3&4 8 0 should include for erection and bolts.

See also Rules C2, C3.

M. 5-7 3-7 *   Erection - site bolts

Divisions   Site bolts are enumerated stating the type in accordance with the Second Division and the diameter as classified in the Third Division.

Rules   -

Generally   If there are not too many different diameters it may be prudent to state the actual diameter.

M. 8 * 0   Off-site surface treatments

Divisions   Off-site surface treatments are measured in square metres stating the type of treatment.

Rules   A8 State the number of applications and the materials for metalspraying, galvanising and painting.

Generally   The steelwork fabricator would pay his supplier in pounds per tonne for some of these items, e.g. pickling, galvanising, etc., and it would seem inappropriate to measure them in square metres. Where the unit of measurement adopted is the tonne a suitable preamble should be incorporated stating the deviation from CESMM 3.

See also Rule M8.

STANDARD DESCRIPTION LIBRARY

Fabrication of main mambers for bridges; steel to B.S.4360

Rolled sections

M. 1 1 1   straight on plan   t

M. 1 1 1.1   straight on plan; tapered   t

Built-up box or hollow sections

M. 1 3 2   curved on plan   t

Fabrication of subsidiary members for bridges; steel to BS4360

Deck panels

M. 2 1 1   straight on plan   t

M. 2 1 2   curved on plan   t

Fabrication of subsidiary members for bridges; steel to BS4360

Bracings

M. 2 2 0	generally	t
M. 2 2 0.1	tapered	t
M. 2 2 0.3	cranked	t

Fabrication of members for frames; steel to BS4360

Columns

M. 3 1 1	straight on plan	t

Beams

M. 3 2 1	straight on plan	t
M. 3 2 1.1	straight on plan; castellated	t

Grillages

M. 3 7 0	generally	t

Anchorages and holding down bolt assemblies

M. 3 8 0	4nr 12mm diameter x 250mm long foundation bolts on 50 x 50 x 2.22kg/m frame as drawing number SM/DWG/1	nr

Erection of members for bridges

Trial erection

M. 5 1 0	generally	t

Permanent erection

M. 5 2 0	generally	t

Site bolts; black

M. 5 3 2	diameter 16-20mm	nr

Erection of members for bridges (cont'd)

Site bolts; HSFG higher grade

M. 5 5 3 diameter 20-24mm nr

Off site surface treatments

Blast cleaning

M. 8 1 0 generally t

Metal spraying; one coat zinc primer

M. 8 5 0 generally t

Chapter 16

CLASS N: MISCELLANEOUS METALWORK

GENERAL TEXT

Principal changes from CESMM 1

The only changes to the divisions in the class were that rectangular frames were to be called miscellaneous framing (N. 1 6 *), a further classification was added for cladding (N. 2 1 0) and the classifications for tanks and tank covers were replaced by the new classification for uncovered tanks and covered tanks (N. 2 7&8 *).

Coverage Rules C2, C3 and C4 made it clear that flooring, weldmesh panelling and duct covers included supporting steelwork unless otherwise stated and that tie rods include concrete, reinforcement and joints.

Additional Description Rule A2 made it necessary to state when ladders have safety loops, rest platforms or returned stringers.

Principal changes from CESMM 2

None

Measurement Rules

M1 All off-site surface treatments including painting are deemed to be included with the items of metalwork themselves and must be described in detail in item descriptions.

M2 This rule applies only to stairways and landings and walkways and platforms because these are the only items of miscellaneous metalwork measured by mass. It is recommended that the mass should not include the weight of weld fillets, rivets, bolts, nuts and washers because it would be impracticable to do so, and that a suitable preamble is included to this effect.

M3 Compares to Measurement Rule M5 in Class M where the criterion for the deduction of holes and notches is 0.1m2. It is considered that this rule should apply only to items such as plate flooring and the superficial items of stairways, landings, walkways and platforms. For sections it is recommended that the criterion should be as Class M ie. 0.2m2.

It is further recommended that no deductions are made for splay cut or mitred ends when calculating the mass of miscellaneous metalwork. A suitable preamble would have to be incorporated stating the deviation from CESMM 3.

M4 An individual ladder may often have a safety cage for part of its length, extended stringers at its top and, if it is exceptionally long, an intermediate rest platform halfway up (see Additional Description Rule A2). None of these items are related to the length of the ladder and it would be inappropriate to measure the ladder in metres but rather by number, stating the length and other relevant details. Measured lengths of ladders, handrails and bridge parapets should only include for lengths which are identical and separate items should be given for components of different cross-sections.

Consideration should also be given to the measurement of gates and safety chains in handrails (it is recommended that these are enumerated) and it should be made clear whether these are measured 'full value' or 'extra over' the handrails in which they occur.

M5 This rule is extracted from Note N5 of CESMM 1, where it applied to rectangular frames. It effectively meant that the frames were to be measured on their overall lengths with no deduction for splay cut or mitred ends. It is thought that this meaning should still apply to CESMM 3, although the external perimeter of miscellaneous framing is not as easy to define (Figure N1).

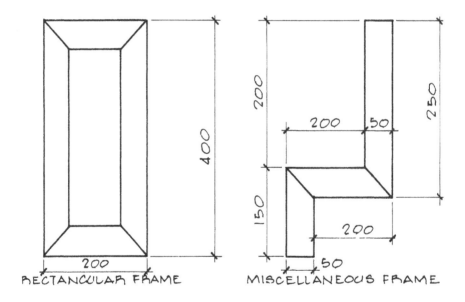

Figure N1

The external perimeter of the rectangular frame is 200 + 400 + 200 + 400 = 1200mm and this the total length of metal from which the frame is fabricated. In the miscellaneous frame the perimeter length is 150 + 200 + 200 = 550mm whereas the overall length from which the frame is made is 150 + 250 + 250 = 650mm. There is a potential problem with this rule and it is recommended that it is amended to say that the lengths measured shall be the gross lengths with no deduction for splay cut or mitred ends.

CLASS N: MISCELLANEOUS METALWORK

Coverage Rules

C1 Care must be exercised with regard to fixing and the provision of fixing components for miscellaneous metalwork. Potential problems arise particularly in respect of the fixing of miscellaneous metalwork to concrete. For example, take a ladder which is fixed to a concrete wall with bolts which are cast in. Class G requires the casting of the bolts to be measured under G. 8 3 2 (inserts). However, this rule states that the item for the ladder measured under Class N includes for fixing the ladder to the concrete, and there would appear to be a confliction. Item descriptions or preambles should make it clear whether or not items have been measured under Class G in connection with the metalwork or whether they should be included with the item (Paragraph 3.3).

C2 These rules mean that the Contractor has to measure the supporting metalwork in
and order to price and include it in his rate for flooring, cladding, panelling or
C3 duct covers. This in turn means that the taker-off must identify separate locations of these items and that the drawings must contain sufficient information and detail to enable the Contractor to measure them. The words 'unless otherwise stated', give the taker-off the opportunity to measure the supports separately and it is recommended that wherever possible this policy is adopted enabling all Contractors to tender on the same basis. This also removes any confusion which may arise whether items of metalwork are supports and therefore deemed to be included, or whether they are miscellaneous framing and measurable under N. 1 6 *.

Additional Description Rules

A1 Reference should be made to the Note at the bottom of the page enabling the taker-off to identify assemblies rather than give detailed description. It is recommended that this alternative is adopted whenever possible.

A2 See Measurement Rule M4.

ITEM MEASUREMENT

N. 1-2 * * Miscellaneous Metalwork Generally

Applicable to all items

Divisions -

Rules   M3 Do not deduct openings or holes from superficial items unless they exceed 0.5m2.

          A1 State the specification and thicknesses of metal, surface treatments and principal dimensions or refer to the drawings or details.
              In many instances it would be extremely difficult for the taker-off to give adequate or clear enough description to enable the estimator to price the work with any reasonable degree of accuracy. It is recommended that

N. 1-2 * * Miscellaneous Metalwork Generally (cont'd)

wherever possible reference is made to the relevant drawing or specification clause.

Generally See also Rules M1, C1.

N. 1 1&2 0 Stairways and landings, walkways and platforms

Divisions Stairways and landings, walkways and platforms are measured in tonnes.

Rules M2 Include the mass of all metal components and attached pieces.

Generally Although the unit of measurement is given as the tonne, it would be very difficult if not impossible for an estimator to price these items on this basis. Stairs are usually composite items comprising sections, flats, treads, bolts, etc., and some of these may be proprietary items. In order to calculate the mass, the taker-off would be required to schedule the individual components so it is recommended that the schedule be presented in the Bill in a format that can be priced. A suitable preamble would have to be incorporated stating the deviation from CESMM 3.

Notwithstanding the above, if these items are measured in tonnes, it is recommended that the mass of weld fillets, rivets, bolts, nuts and washers etc. are not included and item descriptions or preamble should make this clear. It is also recommended that splay cut or mitred ends are not deducted from items and that holes and notches in sections and members are not deducted unless they exceed 0.1m2.

N. 1 3-5 0 Ladders, handrails and bridge parapets

Divisions Ladders, handrails and bridge parapets are measured in linear metres.

Rules M4 Measure ladders along their stringers. Measure handrails and bridge parapets along their top members.

A2 State when ladders have safety loops, rest platforms or extended stringers.

Generally Measure ladders by number if more appropriate.

N. 1 6 * Miscellaneous framing

Divisions Miscellaneous framing is measured in linear metres stating the type of section in accordance with the Third Division.

Rules M5 Measure along the external perimeter (gross lengths).

Generally It would also be appropriate to state the sizes and weights of sections.

N. 1 7&8 0 Plate and open grid flooring

Divisions   Plate and open grid flooring are measured in square metres.

Rules

Generally Measure supporting steelwork separately unless it is clear from the drawing what is to be included.

See also Rule C2.

N. 2 1-3 0 Cladding, weld mesh panelling and duct covers

Divisions   Cladding, weld mesh panelling and duct covers are measured in square metres.

Rules   -

Generally Measure supporting steelwork separately unless it is clear from the drawings what is to be included.

See also Rule C3.

N. 2 4 0 Tie rods

Divisions   Tie rods are enumerated.

Rules   -

Generally See also Rule C4.

N. 2 5 0 Walings

Divisions   Walings are measured in linear metres.

Rules   -

Generally -

N. 2 6 0 Bridge bearings

Divisions   Bridge bearings are enumerated stating the type in accordance with the Third Division.

Rules   -

Generally -

N. 2 7&8 0 Tanks

Divisions Tanks are enumerated stating whether they are covered or uncovered and stating the volume in accordance with the Third Division.

Rules   -

Generally -

STANDARD DESCRIPTION LIBRARY

Galvanised steel to BS2994; metal thickness varying between 4 and 6mm

Stairways and landings; as drawing number MM/DWG/1

N. 1 1 0	overall size 10m x 6m x 4m high	t

Galvanised steel to BS2994

Ladders; as drawing number MM/DWG/2

N. 1 3 0	width 450mm	m
N. 1 3 0.1	width 450mm; with safety cage	m

Handrails; as drawing number MM/DWG/3

N. 1 4 0	height 1.2m	m

Miscellaneous framing; as drawing number MM/DWG/4

N. 1 6 1	angle section	m
N. 1 6 2	channel section	m
N. 1 6 3	I section	m
N. 1 6 4	tubular section	m

Plate flooring; as drawing number MM/DWG/5

N. 1 7 0	overall size 6m x 6m	m2
N. 1 7 0.1	overall 6m x 6m; supporting framework measured elsewhere	m2

Galvanised steel to BS2994 (cont'd)

Tie rods; as drawing number MM/DWG/6

N. 2 4 0	length 2m	nr

Bridge bearings; as drawing number MM/DWG/7

N. 2 6 1	roller	nr
N. 2 6 2	slide	nr
N. 2 6 5	spherical	nr
N. 2 6 6	plain rubber	nr
N. 2 6 7	laminated rubber	nr
N. 2 6 8	rubber pot	nr

Uncovered tanks; as drawing number MM/DWG/8

N. 2 7 1	volume not exceeding 1m3	nr
N. 2 7 3	volume 3-10m3	nr
N. 2 7 8	volume 1050m3	nr

Chapter 17

CLASS O: TIMBER

GENERAL TEXT

Principal changes from CESMM 1

The only change to the divisions was that it was no longer a requirement to state when hardwood was for general or marine use.

Note 07 of CESMM 1 was deleted.

Fittings and fastenings must state the materials, types and sizes (Additional Description Rule A4).

Principal changes from CESMM 2

In the list of exclusion at the beginning of the Class, the wording has been changed to show that carpentry and joinery work is included in the new Class 2.

In item 1-2.7 the word 'stated' is deleted because this requirement is already covered in Additional Description Rule A1.

Measurement Rules

M1   It is not clear whether the classification applies to timber in one length or to the combined effective length of several individual parts which are jointed together. Take for example the case as shown in Figure O1.

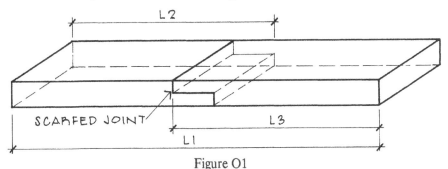

Figure O1

The question arises whether the member should be classified as one item with the length as L1 or as two separate items with the lengths classified as L2 and L3.

It is recommended that the former method is the correct one and that the length for classification should be L1. This is for several reasons:

    a)   the length L1 would provide the estimator with more information with regard to the type of plant he requires for handling

    b)   the Third Division classification extends to 'stated exceeding 20m' and it is difficult to imagine an instance where a piece of timber of this length would be required in one continuous piece or indeed whether it would be obtainable at all

    c)   the Engineer would not generally detail the position or location of the joints anyway.

M2   Open boarded decking would normally be measured without any deduction for the gaps between the boards irrespective of their area, and the preamble should make this clear.

Definition Rules

D1   The nominal size of timber is the size as it is generally purchased in its unworked
and   or unplaned form. If a timber component has a wrought finish then it is necessary
D2   to state the size of the timber out of which it was worked. Alternatively the taker-off could state the finished size providing a suitable preamble is incorporated.

Coverage Rules

C1   Items for timber would also include for any labours on them.

Additional Description Rules

A1   It is taken that 'special surface finishes' means, amongst other things, that it shall be stated when timber is wrought. The grade, species and impregnation requirements would normally be contained in the Specification, to which reference should be made. If the Specification contains different requirements for various components then the item descriptions should differentiate.

A2   The term 'structural use' means defining what function the component fulfils, i.e. beam, column, etc. The extent to which the location of the component is defined would depend upon its cost significance. In most cases it would be sufficient to give the general position of a group of components rather than give the precise location of each individual item.

A3   This rule is exactly the same as Additional Description Rule A1 except that there is no requirement to state the grade of timber. This is thought to be an error and it is recommended that the grade is also stated where this is relevant. As well as stating the thickness of the decking this rule also makes it a requirement to state the cross-sectional dimensions of the components of which the decking comprises. For open type decking it would also be prudent to specify either the spacing of the components or the gap between them.

ITEM MEASUREMENT

O. 1&2 * *   Hardwood and softwood components

Divisions   Hardwood and softwood components are measured in linear metres stating the gross nominal cross-sectional size and the length as classified in the Third Division. It is not necessary to give the size in ranges as classified in the Second Division and these should only be used for coding purposes (Paragraph 3.10).

Rules   M1 Measure the overall length with no allowance for joints.

A1 State the grade or species, any impregnation requirements or special surface finishes.

A2 State the structural use and location of timber components over 3m long.

Generally   It would also be prudent to state when it is an express requirement that a component is to be supplied in one continuous length.

See also Rules C1, D1.

O. 3&4 * 0   Hardwood and softwood decking

Divisions   Hardwood and softwood decking is measured in square metres stating the thickness. It is not necessary to also give the thickness in ranges as classified and these should only be used for coding purposes (Paragraph 3.10).

Rules   M2 Do not deduct the area of openings or holes unless they exceed 0.5m2.

A3 State the grade or species, any impregnation requirements or special surface finishes. State the cross-sectional size of the boards or planks of which the decking comprises.

Generally   For open type decking also state the spacing of the members or the size of the gap between them.
     Although not specifically stated, it would also be appropriate to measure plywood or similar decking under this classification.

See also Rules D2, C1.

O. 5 1-5 0   Fittings and fastenings

Divisions   Fittings and fastenings are enumerated stating the type. The Second Division classifications should be taken as being indicative only and should be added to as required.

Rules   A4 State the materials, types and sizes.

Generally   Strictly speaking nails and screws are fastenings and should be measured separately. In practice this would never be done and a coverage rule stating that nails and screws are deemed to be included should be added.

STANDARD DESCRIPTION LIBRARY

Hardwood components; greenheart

Cross-sectional size 400 x 400mm

O. 1 5 1	length not exceeding 1.5m	m
O. 1 5 2	length 1.5-3m; beams; underside of decking	m
O. 1 5 3	length 3-5m; beams; underside of decking	m
O. 1 5 4	length 5-8m; columns; deck support	m

Softwood components; B.S.4978 grade SS; pressure impregnated with preservative

Cross-sectional size 75 x 75mm

O. 2 1 1	length not exceeding 1.5m	m
O. 2 1 4	length 5-8m; beams; in roof	m

Cross-sectional size 150 x 150mm

O. 2 3 5	length 8-12m; partitions; in external walls	m

Hardwood decking; afromosia; wrought all round

Thickness 19mm

O. 3 1 0	board width 150mm at 200mm centres	m2

Thickness 32mm

O. 3 2 0	board width 200mm	m2

Fittings and fastenings

Straps; aluminium

| O. 5 1 0 | 100 x 4 x 250mm long | nr |

Bolts

| O. 5 4 0 | 12mm diameter x 100mm long; one nut two washers | nr |

Chapter 18

CLASS P: PILES

GENERAL TEXT

Principal changes from CESMM 1

The First and Second Divisions of this class were not changed but the Third Division was different for all types of piles. In CESMM 1 the length of all piles was classified in bands of 5m. In CESMM 2 the length classification was dependent on the type of pile involved (see also Measurement Rules M2, M4, M5 and M6).

Other changes included additional guidance on the calculation of pile lengths (Measurement Rule M1, Definition Rule D3, D4 and D6), further definition of what constitutes a driven cast in place concrete pile and a hollow isolated steel pile (Definition Rules D2 and D5), and a rule covering the disposal of surplus excavated material (Coverage Rule C1).

Additional Description Rule A3 required that the structure to be supported should be identified.

Details of coatings and other treatments, and of driving heads and shoes had to be stated for preformed and timber piles (Additional Description Rules A7 and A8). Coatings and other treatments also had to be specified for isolated steel and interlocking steel piles (Additional Description Rules A7 and A11).

Principal changes from CESMM 2

None

Measurement Rules

M1 Figure P1 shows examples of the calculation of the pile bored and driven depths.
    In the case of driven cast in place piles the depth is taken to the bottom of the casing and no account is taken of shoes although the length of preformed driven piles is taken to the bottom of the toe (therefore the shoe) and this would seem to be inconsistent. No account is taken of enlarged toes because due to their nature it is generally not possible to ascertain the exact depth to the bottom of the toe.
    It is recommended that the depth of driven cast in place piles is taken to the underside of the shoe, and that this also applies to concreted lengths where the concrete fills the shoe as well as the casing.

M2 Careful consideration needs to be given to the definition of the words 'group of items'. The group will obviously contain all piles of the same type, i.e. bored, cast in place concrete, and of the same cross-sectional type and size. Within this type they may be at different locations or under different structures on the same site. Piles in Structure A cannot be classified with the piles in Structure B, even though they are the same type so there would be two separate groups of items for each of these structures, (see also Rule M1).

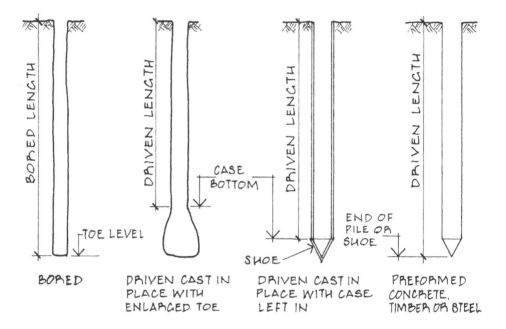

Figure P1

M3 This rule means that the length is determined from the length of the pile which is expressly required without any addition for lengths which are cast above the cut-off level and subsequently removed. Although not specifically stated the toe for bored piles would be the bottom of the bore and for driven cast in place piles it would be the bottom of the casing with no allowance for shoes or enlarged toes. The Preamble should make this clear.

M4 See M2.

M5 See M2.

M6 See M2.

M7 Unlike other piles the driven measurement for interlocking steel piles is square metres and not related to the actual driven depth of the pile. Also the length of piles is treated differently in that the actual length of the pile is not stated but given in very broad bands with no requirement to state the length where it exceeds 24m. Although the function of interlocking steel piles is somewhat different than say preformed concrete piles, the operations involved in driving them are basically very similar. Therefore, the Contractor would still want to know how far the maximum driven depth would be and it is recommended that this is added to the item description. It is also recommended that the actual length of piles exceeding 24m is stated in item descriptions for area of piles.

The undeveloped length of interlocking steel piles is indicated in Figure P2. Although the extra over cost for the supply of special piles is measured separately under P. 8 * 1, this rule means that there is no separate requirement for the driving of special piles and the Contractor's rates for the driving area have to include for driving all types of piles.

UNDEVELOPED LENGTH

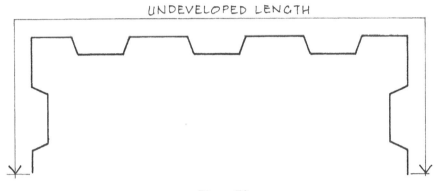

Figure P2

M8 Special piles should always be measured when the Engineer's design specifically calls for them on the drawings or where they are specifically instructed to be installed. Where a run of steel piles meets an abutment on site or where a closure or make up pile is required at changes in direction, the position is not clear. The Engineer's design should wherever possible be such as to make full use of the modular width of interlocking steel piles. If the modular width of a pile is say 500mm, the Engineer would be advised to make a change in direction at say chainage 9500mm rather than 9950mm which may involve the use of a narrow closure pile.

If for some reason the change in direction occurred at 9950mm the question arises whether a closure pile should be measured or not and similarly where the run of piles meets an existing abutment. The prudent taker-off would measure a closure pile in both these instances on the basis that if it is required on site it is already covered in the measurement and if not (it may be that tolerances will allow the 50mm difference to be made up without it) then it is not included in the remeasurement and would be a saving on the Contract.

Definition Rules

D1 The maximum length of the bore or pile to be driven will generally determine the size and type of boring or driving plant which the Contractor will use for the installation of all the piles within the group of items.

D2 The basic types of driven cast in place concrete piles have either concrete or steel casings. With concrete casings, the casing is always left in position and the in situ concrete poured into it. Steel casings can either be left in or withdrawn as the concrete is poured. Permanent casings to cast in place concrete piles are dealt with in Class Q.

D3 The length of the longest pile will determine the size and type of driving plant that the Contractor employs on the site. This is one example of where the measured quantities are the actual gross quantities supplied to the site as opposed to the net quantities finally incorporated in the Works. It should be made clear in Class Q what happens to the surplus cut off lengths. They would either be disposed of by the contractor or remain the property of the Employer (for possible use in extensions). This rule also states that the length of the pile should include the length of the shoe and head which would seem reasonable if they are integral with the pile but unreasonable if they are entirely separate.

D4 See D3.

D6 See D3. This rule would also apply to lengths of special piles.

D7 Any other non-standard pile sections should also be classed as special piles.

Paragraph 5.21

This requires that the Commencing Surface is stated in item descriptions for boring or driving where it is not also the Original Surface and for the Excavated Surface to be stated where it is not also the Final Surface. It is thought that this does not apply to piling operations because Additional Description Rule A3 requires the Commencing Surface to be specified in all cases and because the Excavated Surface (i.e. the bottom of the pile) would always be the Final Surface it would never be stated in item descriptions.

Coverage Rules

C1 It should be made clear either in the Preamble or item descriptions where the excavated material is to be disposed of, i.e. off-site or on-site in stated locations.

Additional Description Rules

A1 This is a very general rule and leaves the scope for item description quite open. For instance, for a preformed prestressed concrete pile it would be necessary to state the specification of the concrete, details of the strands, details of the anchors, whether they are pre or post-tensioned etc. Obviously descriptions may become fairly long and unwieldy and wherever possible reference should be made to drawings or the specification.

A3 Interlocking steel piles do not normally support buildings but act as retaining structures so these piles should be identified by location.
    The identification of the Commencing Surface in piling is very important. The Commencing Surface should be the level at which the boring or driving of the pile itself

commences and not the level at which the boring or driving rig sits. In most instances these two levels will be the same but instances can occur where they are different (see Figure P3).

In the example below the Commencing Surface for driving would be the river bed and not the quay level. Where piling operations are such that the Commencing Surface of the pile is at a different level to the piling rig, it would be prudent to state the two different levels in the item descriptions.

A4 The Second Division gives the diameters of cast in place concrete piles most commonly encountered but other diameters should be added as required. The diameter given should be the outside face of the casing if applicable.

A5 Contiguous bored piles are placed immediately adjacent to each other and are sometimes used as retaining walls (see Figure P4).

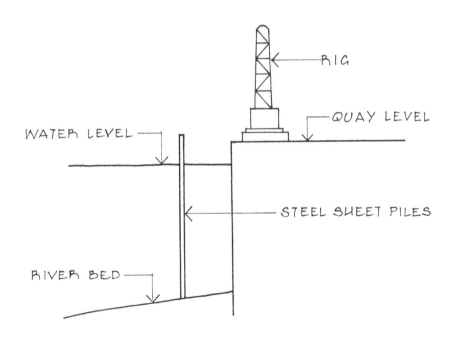

Figure P3

A6 Preformed concrete piles are normally circular but they can also be triangular, square, hexagonal or octagonal. The cross-sectional dimensions for polygonal piles would be the overall or extreme dimensions and item descriptions should so state although the actual classification in the Second Division should be based on the net cross-sectional area of the pile. The diameters or cross-sectional size of circular and square piles would be their external diameter or sizes.

It would also be prudent to state when preformed piles are hollow.

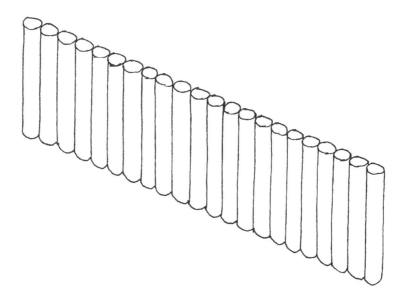

Figure P4 - Contiguous piles

A7 Coatings to preformed piles are applied to prevent down drag on the pile due to ground settlement and the application of a slip coat counteracts this. Coatings are also applied to protect the pile against aggressive ground conditions. Interlocking steel piles are often galvanised or painted with special coatings such as black tar or epoxy resin.

A8 Some proprietary driving heads can be very complicated composite units in which case it would be more appropriate to refer to the manufacturer's catalogue or the specification or drawings.

A9 Although not specifically stated it would also be appropriate to state the cross-section type i.e. circular hollow, H, I, etc.

A10 The section reference to be stated is the manufacturer's reference. The mass per metre is the mass per vertical linear metre of actual pile and not the mass per horizontal metre run of piles.

A11 See A7.

A12 Measurement Rule M7 states that the area of piles is measured through all special piles and these are effectively 'extra over'. Any other piles which are not the standard pile forming the continuous run should also be classed as special piles.

ITEM MEASUREMENT

P. * * * Piling generally

Applicable to all items

Divisions   -

Rules   M1 All depths to be measured along the axis of the pile and from the Commencing Surface. Bored piles to be measured to the toe levels, driven cast in place concrete piles to the bottom of the casings, and other driven piles to the toe levels.

A1 State the materials of which the pile is composed or refer to drawings or specification.

A2 Identify preliminary piles. Identify raked piles and state their inclination ratios.

A3 Identify the structure to be supported or state the location. Identify the Commencing Surface.

Generally   Piling operations are one of the civil engineering activities which can be carried out below water particularly interlocking steel sheet piles. It is recommended that item descriptions for piles installed below a body of open water identified in the Preamble in accordance with Paragraph 5.20 should so state, identifying the body of water.
CESMM 3 cannot cover every variation of pile. For instance jacked piles are sometimes used in underpinning but there is no classification for them. The classifications and rules in Class P should be used for guidance when measuring other types of pile and amended as necessary with suitable preambles.

See also Rule C1.

P. 1&2 * * Cast in place concrete piles

Divisions   Cast in place concrete piles are measured stating the diameter (Additional Description Rule A4) and whether they are bored or driven. For each group of piles separate items are given for the total number of piles, the concreted length and either the maximum bored or driven depth (Measurement Rule M2).

Rules   M3 Measure the concreted lengths from cut-off level to toe level.

    A5 Identify contiguous bored piles.

Generally   Bored piles may be cased or uncased and the casings are generally made from steel or concrete. Concrete casings are invariably left in but steel casings can be either permanent or temporary. It is recommended that item descriptions for bored cast in place concrete piles should state where it is an express requirement that the bore should be temporarily cased together with the type of casing.

    Give details of driving heads and shoes to temporary casings where applicable (driving heads and shoes to permanent casings are dealt with in Class Q).

    See also Rules D1 D2.

P. 3-6 * * Preformed concrete and timber piles

Divisions   Preformed concrete and timber piles are measured stating the cross-section type and dimensions or diameter (Additional Description Rule A6). Preformed concrete piles are further subdivided into preformed, preformed prestressed and preformed sheet piles. For each group of piles separate items are given for the total number of piles stating the lengths of each and the total driven depth (Measurement Rule M4).

Rules   D3 Exclude the length of extension pieces but include the length of heads and shoes.

    A7 Give details of treatments and coatings in items for number of piles.

    A8 Give details of driving heads and shoes in items for number of piles.

Generally   Preformed prestressed concrete piles can be either pre or post-tensioned and it is recommended that item descriptions should state this.

P. 7 * * Isolated steel piles

Divisions   Isolated steel piles are measured stating the mass per metre (Additional Description Rule A9). For each group of piles separate items are given for the total number of piles stating the length of each and the total driven depth

(Measurement Rule M5).

Rules   D4 Exclude the length of extension pieces.

                         A9 State the cross-sectional dimensions.

                         A7 Give details of treatments and coatings in items for number of piles.

Generally   Although not specifically required it would also be prudent to state the cross-section type.

                         See also Rule D5.

P. 8 * *   Interlocking steel piles

Divisions   Interlocking steel piles are measured stating the section modulus (Additional Description Rule A10). For each group of piles separate items are given for the total area of piles supplied, the driven area and the length of special piles which are effectively 'extra over' (Measurement Rule M6).

Rules   M7 Calculate areas from undeveloped lengths and depths as defined in M1 and D6.

                         M8 Measure special piles only where they are expressly required.

                         D6 Exclude the length of extension pieces.

                         A10 State the section reference or mass per metre.

                         A11 Give details of treatments and coatings in items for number of piles.

                         A12 State the type of special piles.

Generally   Interlocking steel piles are the only type of pile which are generally re-usable. They are often used in temporary retaining walls and withdrawn after use. It is recommended that where they are temporary, item descriptions should so state it.
          Also state the driven depth of piles if considered to be appropriate.

                         See also Rule D7.

STANDARD DESCRIPTION LIBRARY

Bored cast in place concrete piles; building C; Commencing Surface Original Surface; sulphate resisting concrete grade C20; reinforced; as specification clause 20.1

Diameter 300mm

P. 1 1 1	number of piles	nr
P. 1 1 2	concreted length	m
P. 1 1 3	depth bored to 30m maximum depth	m

Diameter 300mm; preliminary

P. 1 1 1.1	number of piles	nr
P. 1 1 2.1	concreted length	m
P. 1 1 3.1	depth bored to 30m maximum depth	m

Driven cast in place concrete piles, building D; Commencing Surface underside of topsoil; sulphate resisting concrete grade C20, reinforced; as specification clause 20.2

Diameter 500mm; raked; inclination ratio 1:10

P. 2 3 1	number of piles	nr
P. 2 3 2	concreted length	m
P. 2 3 3	depth bored or driven to 25m maximum depth	m

Preformed concrete piles; building E, Commencing Surface 500mm below Original Surface; concrete grade C30; as specification clause 20.3

Cross-sectional size 300 x 300mm; square section

P. 3 3 1	15m long; coated with bitumastic; integral shoe; steel driving head	nr
P. 3 3 2	depth driven	m

Preformed timber piles; building F; Commencing Surface reduced level; pitch pine; as specification clause 20.4

Cross-sectional size 400 x 400mm; square section; preliminary

P. 6 5 1	20m long; cast iron shoe and driving head	nr
P. 6 5 2	depth driven	m

Isolated steel piles; inlet tower; Commencing Surface underside of general building excavation; grade WR50A; as specification clause 20.5

Mass 85kg/m; cross-sectional size 254 x 254mm

P. 7 4 1	10m long; galvanised	nr
P. 7 4 2	depth driven	m

Interlocking steel piles; located as drawing number P/1; Commencing Surface river bed; section reference E230; as specification clause 20.6

Section modulus 230cm3/m

P. 8 1 1	corner piles	m
P. 8 1 1.1	junction piles	m
P. 8 1 1.2	closure piles	m
P. 8 1 2	driven area	m2
P. 8 1 3	area of piles of length not exceeding 14m	m2
P. 8 1 4	area of piles of length 14-24m	m2

Chapter 19

CLASS Q: PILING ANCILLARIES

GENERAL TEXT

Principal changes from CESMM 1

There were several changes to this class.

The seeming duplication of items for boring through obstructions was resolved and breaking out of obstructions was measured in hours only (Q. 7 0 0). A new unit of measurement was adopted for cutting off surplus lengths of pile (Q. 1-6 7 *) and the classification for pile extensions required them to be described as exceeding and not exceeding 3m (Q. 3-6 5&6 *). Similarly, permanent casings had to be identified as either exceeding or not exceeding 13m (Q. 1 3 &4 *). Reinforcement cages were further subdivided into straight bars and helical bars. The separate classification for the disposal of surplus materials was deleted (Q. 7 0 0 in CESMM 1) and the classification of obstructions was simplified (Q. 7 0 0). The classification of pile tests was amended (Q. 8 * *).

Measurement Rules were included in connection with the measurement of permanent casings and reinforcement (M2, M3 and M4), and also for defining when obstructions were measured (M11).

Definition Rules D1, D3, D4 and D5 made it clear that the Third Division classifications relate to the pile and not to the ancillary item being measured.

Coverage rules were included defining the disposal of surplus materials (C1), shoes and heads to permanent casings (C2), reinforcement (C3), pre-boring (C4) pile extensions (C5) and filling hollow piles (C6).

Additional Description Rule A2 required the materials, thicknesses and treatments to permanent casings to be stated as well as stating when cutting off surplus lengths includes a permanent casing.

It was a requirement to distinguish between tests to preliminary piles and working piles (Additional Description Rule A6), and also where load tests are to raking piles (Additional Description Rule A7).

Principal changes from CESMM 2

In Definition Rule D2 the reference to BS4461 is deleted because it is now incorporated into BS4449:1988.

A new Additional Description Rule A6 is included which requires that the material for pile extensions shall be stated in the item description for the length. The previous A6 and A7 are renumbered to accommodate this new rule and the wording of the new A6 is repeated twice to cover isolated and interlocking steel piles.

Measurement Rules

M1   This rule means that items under this class are not automatically measured except for backfilling empty bores for cast in place concrete piles. It would be necessary for the taker-off to liaise closely with the Engineer about which items and quantities are to be included in this class.

M2   CESMM 3 chooses to deal with permanent casings for cast in place concrete piles under this class rather than Class P. This is considered unusual because the case is left in the pile (particularly if it is the driven variety) is similar to a hollow preformed concrete or steel pile filled with concrete and one might expect the cost of the casing to be included with the items measured under Class P. However, permanent casings to both bored and driven cast in place concrete piles are included here and this rule defines their length as being from the Commencing Surface to the bottom of the casing i.e. measured net. This compares with Class P which defines the lengths of preformed and steel piles as the actual lengths to be supplied i.e. measured gross (Definition Rules D3, D4 and D5 of Class P). Situations can also arise where the permanent casing is required to be left above the Commencing Surface with the surrounding levels subsequently made up and in this case the length of the permanent casing if measured in accordance with this rule would be too short. It is recommended that this rule is amended to suit the situation if it is considered desirable and that where express lengths are required to be supplied these are used for the purpose of the measurement of the permanent casing. It is also interesting to compare this rule with Definition Rule D3 of Class P which states that the measured length of preformed piles to be taken to the underside of the shoe and not the bottom of the casing.

M3   The measurement of reinforcement in piles is one of the examples where the work is not measured in accordance with Paragraph 5.18 i.e. net from the drawings. This rule makes it necessary for the taker-off to measure the additional mass of reinforcement in laps. Potentially the rule is open to abuse by the Contractor in that he could put as many laps as he chooses (within the confines of the specification) and be paid for them all. Also the rule does not impose any constraints with regard to the length of laps measured although the length measured should only be the minimum length and not any length which the Contractor chooses to use over and above the minimum length. This rule would also mean the remeasurement of the reinforcement on site which would be very time consuming in that every lap in the cage would have to be counted before the cage was placed and this could clearly take up more time than the item is worth. It is recommended that this rule is written out by way of a suitable preamble unless there is a strong reason for leaving it in.

M5   The additional driven length of the pile due to the addition of an extension is automatically accounted for in the measurement in Class P. This is because the calculation of the driven length is not based on the length of the actual pile itself but on the difference between the level of the Commencing Surface and the toe or bottom of the casing (See Measurement Rule M1 of Class P).

M6   Under the item for number of pile extensions the Contractor would include for preparing the end of the extension and tying it into the prepared head of the driven pile and also for removing and refixing the pile driving head. The items for length of pile extensions would include only for the supply and delivery of the extensions themselves.

M7   The situation regarding pile extensions which utilize surplus lengths of pile requires some explanation. Pile extensions are generally required for driven piles which are of insufficient length to reach the required 'set'. Most of today's piling systems have special or proprietary jointing techniques for pile extensions enabling the joint to be made quickly and easily. Surplus lengths will require more preparation work to enable them to be used as extension pieces and it is recommended that if it is known or envisaged that surplus lengths will be used as extension pieces then item descriptions for the extensions should so state.

It is interesting to note that there is no facility for the measurement of extensions to driven cast in place piles which have a permanent casing (particularly concrete). Presumably this is because Measurement Rule M2 defines the measurement of permanent casings as being the net length driven into the ground as opposed to the actual length to be supplied and therefore any extensions are deemed to be included. This is not clear and a suitable preamble should be included accordingly.

M8   See Figure Q1 for explanation of this rule.

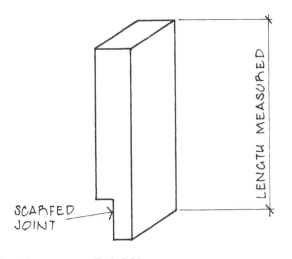

Figure Q1

M9   See Measurement Rule M6.

M10 See Measurement Rule M6.

M11 Items for the removal of obstructions are measured only for bored piles. Driven piles which encounter obstructions are either relocated, in which case the abortive driving would be accounted for in the remeasurement, or the obstruction is removed by other means. In this case the Contractor would have to be paid through remeasurement in other classes of CESMM 3 or through the Contract. There are instances where certain piles can be driven through hard materials, in particular steel H piles, although these are not in common use. Where the Engineer envisages being able to drive through

223

obstructions due to the type of pile then this rule should be amended accordingly.

Obstructions are only measured for those encountered above the founding stratum. Strictly speaking this means that any boring into the founding stratum is not measured as obstructions. As the founding stratum can often be rock, and as it may be an express requirement to bore some way into it, it may be desirable that this also is measured as obstructions. A suitable preamble would have to be incorporated stating the deviation from CESMM 3.

Although not specifically stated the measurement of obstructions could also be applicable for pre-boring and if this is the case then this rule should be amended.

The measurement of obstruction hours is effectively 'extra over' the boring of the pile and no deduction should be made to the length measured under Class P for lengths which are bored through obstructions.

D1,
D3,
D4,
and
D5
These rules prevent any argument whether the classifications in the Third Division apply to the pile or the ancillary item being measured. The classifications are generally given as ranges or bands and there is no requirement to state the actual size or cross-sectional type of the pile. This may make estimating the cost of some ancillary items difficult. It may be desirable that in accordance with Paragraph 5.10 additional description is added as necessary in order to describe more fully the nature of the work concerned. This would involve stating the actual bore of pre-boring, stating the internal size of hollow piles to be filled and stating the actual cross-sectional type and size of the pile or in the case of interlocking steel piles the section reference or mass per metre and the section modulus.

D2
This rule implies that the diameter should be stated in item descriptions for all types of bar reinforcement which is not the case. According to the Third Division straight bars have only two classifications, not exceeding and exceeding 25mm; it is only helical bars for which the actual size has to be stated. This classification of straight bars contrasts with the measurement of bar reinforcement in Class G (Concrete Ancillaries) where the actual size of the bar has to be stated for sizes up to 32mm diameter. As well as assisting the estimator the advantage of the classification in Class G is that it is of more use in post contract control and readily deals with any variations in the quantities of the various bar diameters. Because the various bar sizes would have to be scheduled individually in order for the total mass to be calculated, it is recommended that the classification of straight bar reinforcement as detailed in Class G is adopted in this class. A suitable preamble would have to be included stating the deviation from CESMM 3.

Coverage Rules

C1
It should be made clear in the Preamble or item descriptions where surplus material, particularly excavated material, is to be disposed of, whether off site or on site to stated locations. If the material is to remain on site, then it would be necessary to state the precise location at which the material is to be disposed (or give maximum haul distances) and also how the material is to be treated ie. tipped only, tipped and spread, tipped spread and compacted etc if this is otherwise unclear.

C2   It follows that item descriptions should give details of the driving heads and shoes (see also Additional Description Rule A8 in Class P).

C3   The Engineer must detail the supporting reinforcement comprehensively so that the Contractor cannot come back at a later stage requesting an increase in rate for supporting reinforcement which was not shown on the drawings or inadequately detailed.

C4   Pre-boring is generally only carried out to driven piles in order to ease driving. The pre-bore is usually of smaller size than the actual pile obviating the need for excessive grouting.

C5   This would include exposing the reinforcement of both extensions and pile. It would not be necessary to measure a separate item for the preparation of the head of the pile (Q. 4 8 *) because this would be included in the item of the extension.

Additional Description Rules

Generally   Unlike Class P there are no general Additional Description Rules which apply to the whole of the Piling Ancillaries Class (see A1, A2 and A3 of Class P). For items such as pile extensions it may be desirable to state which materials the extension is composed of (Rule A1), whether it is raking (Rule A2) and its location (Rule A3), although there is no requirement to state these. It is recommended that the additional description in Rules A1, A2 and A3 of Class P should be extended to this class insofar as it may be applicable.

A1   The piling technique commonly used for forming enlarged bases comprises hammering dry concrete or hardcore at the bottom of the pile until a predetermined set is reached. Under these circumstances it is not possible to specify accurately the diameter of the enlarged base or toe because it is dependent solely on the amount of material which is forced in and it is recommended that in this case this rule is suitably amended. It is possible to specify the diameter of enlarged bases for belled piles. These are bored piles which are sunk with fairly complex equipment enabling the lower portion of the bore to be widened but they are not commonly used these days.

A2   Following on from Coverage Rule C2 it would also be necessary to give details of the shoes and heads. Descriptions for cutting off surplus lengths which include permanent casings should also state the nature of the casing.

A3   Materials for reinforcement are normally plain round steel bars or deformed high yield steel bars to B.S.4449. Stainless steel may also be used but is not very common.

A4   Couplers which are used for jointing reinforcement at the Contractor's own volition would be deemed to be included in the rates. Details of couplers would only be included in accordance with this rule when the Engineer specifically required jointing to be carried out in this manner.

A5 The Specification of the concrete would usually comply with the requirements of BS5328 (see also Class F: In situ Concrete).

A6 The materials used for pile extension must be included in the item descriptions for the length.

A7 If the recommendation made earlier with regard to Additional Description and
and Rules A1, A2 and A3 of Class P is adopted then these rules will have already been
A8 complied with in respect of identification of preliminary and raking piles (see beginning of this section - Generally).

ITEM MEASUREMENT

Q. * * * Piling ancillaries generally

Applicable to all items

Divisions -

Rules	M1 Apart from backfilling empty bores for cast in place concrete piles work in this class is only measured where is it is expressly required.
Generally	It is also recommended that the requirements of Additional Description Rules A1, A2 and A3 of Class P are applied here also insofar as they may be applicable.

See also Rule C1.

Q. 1 1-4 * Pre-boring, backfilling empty bore, and permanent casings to cast in place concrete piles

Divisions	Pre-boring, backfilling empty bore, and permanent casing are measured in linear metres. It is recommended that the actual diameter of the pile is stated.
Rules	M2 Measure permanent casings from the Commencing Surface to the bottom of the casing.
	A2 State the materials, thickness and details of treatments and coatings for permanent casings.
Generally	It is recommended that the actual diameter of pre-boring is stated. The length of permanent casings which project above the Commencing Surface should be included in the measured length. Give details of shoes and heads. Ensure that extensions to permanent casings are covered with a coverage rule or measured separately.

See also Rules D1, C2.

Q. 1 5&8 * Enlarged bases and preparing heads to cast in place concrete piles

Divisions   Enlarged bases and preparing heads are enumerated. It is recommended that the actual diameter of the pile is stated.

Rules   A1 State the diameter of enlarged bases where this is considered applicable.

Generally   Do not measure preparation of heads for pile extensions.

See also Rule D1.

Q. 1 7 * Cutting off surplus lengths to cast in place concrete piles

Divisions   Cutting off surplus lengths is measured in linear metres. It is recommended that the actual diameter of the pile is stated.

Rules   A2 Items which include permanent casings should so state. It is recommended that the casing material is also stated.

Generally   Although the unit of measurement given is linear metres, this is not considered particularly relevant to the item involved. A ten metre length of 'cutting off surplus lengths' could either involve ten one metre cut offs or two five metre cut offs or further variations. It is recommended that the number of cut offs is also stated in item descriptions.

See also Rule D1.

Q. 2 1 * Reinforcement to cast in place concrete piles

Divisions   Reinforcement is measured in tonnes, distinguishing between straight bars and helical bars. It is recommended that the sizes given in item descriptions conform to the classification in Class G.

Rules   M3 Include the mass of reinforcement in laps.

A3 State the materials used.

A4 Give details of couplers but only where they are expressly required to be used.

Generally   See also Rules M4, D2, C3.

**Q. 3-5 1-3 * Pre-boring, jetting and filling hollow piles to preformed
concrete, timber and isolated steel piles**

Divisions   Pre-boring, jetting and filling hollow piles are measured in linear metres. It is recommended that the actual cross-sectional type and size or mass of the pile is stated instead of the ranges given in the Third Division.

Rules   A5 State the specification of concrete for filling hollow piles (in accordance with BS5328).

Generally   For pre-boring it is recommended that the actual diameter of the bore is also stated. For filling hollow piles with concrete it is recommended that the internal diameter of the pile is also stated.

See also Rules D3, C4, C6.

Q. 3-6 4&5 * Pile extensions to preformed concrete, timber, isolated steel and interlocking steel piles

Divisions   Pile extensions are measured by enumerating the extensions, together with an item for the total length of the extensions categorised into extensions exceeding and not exceeding 3m long. It is recommended that the actual cross-sectional type and size is stated for preformed concrete, timber and isolated steel piles and that the mass or section modulus of the pile is stated for interlocking steel piles.

Rules   M7 Do not include in the length of extensions the lengths which are obtained from surplus cut off lengths.

M8 Include the length of joints for timber pile extensions.

A6 State material of pile extension.

Generally   If applicable state in item descriptions where extensions are to be used from surplus lengths. Do not measure preparation of heads for pile extensions.

See also Rules M5, M6, M9, D3, D4, D5, C5.

Q. 3-6 7 * Cutting off surplus lengths to preformed concrete, timber, isolated steel piles and interlocking steel piles

Divisions   Cutting off surplus lengths is measured in linear metres. It is recommended that the actual cross-sectional type and size is stated for preformed concrete, timber and isolated steel piles and that the mass or section modulus of the pile is stated for interlocking steel piles.

Rules   -

Generally   See also Q. 1 7 *.

Q. 3-6 8 * Preparing heads to preformed concrete, timber, isolated steel piles and interlocking steel piles

Divisions   Preparing heads are enumerated. It is recommended that the actual cross-section type and size is stated for preformed concrete, timber and isolated steel piles and that the mass or section modulus of the pile is stated for interlocking steel piles.

Rules   -

Generally   Do not measure for pile extensions.

Q. 6 1&2 * Pre-boring and jetting to interlocking steel piles

Divisions   Pre-boring and jetting are measured in linear metres. It is recommended that the actual section modulus be stated.

Rules   -

Generally   It is recommended that the actual bore of the pre-bore be stated.

              See also Rule D5.

Q. 7 0 0 Obstructions

Divisions   Obstructions are measured in hours.

Rules   M11 Only measure for breaking out rock or artificial hard materials
                  encountered above the founding stratum of bored piles.

Generally   If necessary also measure for certain types of driven pile and pre-boring.
                  Measure for boring into the founding stratum if considered to be applicable.

Q. 8 * * Pile tests

Divisions   Pile tests are enumerated stating the type in accordance with the Second Division.

Rules   A7   Identify tests to preliminary piles.

              A8   State the loads for loading tests. Identify loading tests to
                   raking piles.

Generally   -

STANDARD DESCRIPTION LIBRARY

Cast in place concrete piles

Diameter 300mm

Q. 1 1 1	pre-boring	m
Q. 1 2 1	backfilling empty bore with excavated material	m
Q. 1 3 1	permanent casings; each length not exceeding 13m	m
Q. 1 5 1	enlarged bases; diameter 750mm	nr
Q. 1 7 1	cutting off surplus lengths	m
Q. 1 8 1	preparing heads	nr

Reinforcement; deformed high yield steel bars to BS4449

Q. 2 1 1	straight bars; nominal size not exceeding 25mm	t
Q. 2 1 3	helical bars; nominal size 10mm	t

Preformed concrete piles

Cross-sectional area 0.05-0.1m2

Q. 3 1 3	pre-boring	m
Q. 3 2 3	jetting	m
Q. 3 3 3	filling hollow piles with concrete grade C15	m
Q. 3 4 3	number of pile extensions	nr
Q. 3 6 3	length of concrete pile extensions; each length exceeding 3m	m
Q. 3 7 3	cutting off surplus lengths	m
Q. 3 8 3	preparing heads	nr

CLASS Q: PILING ANCILLARIES

Timber piles

Cross-sectional area 0.15-0.25m2

Q. 4 1 6	pre-boring	m
Q. 4 2 6	jetting	m
Q. 4 4 4	number of pile extensions	nr
Q. 4 4 6	length of timber pile extensions; each length exceeding 3m	m
Q. 4 7 6	cutting off surplus lengths	m
Q. 4 8 6	preparing heads	nr

Isolated steel piles

Mass 60-120kg/m

Q. 5 1 4	pre-boring	m
Q. 5 2 4	jetting	m
Q. 5 3 4	filling hollow piles with concrete grade C15	m
Q. 5 4 4	number of pile extensions	nr
Q. 5 6 4	length of steel pile extensions; each length exceeding 3m	m
Q. 5 7 4	cutting off surplus lengths	m
Q. 5 8 4	preparing heads	nr

Interlocking steel piles

Section modulus 3000-4000cm3/m

Q. 6 1 6	pre-boring	m
Q. 6 2 6	jetting	m
Q. 6 4 6	number of pile extensions	nr
Q. 6 6 6	length of interlocking steel pile extensions; each length exceeding 3m	m
Q. 6 7 6	cutting off surplus lengths	m

Interlocking steel piles (cont'd)

Q. 6 8 6	preparing heads	nr

Obstructions

Bored piles

Q. 7 0 0	rock or artificial hard material above founding stratum	h

Pile tests

Constant rate of penetration

Q. 8 2 2	test load 150t; preliminary piles	nr

Non-destructive integrity

Q. 8 4 0	generally	nr

Chapter 20

CLASS R: ROADS AND PAVINGS

GENERAL TEXT

Principal changes from CESMM 1

The only change to the divisions in this class was that geotextiles has been added as an additional classification (R. 1 7 0). Other bar reinforcement became high yield bar reinforcement (R. 4 7 *) and the size classification was changed in that the largest classification is now 32mm or greater.

As with Class G (Concrete Ancillaries) the mass of reinforcement includes the mass of steel supports to top reinforcement. (Measurement Rule M4).

Principal changes from CESMM 2

The references in the BS numbers for reinforcing bars have been amended: BS4461 is deleted and those to BS4449 updated.

In the Second Division the wording of item R. 4 5 * has been changed to require that the material must be stated where reinforcement other than BS4483 is used. This now ties in with Additional Description Rule A4.

There is also a change of wording in the Second Division item R. 6 1-5 * where the references to BS have been updated.

A new measuremnt Rule M2 has been inserted which repeats the instruction in M24 of Class E, stating that the area contained in the laps of geotextiles shall not be measured. This means that Rule M2 to M9 are renumbered M3 to M10. Additional Description Rule A4 has been amended to confirm to the change to reinforcement of material other than steel, in item R. 4 5 *.

Measurement Rules

M1   On the face of it this rule would seem to be fairly simple but it does require detailed explanation.

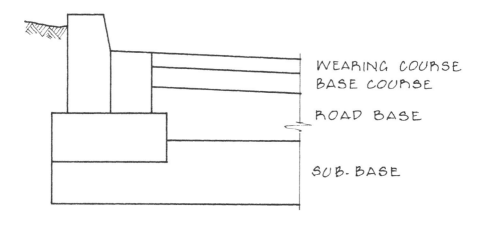

Figure R1

The measurement of the various courses in Figure R1 poses several measurement questions.

Sub-base   Is the top surface as defined in this rule the width up to the kerb foundation or does it extend under the kerb foundation as well, or is the width of sub-base under the kerb foundation classed as filling to kerbs and measured under Class E?

    The answer is not clear but it is thought that as the sub-base under the foundation would be laid at the same time as the main body of the layer, to class it as filling to kerbs would be inappropriate. It is also recommended that in this case the top surface is taken as being the full width of the layer extending under the kerb.

Road base   Is the top surface the full width up to the inside face of the channel or up to the inside face of the foundation where the main body of the layer terminates?

    It is recommended that the width measured is taken as being the maximum width.

Wearing   Is the top surface the net plan area or the gross surface area taking into
course   account the additional area due to cambers and falls?

    The additional area due to the camber would be negligible on most contracts but could have some effect on major motorway schemes. It is impractical for the taker- off to make allowance for cambers and it is recommended that the area measured is the net plan area.

All the above should be made clear in the Preamble.

M2   The area of geotextiles shall be measured ignoring any laps.

M5   See Measurement Rule M8 of Class G.

M7   The words 'at locations where they are expressly required' are extremely important. Paragraph 1.6 defines the words expressly required as being '...described in the Specification'. Although there would not be any express requirement for them to be at specific locations the Engineer would normally want the Contractor's own day construction joints to be carried out in a particular way and would include details of their construction in the Specification; it could be construed therefore that they are expressly required. The addition of the words 'at locations' solves this potential problem.

M8   This rule means that excavation and filling for kerbs, channels and edgings has to be measured separately and normally this would be included in the overall excavation for the road construction. Filling in this case would not usually include volumes of sub-base carried under the kerb as part of the normal sub-base layer (see Measurement Rule M1) but would generally only be the backfilling required behind the kerb line (see Figure R2).

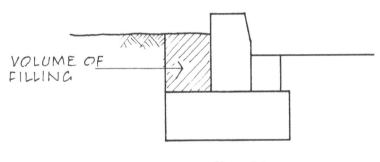

Figure R2

M9   In other words the length measured is the net length as laid.

M10 The taker-off should use his discretion whether supporting structures to traffic signs should be measured separately. If he elects to do so, then they must be measured in detail from the excavation of the foundation to the painting of the members.

Definition Rules

D1   The Specification for Road and Bridge Works also has subsidiary publications which are associated with it (such as the Notes for Guidance) and these should all be read in conjunction with each other.

Coverage Rules

C1   It follows that the drawings and/or specification must state precisely the finishes to the concrete. Concrete items which have different finishes must be measured separately.

C2   All supports and tying wire except supports to top reinforcement are deemed to be included in the Contractor's rates.

C3   This rule now makes it perfectly clear that the rates for kerbs are to include for all the ancillary items that are associated with them. The rule specifically mentions beds, backings, reinforcement and joints but any other items such as dowels would also be included and this should be made clear in item descriptions or the Preamble.

C4   The problem in trying to list all the work to be included in a measured item is the danger of omission. This rule attempts to do just this but does not list every item which may be required. For instance what if the sign has a bolted down grouted base plate? This rule makes no mention of bolts or grouting. It is clear that items for traffic signs measured in accordance with this rule are fully inclusive items and the rule should be expanded to cover any further items not already mentioned or reference should be made to the relevant drawing or specification clause.

Additional Description Rules

A1   By specifying the actual depth of the course, it is unnecessary to state the ranges of thickness as given in the Divisions and these should be used for coding purposes only, (Paragraph 3.10).

A2   The angle referred to in this rule would generally be that along the longitudinal axis of the road. It would be a rare occurrence where the angle to the horizontal across the width of the road exceeded 10 degrees.

A5   Waterproof membranes are often specified by their grade or gauge as opposed to their thickness and it would be acceptable to use this as an alternative.

A6   Class G (Concrete Ancillaries) measures joints in concrete structures divided into their constituent parts. Class R requires that joints in concrete pavements to be measured as all encompassing items with details of all the constituent parts given in the item description of the joint. One reason for this approach could be that the joints in concrete pavements will be of uniform width and specification throughout their length, whereas joints in concrete structures can vary considerably. However, should the joints not be uniform then it may be prudent to measure them in accordence with Class G.

A7   Any other salient features should also be stated or reference made to the relevant drawing or specification clause.

Note to Class R

The note to Class R states that the earthworks for kerbs, channels and edgings may be included in the item descriptions. Generally the excavation for the kerb would form part of the overall excavation for the road itself and would not therefore warrant separate measurement.

ITEM MEASUREMENT

R. * * *   Roads and pavings generally

Applicable to all items
Divisions -

Rules   A1   State the material and depth of the course or slab and the spread rate of applied surface finishes.

           A2   State when work exceeds 10 degrees to the horizontal.

Generally   See also Rules D1, C1.

R. 1-3 * *   Sub-bases, flexible road bases and surfacing

Divisions   Sub-bases, flexible road bases and surfacing are measured in square metres except for additional depths of stated material (measured in cubic metres) and regulating courses (measured in tonnes). For all items except geotextiles, additional depth of stated material and regulating courses, the thickness of the course has to be stated.

Rules   M1   Measure the width of each course along its top surface. Do not deduct the area of intrusions unless they exceed 1m2.

           A3   State the type and grade of material for geotextiles.

Generally   It is not clear whether items for additional depth of stated material should include for any additional excavation required. This item would primarily be measured for soft spots in a similar manner to item E. 5 6 0, and because this includes for excavation it is recommended that additional depth of stated material should be treated in the same way. The item description or preamble should state that additional excavation is deemed to be included and it should also be made clear where the surplus material is to be disposed of.

           For items such as slurry sealing, surface dressing and bituminous spray it would be difficult to state the actual thickness of the layer. In these cases the thickness should be substituted with the rate of application.

It should be made perfectly clear in the Preamble under what circumstances the laying of regulating courses will be paid for. It should not be paid for when it is repairing damage caused by the Contractor's plant using the surface as a site road. It should only be paid when the regulating course is required because of circumstances which are beyond the control of the Contractor.

The items listed in the Second Division are more often than not the generic term for several types of that material. It is necessary to state the precise nature of the material in item descriptions.

R. 4 1-3 * Concrete pavements

Divisions — Concrete pavements are measured in square metres stating the quality of the concrete and the depth.

Rules — -

Generally — -

R. 4 4&5 * Fabric reinforcement

Divisions — Fabric reinforcement is measured in square metres.

Rules — M2 Exclude the additional area of fabric in laps.

A4 State the reference of fabric reinforcement which is to B.S.4483. State the materials, sizes and nominal mass per square metre for other types of fabric. It would not be necessary to state the mass of reinforcement for fabric which is to the British Standard.

Generally — -

R. 4 6&7 * Bar reinforcement

Divisions — Bar reinforcement is measured in tonnes stating the type of material, relevant British Standard and the nominal size.

Rules — M4 Take the mass of reinforcement as being 0.785kg/100mm2 of cross-sectional area.

M5 Measure the mass of steel supports to top reinforcement.

Generally — It is interesting to note that the measurement of bar reinforcement in this class differs from Class G. Class G has two further classifications for bar reinforcement viz: stainless steel bars of stated quality and reinforcing bars of other stated quality (G. 5 3&4 *). Also Class G has a classification for the measurement of special joints (G. 5 5 0). It is recommended that if the need arises then these items should be incorporated into this class.

See also Rules D2, C2.

R. 4 8 0 Waterproof membranes

Divisions   Waterproof membranes are measured in square metres.

Rules   M6 Exclude the additional area of fabric in laps.

             A5 State the materials and thickness or gauge.

Generally   -

R. 5 * * Joints in concrete pavements

Divisions   Joints in concrete pavements are measured in linear metres stating the type of
             joint and the depth.

Rules   M7 Measure construction joints only where they are expressly required at
                 stated locations.

             A6 State the dimensions, spacing and nature of sealed grooves and rebates,
                 waterstops, dowels and other components.

Generally   It is not clear whether the actual depth of the joint has to be stated or whether
             the depth as classified in the Third Division is sufficient. Because the actual
             depth is more meaningful it is recommended that this is stated.

R. 6 * * Kerbs, channels and edgings

Divisions   Kerbs, channels and edgings are measured in linear metres. Quadrants, drops
             and transitions are enumerated and are full value items and therefore the
             kerbs, channels and edgings in which they occur are not measured through
             them.
                 Items measured in linear metres have to distinguish between those which
             are straight or curved exceeding 12m and those curved not exceeding 12m.

Rules   M8 Measure excavation and filling to kerbs, channels and edgings in Class E
                 or include it in the items themselves either by way of preamble or
                 additional item desciption.

             A7 State the sizes and materials including beds and backings in item
                 descriptions.

Generally   See also Rule C3.

R. 7 * *   Light duty pavements

Divisions   Light duty pavements are measured in square metres. State the precise nature of the material and the actual depth.

Rules   M1 Measure the course at the top surface. Do not deduct the area of intrusions unless they exceed 1m2.

Generally   -

R. 8 1 *   Traffic signs

Divisions   Traffic signs are enumerated stating whether they are illuminated or non-illuminated.

Rules   M10 Measure substantial supporting structures in detail in accordance with other classes.

A8 State the materials, size and diagram number taken from 'Traffic signs, regulations and general directions' issued by the Department of Transport.

Generally   Although not specifically stated, it should be made clear of the extent, if any, of electrical work which is to be included with the items of illuminated signs.

See also Rule C4.

R. 8 2 1-3   Road studs, letters and shapes

Divisions   Road studs, letters and shapes are enumerated. State whether studs are reflecting or non-reflecting.

Rules   A8 State the materials, size and diagram number taken from 'Traffic signs regulations and general directions' issued by the Department of Transport.

A9 State the shape and colour of aspects for studs.

Generally   -

R. 8 2 4&5   Line markings

Divisions   Line markings are measured in linear metres.

Rules   M8 Do not include the gaps in the measurement of intermittent markings.

A8 State the materials, size and diagram number taken from 'Traffic signs regulations and general directions' issued by the Department of

Transport.

Generally   -

STANDARD DESCRIPTION LIBRARY

Sub-bases; flexible road bases and surfacing

Granular material; DTp Specified type 1 sub-base

R. 1 1 7	depth 300mm	m2

Soil cement; DTp Specification clause 805 road base

R. 1 3 4	depth 150mm	m2

Hardcore road base

R. 1 6 8	depth 350mm	m2

Geotextiles

R. 1 7 0	Terram	m2

R. 1 7 0.1	Terram; inclined at an angle exceeding 10 degrees to the horizontal	m2

Additional depth of material

R. 1 8 0	granular material; DTp Specified type 1 sub-base	m3

R. 1 8 0.1	hardcore	m3

Wet mix macadam; DTp Specification clause 808 base course

R. 2 1 2	depth 60mm	m2

Dense bitumen macadam; DTp Specification clause 908 wearing course; 14mm nominal size aggregate

R. 2 3 2	depth 40mm	m2

Open textured tarmacadam; DTp Specification clause 913 base course; 14mm nominal size aggregate

R. 2 6 2	depth 60mm; inclined at an angle exceeding 10 degrees to the horizontal	m2

Cold asphalt; DTp Specification clause 910 wearing course; 10mm nominal size aggregate

| R. 3 1 2 | depth 50mm | m2 |

Removal of flexible surfacing

| R. 3 6 1 | depth 25mm | m2 |

| R. 3 6 3 | depth 100mm; inclined at an angle exceeding 10 degrees to the horizontal | m2 |

Regulating course

| R. 3 8 0 | dense tarmacadam | t |

Concrete pavements

Carriageway slabs of DTp specified paving quality

| R. 4 1 4 | depth 125mm | m2 |

| R. 4 1 4.1 | depth 125mm; inclined at an angle exceeding 10 degrees to the horizontal | m2 |

Other carriageway slabs; grade C20

| R. 4 2 4 | depth 125mm | m2 |

Steel fabric reinforcement to B.S.4483

| R. 4 4 2 | reference A142 | m2 |

| R. 4 4 5 | reference B503 | m2 |

Plain round steel bar reinforcement to BS4449

| R. 4 6 1 | nominal size 6mm | t |

| R. 4 6 8 | nominal size 32mm or greater | t |

Waterproof membranes below concrete pavements

| R. 4 8 0 | plastic sheet; 1200 gauge | m2 |

Joints in concrete pavements

Longitudinal joints; 10mm diameter x 750mm long mild steel
dowels at 750mm centres; 12 x 20mm groove sealed with
polysulphide sealant

R. 5 1 4   depth 125mm   m

Kerbs, channels and edgings

Precast concrete kerbs to BS7263; Part 1; figure 1(b)

R. 6 3 1   straight or curved to radius exceeding 12m;
as drawing number RP/DWG/1   m

Precast concrete edgings to BS7263; Part 1; figure 1(n)

R. 6 6 2   curved to radius not exceeding 12m; as drawing
number RP/DWG/2   m

Quadrants; B.S.340

R. 6 1-8 3   455 x 455mm; as drawing number RP/DWG/3   nr

Light duty pavements

Granular base; DTp Specified type 1

R. 7 1 3   depth 100mm   m2

Precast concrete flags to B.S.368; 600 x 450mm

R. 7 8 2   depth 50mm   m2

Ancillaries

Traffic signs

R. 8 1 1   non-illuminated; steel; diagram number__;
size__   nr

Surface markings

R. 8 2 3   letters and shapes; polyester; diagram number___;
size___   nr

R. 8 2 4   continuous lines; rubber paint; 100mm wide   m

Chapter 21

CLASS S: RAIL TRACK

GENERAL TEXT

Principal changes from CESMM 1

Class S has had a major overhaul.

The divisions to the class were rationalised in their format and most of the original descriptive features are contained in the rules. Additional classifications were created such as waterproof membranes (S. 1 5 0), taking up track (S. 2 * *), and lifting, packing and slewing (S. 3 * *). The notes were extensively amplified and the class wasnow much more definitive.

Principal changes from CESMM 2

The Second Division item 4-5.1* 'Switches and Crossings' has been replaced by two items from the Third Division - 'Turnouts' and 'Diamond Crossings'. This change in terminology also affects the wording in rules M3, M4, M5, C1, C6, C9, A10 and A16.

Measurement Rules

M2 This rule is not really necessary (see Paragraph 5.18).

M3 An item measured for taking up one linear metre of plain track includes for taking up two metres of rail. It may be necessary to define on the drawings or in the Specification exactly at what point plain track is deemed to become part of a turnout or diamond crossing in order to prevent arguments at a later stage. This applies to all items involving the measurement of turnouts and diamond, crossings, etc. Figure S1 shows a turnout which is defined by the dotted lines. All track outside these lines would be measured as plain track.

M4 Taking up check and guard rails are measured separately only where they are located outside turnouts and diamond crossings, etc., i.e. normally on plain track which is curved.

M5 The enumerated item for twist rails is 'extra over' the mass of the rail of which it is made.

M6   Fishplates are used for jointing the ends of two rails and are placed on either side of the rail and bolted through. The unit inserted in the unit column could be nr but in order to save any confusion it is recommended that pr (pair) is used. This does not comply with Paragraph 5.19 so a suitable preamble should be incorporated.

M8   See Measurement Rule M3.

M9   See Measurement Rule M4.

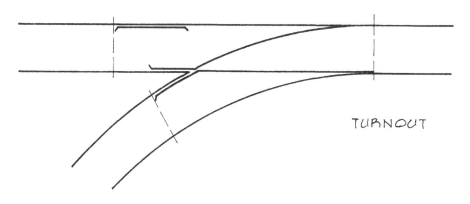

Figure S1

Definition Rules

D3   See Measurement Rule M3.

Coverage Rules

C1   The items for taking up turnouts and diamond crossings include for all items within the plan area of the turnout or crossing (see Measurement Rule M3).

C3   Although not specifically stated, delivery to the site would also normally include offloading and storing.

C4   The Specification should make it perfectly clear which fittings are to be supplied with the sleepers in order to avoid any confusion.

C5   It is only the main fittings which are measured. All the ancillary items which are associated with them are deemed to be included.

C6   Items for turnouts and diamond crossings are fully inclusive and include for all the constituent parts within the plan area of the switch or crossing, apart from track foundations. It follows that sufficient information must be given to enable the estimator to ascertain the total amount of work involved.

C8   Location in this context is the location where the item is to be laid.

Additional Description Rules

A1   The materials would normally be imported. If materials are to be used from other sources, i.e. from excavations, then item description should state this.

A2   Waterproof membranes are often specified by grade and this would be acceptable instead of thickness.

A3   Materials which are to remain the property of the Employer would be better described as being 'carefully' taken up. The description for dismantling should state what is to be left at the end of the dismantling process rather than how it is to be dismantled.

A4   Where it is not possible to state the approximate mass then it would be prudent to state the location.

A6   Item descriptions should also state the type of ballast.

A7   This would normally be done by reference to the Specification or drawings rather than by detailed description.

A8   Fittings identified as being attached to sleepers are not measured separately.

A10 It would be prudent to refer to the relevant drawing.

A13 Location in this context is the location where the item is to be laid. It would also be prudent to state where the Contractor is to obtain the materials.

A15 It is necessary to state the type of joint for the laying of rails. Where the joint is made with fishplates, the cost of making the joint is included with the rate for laying the rail. Where the joints are welded, these have to be enumerated separately as well as describing the type of joint in the rail.

ITEM MEASUREMENT

S. 1 * 0   Track foundations

Divisions   Bottom ballast and top ballast are measured in cubic metres. Blinding, blankets and waterproof membranes are measured in square metres.

Rules   M1 Do not deduct the volume of sleepers from top ballast.

   M2 Do not measure the area of membranes in laps.

   A1 State the material.

   A2 State the thickness or grade of blinding, blankets and membranes.

CLASS S: RAIL TRACK

Generally See also Rules D1, D2.

S. 2 1-5 * Taking up rails

Divisions Taking up plain track is measured in linear metres. Taking up turnouts and diamond crossings are enumerated. The type of rail involved must be stated in accordance with the Second Division.
Taking up check, guard and conductor rails are measured in linear metres.

Rules M3 Measure plain track along the centre line excluding turnouts and diamond crossings.

M4 Measure check guard and conductor rails along the rail excluding turnouts and diamond crossings.

A3 State the extent of the dismantling and of disposal details. Give details of the rail type, sleeper and joint.

Generally See also Rule C1.

S. 2 8 * Taking up sundries

Divisions Taking up sundries is enumerated stating the item to be taken up.

Rules A4 State the approximate weight and construction of buffer stops.

Generally -

S. 3 * 0 Lifting, packing and slewing

Divisions Lifting, packing and slewing is enumerated stating the item concerned in accordance with the Second Division.

Rules D3 Measure along the centre line of the track including lengths over roads. Measure switch roads from the toes of switches.

A5 State the length of track, maximum distance of slew and maximum lift.

A6 State when extra ballast is required. It is also recommended that the type of ballast is stated.

Generally See also Rule C2.

CLASS S: RAIL TRACK

S. 4 1-6 * Supplying rails

Divisions The supply of rails is measured in tonnes. A further enumerated item has to be given for the number of twist rails and these are 'extra over'. State the type of rail in accordance with the Second Division and the mass per metre (Additional Description Rule A9). It is not necessary to also give the mass in the ranges listed in the Third Division (Paragraph 3.10).

Rules M5 Include the mass of twist rails in the mass measured.

A9 State the section reference or cross-sectional dimensions.

Generally See also Rule C3.

S. 4 7 * Supplying sleepers

Divisions The supply of sleepers is enumerated stating whether timber or concrete.

Rules A7 State the type.

A8 State the size and identify fittings attached by the manufacturer.

Generally See also Rules C3, C4.

S. 4 8 * Supply of fittings

Divisions The supply of fittings is enumerated stating the type in accordance with the Third Division.

Rules M6 Measure fishplates in pairs.

A7 State the type.

Generally See also Rule C3, C5.

S. 5 1-2 * Supplying turnouts and diamond crossings

Divisions The supply of turnouts and diamond crossings is enumerated.

Rules A10 State the type.

Generally See also C3, C6.

S. 5 8 * Supplying sundries

Divisions The supply of sundries is enumerated stating the item required in accordance with the Third Division.

Rules   M7   Measure conductor rail guard boards on both sides of the rail.

           A11 State the type.

           A12 State the approximate weight of buffer stops.

Generally See also Rule C3, C7.

S. 6 1-4 1-3 Laying plain track

Divisions Laying plain track is measured in linear metres stating the type of rail in accordance with the Second Division. Additional items must be given for laying the track to a curve distinguishing between exceeding and not exeeding 300 metres radius. The item for forming the curve is 'extra over'. Further items must be given for any welded joints (S. 6 1-4 7) and spot re-sleepering (S. 6 1-4 8).

Rules   M8   Measure along the centre line excluding lengths occupied by turnouts and diamond crossings.

           A13 For track which is not supplied by the Contractor state the form in which it is to be supplied and the location.

           A14 Identify prefabricated lengths.

           A15 State the type and mass per metre and the type of joint and sleeper.

Generally See also Rules C8, C9.

S. 6 1-4 4&5 Laying turnouts and diamond crossings

Divisions Laying turnouts and diamond crossings are enumerated stating the type of rail in accordance with the Second Division. Further items must also be given for any welded joints (S. 6 1-4 7) and spot re-sleepering (S. 6 1-4 8).

           A13 For track which is not supplied by the Contractor state the form in which it is to be supplied and the location.

           A15 State the type and mass per metre and the type of joint and sleeper.

           A16 State the type and length.

Generally See also Rules C8, C9, C10.

S. 6 5-7 1-3 Laying check, guard and conductor rails

Divisions   Laying check, guard and conductor rails are measured in linear metres. Additional items must be given for length ends and side ramps. Welded joints are enumerated separately (S. 6 5-7 7).

Rules   M9   Measure along the centre line.

          A13 For track which is not supplied by the Contractor state the form in which it is to be supplied and the location.

          A15 State the type and mass per metre and the type of joint and sleeper.

Generally   See also Rules C8, C9.

S. 6 1-4 7 and S. 6 5-7 7 Welded joints

Divisions   Welded joints are enumerated.

Rules   A17 State the rail section and weld type.

Generally   -

S. 6 1-4 8 Spot re-sleepering

Divisions   Spot re-sleepering is enumerated.

Rules   -

Generally   Although not specifically stated, it is recommended that the type and size of sleeper together with its fixing details are stated in item descriptions.

          See also Rule D4.

S. 6 8 * Laying sundries

Divisions   Laying sundries are enumerated except for conductor rail guard boards which are measured in linear metres. State the item to be laid in accordance with the Third Division.

Rules   M7   Measure conductor guard rail boards to both sides of the rail.

          A18 State the approximate weight of buffer stops.

Generally   See also Rules C8, C9.

CLASS S: RAIL TRACK

STANDARD DESCRIPTION LIBRARY

Track foundations

Bottom ballast

S. 1 1 0	granular material as specification clause 23.1	m3

Blinding

S. 1 3 0	sand; 25mm thick	m2

Waterproof membranes

S. 1 5 0	uPVC sheet; 1200 gauge	m2

Taking-up complete installation including ballast; rails and fittings to remain the property of the Employer; all other materials to be disposed of off-site

Bullhead rail

S. 2 1 1	plain track; timber sleepers, fishplate joints	m
S. 2 1 4	turnouts; timber sleepers; fishplate joints	nr

Flat bottom rails

S. 2 2 1	plain track; concrete sleepers; welded joints	m
S. 2 2 5	diamond crossings; concrete sleepers; welded joints	nr

Check and guard rails

S. 2 4 0	generally	m

Conductor rails

S. 2 5 0	generally	m

Sundries

S. 2 8 1	buffer stops; approximate weight kg	nr
S. 2 8 2	retarders	nr
S. 2 8 3	wheel stops	nr

Taking-up complete installation including ballast; rails and fittings to remain the property of the Employer; all other materials to be disposed off-site (cont'd)

S. 2 8 4	lubricators	nr
S. 2 8 5	switch heaters	nr
S. 2 8 6	switch levers	nr

Lifting packing and slewing

Bullhead rail track

S. 3 1 0	track length 10m; maximum slew 2m; maximum lift 5m	nr

Flat bottom rail track

S. 3 3 0	track length 15m; maximum slew 3m; maximum lift 6m; additional ballast as specification clause 23.2	

Supplying

Bullhead rails

S. 4 1 2	mass 20-30kg/m; section reference	t

Dock and crane rails

S. 4 3 3	mass 30-40kg/m; size	t

Sleepers

S. 4 7 1	timber; size 250 x 450 x 2400mm long; no fittings to be supplied	nr
S. 4 7 2	concrete; size 250 x 450 x 2400mm long; supplied with bolts and spring clips	nr

Fittings

S. 4 8 1	chairs	nr
S. 4 8 2	base plates	nr
S. 4 8 4	plain fishplates	pr

Supplying (cont'd)

S. 4 8 5	insulated fishplates	nr
S. 4 8 6	conductor rail insulators	nr
S. 4 8 7	conductor rail side ramps	nr
S. 5 1 0	Turnouts	nr
S. 5 2 0	Diamond crossings	nr

Sundries

S. 5 8 1	buffer stops; approximate weight kg	nr
S. 5 8 2	retarders	nr
S. 5 8 3	wheel stops	nr
S. 5 8 4	lubricators	nr
S. 5 8 5	switch heaters	nr
S. 5 8 6	switch levers	nr
S. 5 8 7	conductor rail guard boards	m

Laying

Bullhead rails

S. 6 1 1	plain track; mass kg/m; fishplate joints; timber sleepers	m
S. 6 1 2	form curve in plain track; radius not exceeding 300m	m
S. 6 1 4	turnouts; length 50m	nr
S. 6 1 8	spot re-sleepering	nr

Check rails; mass   kg/m; concrete sleepers

S. 6 5 1	rail	m
S. 6 5 2	length ends	nr
S. 6 5 3	side ramps	nr

CLASS S: RAIL TRACK

S. 6 5 7	welded joints	nr
	Sundries	
S. 6 8 2	retarders	nr
S. 6 8 3	wheel stops	nr
S. 6 8 5	switch heaters	nr
S. 6 8 7	conductor rail guard boards	nr

Chapter 22

CLASS T: TUNNELS

GENERAL TEXT

Principal changes from CESMM 1

There were only one or two minor changes to the divisions of this class apart from a minor re-shuffling of the items.

Formwork to in situ linings must state the finish (T. 2-3 5 *). In accordance with Paragraph 3.2 the item description for face packers was simplified (T. 8 3 2). The item for standby driving or sinking operations was deleted (T. 8 5 0 in CESMM 1) and the Contractor had to seek payment for standing idle due to emergency supporting or stabilization operations, through the Contract and not through measurement.

Definition Rule D1 defines 'other cavities' as being work executed outside the normal profiles. Measurement Rule M3 defines the minimum piece size before the excavation of rock has to be measured separately. The diameter used for the classification of excavation became the external diameter as opposed to the internal diameter (Definition Rule D2). The disposal of surplus materials was not dealt with at all in CESMM 1 and this was rectified by the inclusion of Coverage Rule C1 and Additional Description Rule A4, whilst Additional Description Rule A2 requires the excavation and lining of other cavities to identify the cavity. Measurement Rules M6, M7 and M10 further defined the measurement of in situ concrete linings, packing and packers respectively. Definition Rules D3 and D4 clarify the position with respect to reinforcement to sprayed concrete. Coverage Rules C2, C3 and C4 list the work to be included for items for linings, preformed linings and face packers.

Principal changes from CESMM 2

A new Definition Rule D4 has been included which states that where the internal diameter is given in the item description, it refers to the dimeter of the lining. The old D4 is now re-numbered to become D5.

A new Additional Description Rule A12 is included which states that the internal diameter of the lining shall be stated in the description for lining ancillaries.

A further Additional Description Rule A13 is inserted which covers situations where the linings do not have a circular cross-section. In these cases the maximum internal dimensions of the cross-sections should be used instead of the internal diameter and stated in the item description. This rule now conforms to rule A6.

Additional Description Rules A12 to A15 are re-numbered A14 to A17 respectively.

CLASS T: TUNNELS

Measurement Rules

M1 The term 'cut and cover' means the installation of a tunnel for which the excavation is carried out from the surface. Tunnels measured under Class T are those for which the excavation work is carried out without disturbing the original ground surface. It is used where the excavation for a cut and cover tunnel would be too deep to be economically viable or where it is not physically possible to construct it from the surface i.e. in built up areas or under rivers, etc.

One form of tunnel construction now in fairly common use is the submersible tube for tunnels which are under water. In simple terms this involves the construction and floating of the tube on top of the body of water and then sinking it into its final position. CESMM 3 does not deal specifically with this mode of construction and should this be encountered then suitable additional rules would have to be drawn up accordingly.

M2 The volume of excavation measured is the net volume based on either the payment line or the net volume occupied by the permanent works if no payment line is shown. The payment line is the limit to which the payment for excavation work will be made. It is not intended as being the limit of excavation or excavation overbreak and the Contractor must allow for any additional excavation and filling beyond the payment line in his rates. In most instances, the payment line should correspond to the outside face of the permanent works.

M4 The area of the payment surface is that area of surface on the payment line or if no payment line is shown, the area outside the area of the permanent works.

M5 The position of the payment line should not only affect the thickness of the lining but also the area measured. Cast in situ linings are measured by volume (see Measurement Rule M6) but sprayed linings are measured in square metres. The area of sprayed linings will be calculated on the basis of the centre of the lining (see Figure T1).

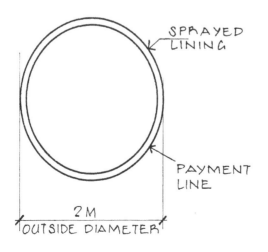

In the example the lining is a sprayed lining of 50mm minimum thickness. The diameter upon which the area of the lining is calculated is 2000 - ((50/2) x 2) = 1950mm.

Figure T1

M6 Specifically refers to Measurement Rules M1 and M2. (For discussion of both these Rules see Class F).

M7 The measurement of circumferential packing is by the number of completed rings and not the number of individual pieces in the ring. Packing is the filling of the joints between the individual rings forming the tunnel. Tapered circumferential packings are used in order to 'bend' the tunnel. Packings between the individual segments of a ring (longitudinal packings) are included with the item for the ring itself.

M8 This rule caused much discussion in CESMM 1 and 2 but it remains unchanged. In effect it means that the Contractor will be paid for all the temporary support and stabilisation which he decides is necessary, whether or not it is at the Engineer's instruction which means that the Employer bears all the risk.
  In today's environment, many client bodies may not wish to bear this risk and it may be necessary to amend this rule accordingly.

M9 Provided always that temporary support is measured then this rule details how to calculate the weights, volumes and areas of support. This in turn means that the Engineer should detail on the drawings such temporary support although in practice he may well wish to leave this to the Contractor's discretion. If this is the case the quantities inserted for temporary support would, at best, be an approximate estimate, and it may be more appropriate to include a provisional sum instead.

M10 In other words one packer for each injection made.

M11 The measurement of grout and injection of grout differs quite markedly between this class and Class C. In Class C materials and injection are measured separately and the measurement of material is further subdivided into the constituent parts of the grout. In this class the injection of the grout is the only item measured and the supply of all the grout materials as well as the injection itself is deemed to be included. The mass measured for injection of grout should include the mass of all materials except for mixing water.

Definition Rules

D1 The critical words in this rule are 'outside the normal profiles'. Work would only be classed as other cavities where work was expressly required to be done beyond the external face of the tunnel or shaft (see Figure T2).

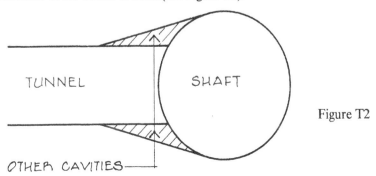

Figure T2

D2   The external diameter would be the diameter taken to the payment line if one was shown or otherwise the net diameter of the permanent works.

D3   The main type of reinforcement under this rule would be glass fibre or polyester.

D4   The internal diameter stated in the item description refers to the lining.

D5   This would be measurable under T. 8 2 7.

Coverage Rules

C1   Disposal would normally be to tips provided by the Contractor unless otherwise stated. Where material is to be disposed of on or about the site item descriptions should so state. Specific areas of filling which are designated for various types of material should be described as such. It may even be necessary to create a separate disposal classification and it is recommended that this should follow the same lines as Class E.

C2   It follows that details, sizes, spacing, etc., of joints and finishes should be given either on the drawings or in the Specification.

C3   Finishes on concrete would also be included.

Additional Description Rules

A1   It is recommended that tunnelling work to be carried out under compressed air is collected under a general heading which defines the gauge pressure in accordance with this rule. Separate quantities have to be given for working under different pressures and these start at not exceeding 1 bar, 1-1.4 bars, 1.4-1.8 bars and so on in increments of 0.4 bars.
    Whilst the last sentence of this rule specifically mentions provision and operation of plant and equipment, it does of course mean establishment, continuing operation and maintenance and removal.

A3   It is not necessary to distinguish between horizontal tunnels and tunnels which are inclined at an angle less than 1 in 25 and they may be grouped together for the purposes of measurement.

A4   See Coverage Rule C1.

A5   Descriptions should not try and dictate to the Contractor how to fill the voids but merely the material which is to be used for the filling. The amount of overbreak is dependent upon several factors. Loose ground will produce large overbreak whereas firm ground or rock will produce little. Drill and blast techniques will increase the amount of overbreak that would be experienced in tunnel excavation carried out by machine. The amount of care the Contractor takes in the excavation will be a factor which could be influenced by the quality of the Bill preparation.

The backgrouting of overbreak is deemed to be included in the rate for excavated surfaces, whereas the grouting of the surrounding ground for support and stabilization is measured in tonnes under T. 8 3 6. However, it is often difficult to distinguish between the two.

Whilst grouting for stabilization the Contractor may be filling all or part of the overbreak for which he is responsible and effectively be paid twice for the same operation. Alternatively the back grouting of overbreak could be removed from the excavated surfaces item and paid for on a tonneage basis, but the Contractor would then have little incentive to be economic in the amount of overbreak he excavates so this course of action is not recommended.

There is no easy answer to this problem. One possible solution which is really an engineering consideration is to require the Contractor to back grout the overbreak first, by inserting the injection pipes into the overbreak volume only. He would not be paid for this initial grouting operation as this would be deemed to be included in his rates for excavated surfaces. Subsequent drilling and grouting would be paid for under T. 8 3 *. An alternative but not particularly practical solution would be to try and measure the volume of overbreak as the excavation is actually executed. From this volume the mass of grout which has theoretically been used in filling the overbreak could be calculated and deducted from the total mass injected.

Whatever procedure is used to try and overcome this problem, careful post contract administration is required and the Contractor should be informed of the procedures to be adopted in the tender documentation.

A6   This rule means that the external cross-sectional dimensions must be stated in item descriptions but the largest dimension must be used for the coding/classification in the Third Division.

A7   See Additional Description Rule A3.

A8   The concrete specification would normally be in accordance with BS5328. The reinforcement must be measured separately under Class G. The measurement of in situ concrete under this class applies only if the concrete lining is of a very simple nature (see the footnote to the rules in Class G). In fact there is no provision in this class for measuring any concrete ancillaries apart from formwork, and these should be measured separately in accordance with Class G (as with reinforcement) except where provided for elsewhere in this class, (see Coverage Rule C2 for the treatment of joints and finishes).

A9   Rings are normally purchased from the supplier as a complete unit or package. It may be simpler to refer to the manufacturer's catalogue or literature making sure that any options or optional extras are dealt with in the item description.

A12   This rule requires that the internal diameter of the lining shall be stated in the description for the lining ancillaries.

A14   Rock bolts are simply bars or rods, usually of steel, which are grouted into predrilled holes set in an epoxy resin or similar material. The other end has a plate or large washer, and they are used to support fragmented or loose material.

A16 The concrete specification would usually be in accordance with BS5328. Unlike in situ linings measured under T. 2-4 * *, reinforcement for internal support is measured here rather than in Class G.

A17 This would be not exceeding 5 metres; 5-10 metres; 10-15 metres, etc.

ITEM MEASUREMENT

T. 1-7 * * Excavation and lining to tunnels generally

Applicable to all items

Divisions   -

Rules   A1 State when work is done under compressed air in the stages of pressure: not exceeding 1 bar; 1-1.4 bars; 1.4-1.8 bars and so on in increments of 0.4 bars. Give items in the General Items (specified requirements) for provision, ongoing operation and maintenance and removal of plant.

            A2 Identify the cavity for excavation of other cavities.

Generally   It would be normal practice to measure the items relating to a particular location of tunnel or relating to a particular mode of construction under a general heading, i.e. all excavation and lining done under compressed air would be measured together or all items relating to tunnel between termination points A and B would be grouped together.
            Furthermore, although it is not a specific requirement to state whether work is in tunnels, shafts or other cavities in all cases (for example excavated surfaces) it would be sensible to include this classification also as under the general heading.

            See also Rules M1, D1.

T. 1 1-6 * Excavation

Divisions   Excavation in tunnels is measured in cubic metres, stating whether in tunnels, shafts or other cavities; whether in rock or other stated material, and stating the actual external diameter of the excavation cross-section. It is not necessary to state the diameter in the ranges given in the Third Division and these should be used for coding purposes only (Paragraph 3.10).

Rules   M2 Calculate volumes from the payment lines or from net dimensions if no payment lines are shown.

            A3 State whether straight, curved, or tapered. Tunnels inclined exceeding a gradient of 1 in 25 shall be so described stating the gradient. Inclined shafts shall be so described stating the inclination to the vertical irrespective of their angle.

A4 State the location of the disposal areas for material to be disposed of on site. Item descriptions shall describe excavated material to be used as filling.

Rules   A6   State the external cross-sectional dimensions for non-circular tunnels, shafts and other cavities.

Generally   The classification of material for excavation differs in Class T from other classes in CESMM 3. There are two classifications listed, 'rock', and 'stated material other than rock'. More than any other class the definition of what constitutes rock in accordance with Paragraph 5.5 must be given careful consideration.

The classification of 'stated material other than rock' also requires careful management. The phrase covers a multitude of different material types and the question arises as to how far one should go in measuring them separately.

For small jobs it may be sufficient to have just three classifications. These would be:

a)   rock

b)   natural material other than rock or artificial hard material

c)   stated artificial hard material

In practice the likelihood of encountering artificial hard material at the depths involved would be remote and therefore the number of classifications would effectively be two.

However, for large contracts, it may be desirable to reduce the risk to the Contractor, and hence the total cost of the contract by subdividing the second classification into various grades of material. In most cases, a detailed site investigation will have been carried out and most of the various material types should be known. If these vary widely over the length of the tunnel it may be possible to classify them individually, or if it is considered that there are too many by grouping similar types together.

Whatever policy is adopted it is necessary to inform the Contractor in detail about it in the Preamble.

See also Rules M3, D2, C1.

T. 1 7&8 * Excavated surfaces

Divisions   Excavated surfaces are measured in square metres stating whether in rock or other stated material. It is not necessary to state whether the surface is in tunnels, shafts or other cavities.

Rules   M4   Measure at the payment line or from the net dimensions if no payment line is shown.

A5   Give details of the filling to overbreak.

Generally   Whilst it is not a specific requirement to identify whether excavated surfaces are in tunnels, shafts or other cavities, they should be kept separate if it is considered that they have different cost consequences, or if the individual treatment to each varies.

See also Generally under T. 1 1-6 *.

T. 2-4 1-4 * In situ linings

Divisions   Sprayed concrete linings are measured in square metres stating whether in tunnels, shafts or other cavities and whether primary or secondary together with the actual diameter.

  Cast concrete linings are measured in cubic metres stating whether in tunnels, shafts or other cavities and whether primary or secondary together with the actual internal diameter of the lining.

  In both cases it is not necessary to state the range of diameters as stated in the Third Division.

Rules   M5 Measure to the payment lines or from the net dimensions if no payment lines are shown.

M6 Calculate concrete volumes in accordance with Class F (specifically Measurement Rules M1 and M2).

A7 State whether tunnels and shafts are straight, curved or tapered. If tunnels slope more than 1 in 25, items descriptions must so state giving the gradient. Inclined shafts must be always described as such stating the angle of inclination to the vertical.

A8 State the concrete specification and whether it is reinforced (measure the reinforcement under Class G). Identify separate components i.e. headwalls, shaft bottoms, etc. Alternatively measure in Class F.

  State the minimum thickness of sprayed concrete.

See also Rules D3, C2, C3.

T. 2-4 5 * Formwork to in situ linings

Divisions   Formwork is measured in square metres stating whether in tunnels, shafts or other cavities and the actual internal diameter of the lining. It is not necessary to state the range of diameters as stated in the Third Division.

Rules   A7 State whether tunnels and shafts are straight, curved or tapered. If tunnels slope more than 1 in 25, item descriptions must so state giving the gradient. Inclined shafts must be always described as such stating the angle of inclination to the vertical.

Generally   Although not specifically stated it is recommended that the diameter used for classification is the actual diameter of the formwork face. It would also be prudent to state the type or grade of formwork.

It is further recommended that unless the formwork is of a very simple nature with little or no intrusions or projections that it is measured in accordance with Class G. A suitable preamble would have to be incorporated stating the deviation from CESMM 3.

T. 5-7 1-6 *   Preformed segmental linings

Divisions   Preformed segmental linings are measured by the number of complete rings, stating the type of ring in accordance with the Second Division; whether to tunnels, shafts or other cavities together with the actual diameter. It is not necessary to state the diameter in the ranges given in the Third Division.

Rules   A7   State whether tunnels and shafts are straight, curved or tapered. If tunnels slope more than 1 in 25, item descriptions must so state giving the gradient. Inclined shafts must be always described as such stating the angle of inclination to the vertical.

            A9   State the components of each ring and the nominal ring width and maximum piece weight.

            A10 Identify when in pilot tunnels. Rings in pilot tunnels which are to remain the property of the Employer shall so state.

            A11 State whether concrete rings are flanged or solid. State whether metal rings have machined abutting faces.

Generally   Although not specifically stated the diameter used for classification should be the internal diameter of the ring because this is how they are normally specified. The external diameter could be used providing it is made clear either in the item description or Preamble.

            See also Rules C2, C3.

T. 5-7 7 *   Lining ancillaries

Divisions   Parallel and tapered circumferential packing and stepped junctions are enumerated. Caulking is measured in linear metres stating the type of material. It is necessary to state whether the ancillaries are in tunnels, shafts or other cavities.

Rules   M7   Packing shall be measured by the number of rings of segments packed.

             A7   State whether in tunnels and shafts are straight curved or tapered. If tunnels slope more than 1 in 25, item descriptions must so state the gradient. Inclined shafts must be always described as such stating the angle of inclination to the vertical.

             A9   Whilst lining ancillaries appears under the general heading of
and   preformed segmental linings it is considered that these rules do not
A11   apply.

             A10   Identify when in pilot tunnels. Materials in pilot tunnels to remain the property of the Employer shall so state.

Generally   Although not a specific requirement it would be prudent to also state the nature of the packing material.

T. 8 * * Support and stabilization generally

Applicable to all items

Divisions   -

Rules   A1   State when work is carried out under compressed air in the stages of pressure: not exceeding 1 bar; 1-1.4 bars; 1.4-1.8 bars and so on in increments of 0.4 bars. Give items in the General Items (specified requirements) for provision, ongoing operation and maintenance and removal of plant.

             M8   Measure both temporary and permanent support and stabilization.

Generally   Although it is not necessary to distinguish between support and stabilization in tunnels, shafts and other cavities it would probably be sensible to include it under the relevant general heading for each rather than collect all the items together under one heading of support and stabilization for all tunnels, shafts and other cavities regardless of location or method of working.

           See also M1, D1.

T. 8 1 * Rock bolts

Divisions   Rock bolts are measured in linear metres stating the type in accordance with the Third Division.

Rules   A12   State the size, type, shank detail and maximum length.

Generally   It would also be prudent to state the grout type if this would be otherwise
             unclear. The maximum length is the maximum length of the longest bolt. It is
             considered that it would be more helpful to the estimator if the bolts were
             grouped together in lengths given in increments, i.e. not exceeding 1m; 1-2m; 3-
             4m, etc. The increments could be shorter or longer depending on the situation.
             If this approach is adopted then a suitable preamble would have to be incor-
             porated stating the deviation from CESMM 3.

T. 8 2 *   Internal support

Divisions   Steel arch supports are measured in tonnes distinguishing between supply and
             erection. Include the mass of ribs. Timber supports are measured in cubic
             metres distinguishing between supply and erection. Lagging, sprayed concrete
             and mesh and link are measured in square metres.

Rules   M9   Calculate the weight of steel arches as set out in Class M. Calculate the
                  volume of timber as set out in Class O. Calculate the area of sprayed concrete
                  at the payment line or from the net dimensions if no payment line is
                  shown.

             A13 State the nature of the materials used for lagging and packing or grouting
                  behind lagging.

             A14 State the specification of concrete for sprayed concrete support, whether
                  it is reinforced and its minimum thickness. State the size and mass of
                  mesh or link fabric.

Generally   See also Rules D3, D4.

T. 8 3 *   Pressure grouting

Divisions   Several items must be measured for pressure grouting. Sets of drilling and
             grouting plant, face packers, and deep packers of stated size are enumerated.
             Drilling and flushing to stated diameter and redrilling and flushing are
             measured in linear metres. Injection of grout is measured in tonnes stating the
             composition.

Rules   M10 Measure the number of face packers and deep packers as the number of
                  injections.

             M11 Do not include the mass of mixing water for injection of grout.

             A17 State the length of holes in stages of 5 metres in items for drilling and
                  redrilling.

Generally The treatment of pressure grouting in Class T is another example of how similar items of work are treated quite differently. Whilst it is appreciated that the type of plant will be different, the actual grouting techniques may be very similar. There may be staged drilling, the material drilled may be rock or other material and the injection pipes may be driven. It is also possible that the injection of materials is done in stages, either ascending or descending, and that the angle of the grout hole has a bearing on cost. It is recommended that careful consideration is given to the measurement of pressure grouting in Class T, and that if it is considered appropriate, that the rules of Class C are adopted complete or in an amended form. A suitable preamble would have to be incorporated in this case.

The separate measurement of grouting plant could also pose some problems. This is possibly the only time CESMM 3 measures numbers of sets of plant which are to be used for installing the permanent works and the inherent dangers in doing this are obvious. It is considered more prudent if, as with all other work, the Contractor is left to decide himself how much drilling and grouting plant he requires in order to maintain his programme.

See also Rule C4

T. 8 4 0 Forward probing

Divisions Forward probing is measured in linear metres.

Rules A17 State the length of holes in stages of 5 metres.

Generally See Generally under T. 8 3 *.

STANDARD DESCRIPTION LIBRARY

Work carried out under compressed air; gauge pressure 1-1.4 bars; items are given under Class A - specified requirements for provision and maintenance of plant

Excavation

Tunnels in rock; straight

T. 1 1 4   diameter 4.57m   m3

Tunnels in clay; straight

T. 1 2 4   diameter 4.57m   m3

Shafts in rock; inclined at 5 degrees to vertical

T. 1 3 5   diameter 5.94m   m3

Shafts in mudstone; inclined at 5 degrees to vertical

T. 1 4 5	diameter 5.94m	m3

Other cavities in rock; intersection I1

T. 1 5 4	diameter 4.77m	m3

Other cavities in material as specification clause 24.1
intersection I1

T. 1 6 4	diameter 4.77m	m3

Excavated surfaces in rock

T. 1 7 0	filling voids with grout as specification clause 24.2	m2

Excavated surfaces in clay

T. 1 8 0	filling voids with grout as specification clause 24.3	m2

Excavated surfaces in mudstone

T. 1 8 0.1	filling voids with grout as specification clause 24.4	m2

Excavated surfaces in material as specification clause 24.1

T. 1 8 0.2	filling voids with grout as specification clause 24.6	m2

**In situ lining to tunnels; straight; concrete grade
C20; unreinforced**

Sprayed concrete primary; tunnel walls; minimum thickness 50mm

T. 2 1 3	lining internal diameter 4m	m2

Sprayed concrete secondary; tunnel walls; minimum thickness 75mm

T. 2 2 3	lining internal diameter 3.9m	m2

**In situ lining to shafts; inclined at 5 degrees to
vertical; concrete grade C20; reinforced**

Cast concrete primary; shaft bottoms

T. 3 3 5	diameter 5.94m	m3

**In situ lining to shafts; inclined at 5 degrees to
vertical; concrete grade C20; reinforced (cont'd)**

Cast concrete primary; head walls

T. 3 3 5.1   diameter 5.94m   m3

Cast concrete secondary; shaft bottoms

T. 3 4 5   diameter 5.94m   m3

Cast concrete secondary; head walls

T. 3 4 5.1   diameter 5.94m   m3

**In situ lining to other cavities; intersection I1;
concrete grade C20; reinforced**

Cast concrete primary; walls

T. 4 3 4   diameter 4.93m   m3

Cast concrete secondary; walls

T. 4 4 4   diameter 4.93m   m3

**Preformed segmental lining to tunnels; sloping at a
gradient of 1 in 20**

Precast concrete bolted flanged rings; nominal ring width 610mm;
maximum piece weight 0.25t; comprising 4nr 0 segments, 2nr T
segments and 1nr key segment; 28nr 20 x 165mm long circle and
cross bolts and 2nr 20 x 265mm long key bolts; 3mm thick
bituminous felt packings to longitudinal joints

T. 5 1 3   diameter 3.35m   nr

T. 5 1 3.1   diameter 3.35m; to pilot tunnels; to remain the
                     property of the Employer   nr

Lining ancillaries

T. 5 7 7.1   parallel circumferential timber packing   nr

T. 5 7 2   tapered circumferential precast concrete packing;
                     for 20 metre radius   nr

T. 5 7 7.4   caulking with material as specification clause 24.7   m

Support and stabilization

Rock bolts; galvanised steel; mechanical grouted with epoxy mortar; maximum length 2m; round

T. 8 1 2	diameter 20mm	nr

Internal support

T. 8 2 1	steel arches; supply	t
T. 8 2 2	steel arches; erection	t
T. 8 2 3	timber supports; supply	m3
T. 8 2 4	timber supports; erection	m3
T. 8 2 6	sprayed concrete; grade C20; unreinforced; minimum thickness 100mm	

Pressure grouting

T. 8 3 1	sets of drilling and grouting plant	nr
T. 8 3 2	face packers	nr
T. 8 3 4	drilling and flushing to 100mm diameter; not exceeding 5m	m
T. 8 3 5	re-drilling and flushing; not exceeding 5m	m
T. 8 3 6	injection of grout as specification clause 24.9	t

Forward probing

T. 8 4 0	length not exceeding 5m	m
T. 8 4 0.1	length 5-10m	m
T. 8 4 0.2	length 10-15m	m

Chapter 23

CLASS U: BRICKWORK, BLOCKWORK AND MASONRY

GENERAL TEXT

Principal changes from CESMM 1

The separate classifications for brick, block and stone walls were brought together (U. 1-8 * *). The classification of material thicknesses were standardised for all types of work and it became necessary to state the nominal thickness of walls, facings to concrete and casings to metal sections (Additional Description Rule A5).

The Third Division classification of walls has been rationalised (U. * 1-5 *) and columns and piers were given their own Second Division classification (U. * 6 0).

The ancillaries section (U. * 8 *) had new classifications added and these were movement joints, fixings and ties. The cross-sectional areas for the classification of built-in pipes and ducts were changed and the item in CESMM 1 for centering to arches were deleted.

Principal changes from CESMM2

The wording of Measurement Rule M6 has been amended to clarify that it is the smaller of the two areas that is measured when two areas of brickwork, blockwork or masonry are tied or fixed together.

Measurement Rules

M1 A composite wall is one which is composed of different materials e.g. a block wall faced with masonry. Separate skins should only be measured when they are quite separate notwithstanding that they may be tied together with ties. For instance, a two brick thick wall built in commons with one face only constructed in facings would not warrant the measurement of two separate skins. In this case the actual construction of the entire thickness would be given in the item description.

 Note that there is no need to measure the forming of the cavity in cavity walling although the ties are now measured separately under U. * 8 6.

M2 Although not specifically stated, this rule would also apply to linear items of columns and piers measured under U. * 6 0. Calculations should be made overall, including the space occupied by joints, but excluding the space occupied by copings and sills. This means that the copings and sills measured under U. * 7 1 are 'full value'. No deduction or additions are made for intruding surface features, and that projecting surface features are also 'full value' items whereas intruding surface features are 'extra over'.

M3 All dimensions taken should be mean but common sense must be applied. In Figure U1 the two walls must be measured separately and not weighted into a mean thickness.

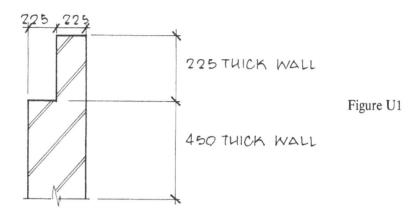

225 THICK WALL

Figure U1

450 THICK WALL

M4 Fair facing need only be measured where expressly required. This requirement would most commonly arise when common brickwork is exposed to view. The instruction from the Engineer would probably not appear on the drawings but would be included in the Specification which is another reason the taker-off should have access to this document before completing the Bill of Quantities (see also Coverage Rule C1).

M5 It is difficult to see why this rule has been included because all measurements in CESMM 3 are taken net unless otherwise stated, so drawing attention to the fact that the laps are not to be included for joint reinforcement and damp proof courses seems superfluous.

Definition Rules

D1 Walls or facing which have either one or both sides battered are classed as battered. One important reason why these are itemised separately is to allow the estimator the opportunity to include the extra cost of scaffolding to a non-vertical face.

D2 If a wall is isolated it shall be described as a pier unless its length is more than four times its thickness. (See Figure U2).

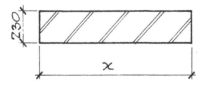

Where x is up to 4 x 230mm - Pier
Where x is over 4 x 230mm - Wall

Figure U2

D3   Figure U3 gives examples of the classification of wall thicknesses for walls which have surface features.

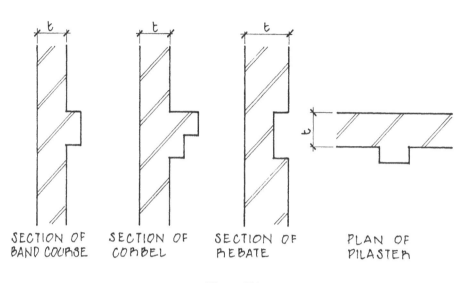

Figure U3

Unfortunately CESMM 3 does not give any definition where a pilaster becomes a wall as it gets wider. We can take guidance from Definition Rule D2, however, and say that the pilaster becomes a wall when its length exceeds four times its width measured from the face of the wall in which it occurs (see Figure U4).

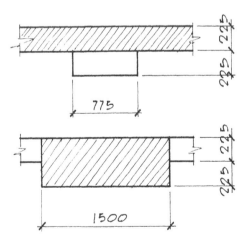

Shaded portion measured as 225mm thick wall. Unshaded portion measured as pilaster 225 x 775mm because its length does not exceed 4 times its projecting width.

Shaded portion measured as 450mm thick wall because its length exceeds 4 times its projecting width.

Figure U4

275

D4   See Definition Rule D1.

D5   See Definition Rule D3.

Coverage Rules

C1   The question of what constitutes fair facing requires clarification. This rule means that there is no requirement to measure fair facing to masonry work or, put another way, fair facing has to be measured to all other work where it is specifically required. It is generally accepted throughout the construction industry that the term fair facing means the additional work involved in sorting common bricks or blocks, selecting the better ones for use, constructing them in a neat and smooth manner and pointing. Fair facing in CESMM 3 applies to all types of brickwork and blockwork, even those elements constructed entirely of facings. It is necessary to measure fair facing to all items of brickwork and blockwork which have a neat and uniform appearance regardless of the material involved. Alternatively, this rule could be extended to include elements constructed entirely of facings.

C2   The cost significance of building-in pipes is very small, particularly smaller diameter pipework whose supply is measured elsewhere, but CESMM 3 elects to measure them separately. If it is felt that in comparison with the cost of the overall scheme they are cost insignificant and their measurement classification could be written out in suitable preamble.

    The supply of the pipes and ducts to which the building-in applies will in most instances have been catered for in Class I and item descriptions must contain the words 'supply measured elsewhere' or similar words to that effect.

Additional Description Rules

Generally   Additional Description Rules A1-A5 inclusive mean that a considerable amount of additional information must be given than would be assumed from a superficial examination of the Divisions. This may either be done in the Specification to which reference should be made or in the item descriptions themselves. If the Specification is referred to, then the taker-off must ensure that it complies in all respects with these rules and that where different requirements are specified on the drawings, that the Specification deals with those quite separately so that the references which appear in the take-off and billing will be unique for each requirement.

    These rules only apply to walls, facing to concrete, casing to metal sections and columns and piers, because these are the classification in the Divisions which they appear opposite. However, there are other items in the class to which at least part, if not all, of these rules would equally apply. The main ones would be copings and sills, cornices, band courses, corbels, pilasters and plinths, and Rules A1-A5 should also cover these items insofar as they apply.

CLASS U: BRICKWORK, BLOCKWORK AND MASONRY

A1   This rule covers the specification of brick, block and stone. Specification of materials would either be by manufacturer's reference, reference to the relevant British Standard and type, or, if at the time of bill preparation the actual type is not known, by the insertion of a PC (Prime Cost) sum per unit of number (normally per thousand or for specials per hundred). Specifying brickwork by way of a PC sum is not particularly desirable and should be used only as a last resort. If a PC sum is used then several important items must be made clear either in the item description or the preamble. These are:

   a) the type of material i.e. facing brick, dense concrete block, etc. The Contractor will need to know this in order to assess the labour aspect of laying them

   b) what the P.C. sum is to include for i.e. supply, delivery to site, unloading, etc. This will inform the Contractor whether he must allow for any items in his rate for laying apart from labour and jointing materials.

   c) whether the P.C. sum includes any percentage for Main Contractor's discount or not. It is also usual to include a statement to the effect that the sum is exclusive of Main Contractor's overheads and profit and that the Contractor must allow for waste.

The nominal sizes of bricks and blocks are their formatted sizes i.e. the actual dimensions of the brick or block plus half the joint width all round. However, with some types of stonework it would be difficult to give the nominal size because of the variation in the joint widths. Similarly, it would be difficult to give the sizes of stones which comprise random rubble walls. In these cases it would be more usual to refer to the Specification which would give the sizes to which the stones must comply.

A2   This rule only applies to masonry because the finish on brickwork and blockwork is non-optional, i.e. red rustic facings will have a red rustic finish. Masonry can have various finishes and these include tooled, dragged, straight cut and pitched finishes.

A3   The specification of the bonding pattern is relatively simple for brickwork and blockwork, but can become quite involved for stonework. Masonry may have a flemish type face, be uncoursed, built to courses or regular coursed. In addition the courses may be of diminishing thickness or of alternating thickness and all this information must be given.

   Jointing is the method by which the bricks, blocks or stones are joined together. Pointing is the term given to the work involved in treating the exposed part of the joint to give it a neat finish and this is done by either treating the joint mortar or by applying mortar of a different colour or type into the exposed face of the joint. It follows that pointing is only applicable for items of fair facing, and because it is measured separately for all work except masonry. Item descriptions for brickwork and blockwork walls, facing to concrete, casing to metal sections and columns and piers would not contain the specification of the pointing because this would be included in U. * 7 8. Item descriptions for masonry would require the pointing to be specified as these items are deemed to include the fair face.

Jointing is usually done with mortar and it is necessary to state the type and mix. Jointing of masonry can take several different forms and these should be described in detail.

A4   See Measurement Rule M1.

A5   The thickness of battered walls would be best described as being the mean or average thickness.

A6   There are standard special bricks and blocks which can be bought or ordered 'off the shelf' and BS4729 gives details of the types of special bricks available. The taker-off should not decide whether bricks, blocks or stones have to be cut, but give as much information about the surface feature as possible and let the Contractor decide. In doing so it may be sometimes necessary to give the cross-sectional sizes of surface features where they are less than 0.05m2 even though this contravenes Additional Description Rule A7. The departure should be noted in the preamble. It may also be prudent to refer to a drawing.

A7   It is necessary to state the size of surface features only when their cross-sectional size exceeds 0.05m2. In effect this means that a group of one type of surface feature below a size of say 300 x 165mm would be measured and billed together even if their sizes differed and this is not desirable. It is recommended that the taker-off uses some discretion regarding this rule and states the sizes of all surface features if he considers it to be applicable. A preamble would have to be incorporated stating the deviation from CESMM 3.

A8   This should be regarded as a general rule for the describing of joint reinforcement, damp proof courses, infills, fixing and ties. Sufficient information should always be given in the item descriptions to enable the estimator to price the work with a reasonable degree of accuracy.

A9   It would more prudent to refer the Contractor to the relevant drawing rather than become involved in lengthy bill descriptions. Joints in brickwork are measured as completely composite items and the rates should include for all the components and work involved in forming the joint.

A10 Where the length of built-in pipes and ducts exceeds 1m, it is possible that jointing of the ducts may be involved. It is recommended that any ducts which are measured in this class for which the supply is included are only those of a very simple nature and anything which may require jointing or is more specialised is measured under Classes I and J.

ITEM MEASUREMENT

U. * * * Brickwork, blockwork and masonry generally

Applicable to all items

Divisions   State the type of brickwork, blockwork or masonry in accordance with the First Division.

Rules   M1 Measure each skin of brickwork, blockwork or masonry which is in cavity or composite construction.

M2 Include the volume and areas of joints; exclude the volume of copings and sills. Do not adjust the measurement for intruding or projecting surface features. Do not deduct the measurement for holes or openings unless they exceed 0.25m2.

Generally   Additional Description Rules A1-A5 inclusive should also be taken as being applicable to surface features insofar as they apply.

U. * 1-5 * U. * 6 0 Walls, facing to concrete, casing to metal sections and columns and piers

Divisions   Walls, facing to concrete and casing to metal sections are measured in square metres for walls less than one metre thick and in cubic metres for those walls over one metre thick. State whether vertical, battered, straight or curved in accordance with the Third Division. Columns and piers are measured in linear metres stating the cross-sectional dimensions.

Rules   M3 Use mean dimensions for the purposes of measurement.

D3 Ignore the presence of surface features when determining the thicknesses of walls.

D4 Take the thickness of battered walls as the mean thickness.

A1 State the materials, nominal dimensions and types of brick, block or stone or give the relevant British Standard specification.

A2 State the surface finish to masonry work.

A3 State the bonding pattern and method of jointing. State the type of pointing for masonry work.

A4 State when in cavity or composite construction.

A5 State the nominal thickness of walls, facing to concrete and casing to metal sections. It is not necessary to state the thickness in ranges in accordance with the Second Division (Paragraph 3.10).

Generally The Third Division classification for walls and facing to concrete should not be taken as being definitive. For instance a facing which is described as being battered could be either of the two examples shown in Figure U5.
  However, there would be a considerable difference in constructing the two walls in that example B would involve much more cutting and waste of material than example A which is of a uniform thickness. Also, further classifications are possible i.e. curved facing to concrete. It is recommended that the taker-off uses his discretion when measuring brickwork, blockwork and masonry and invokes Paragraph 5.10 with regard to additional description should he consider it necessary.

See also Rules D1, D2, C1.

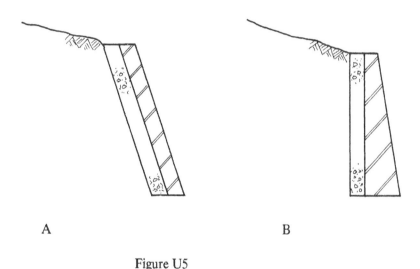

A   B

Figure U5

U. * 7 * Surface features

Divisions Surface features are measured in linear metres except for fair facing which is measured in square metres. State the type of surface feature.

Rules M4 Lengths measured shall be the mean lengths. Fair facing is measured at the face.

A6 Give sufficient detail to identify special masonry and special or cut bricks or blocks. State the spacing of intermittent surface features.

A7 State the cross-sectional dimensions where the cross-sectional area exceeds 0.05m2.

Generally The Third Division classifications should not be taken as being definitive. In particular there are no enumerated items such as air bricks, projecting headers, or special bricks at corners of copings and the like. Again it is recommended that the taker-off use his discretion in these matters.

See also Rule D5.

U. * 8 * Ancillaries

Divisions Ancillaries are measured according to their type. Reinforcement, damp proof courses and movement joints are in linear metres; bonds to existing work, infills of stated thickness and fixings and ties are in square metres; built in pipes are enumerated.

Rules M5 Do not measure the additional material in laps.

M6 Measure the area of fixings and ties as the smaller area of two areas of brickwork, blockwork or masonry involved.

C2 State when built in pipes exclude the supply.

A8 State the materials and dimensions of reinforcement and damp proof courses. State the materials of infills. State the type and spacing of fixings and ties.

A9 State the dimensions and nature of components including face or internal details for movement joints.

A10 State the length of built in pipes where they exceed 1m.

Generally There are instances when it may be more desirable to measure ties in linear metres or by number, for instance where the end of a wall is tied into a column. Preambles should state the deviation from CESMM 3.

STANDARD DESCRIPTION LIBRARY

**Common brickwork; BS3921; size 225 x 112.5 x 65mm;
jointing in cement mortar (1:3); stretcher bond**

Thickness 102.5mm

U. 1 1 1	vertical straight walls	m2
U. 1 1 1.1	vertical straight walls; cavity construction	m2

Common brickwork; BS3921; size 225 x 112.5 x 65mm; jointing in cement mortar (1:3); stretcher bond (cont'd)

U. 1 1 2	vertical curved walls	m2
	Columns and piers	
U. 1 6 0	215 x 112.5mm	m
	Surface features	
U. 1 7 8	fair facing; flush pointing	m2
	Ancillaries	
U. 1 8 2	damp proof courses; pitch polymer; 112.5mm wide	m
U. 1 8 6	cavity ties; butterfly twist type	m2

Facing brickwork; red rustic; size 225 x 112.5 x 65mm; jointing in cement lime mortar (1:1:6); pointing in coloured cement lime mortar (1:1:6); Flemish bond

Thickness 215mm

U. 2 2 3	battered straight walls	m2
U. 2 2 4	battered curved walls	m2
	Surface features	
U. 2 7 1	coping 215 x 102.5mm; headers on end	m
U. 2 7 8	fair facing; weather struck pointing	m2
	Band courses	
U. 2 7 4	112.5 x 50mm; projecting	m

Engineering brickwork; BS3921 type A; size 225 x 112.5 x 65mm; jointing in cement mortar (1:3); English bond

Thickness 215mm

U. 3 2 1	vertical walls	m2
U. 3 2 2	vertical curved walls	m2

U. 3 2 7	casing to metal sections	m2

Columns and piers

U. 3 6 0	450 x 215mm	m

Surface features

U. 3 7 6	pilasters; 375 x 112.5mm	m
U. 3 7 8	fair facing; flush pointing	m2

Ancillaries

U. 3 8 7	built-in pipes and ducts; cross-sectional area not exceeding 0.05m2	nr

Blockwork; BS6073 part 1; 7N/mm2 solid; 450 x 225 x 100mm; jointing in cement mortar (1:3); stretcher bond

Thickness 100mm

U. 5 1 1	vertical straight walls	m2
U. 5 1 1.1	vertical straight walls; cavity construction	m2
U. 5 1 5	vertical facing to concrete	m2

Ashlar masonry; granite; size 400 x 200 x 200mm jointing and pointing in cement mortar (1:3); flemish face built to courses; both faces hand dressed

Thickness 200mm

U. 7 2 1	vertical walls	m2
U. 7 2 5	vertical facing to concrete	m2

Rubble masonry; sandstone; minimum size 200 x 100 x 100mm; maximum size 400 x 200 x 200mm; jointing and pointing in cement lime mortar (1:1:6); built to courses; one face quarry dressed, other face hammer dressed; courses of diminishing thickness

Average thickness 300mm

U. 8 3 3	battered straight walls	m2
U. 8 3 4	battered curved walls	m2

Surface features

U. 8 7 1	coping; 200 x 150mm	m

Chapter 24

CLASS V: PAINTING

GENERAL TEXT

Principal changes from CESMM 1

There are no significant changes to this class except that preparation of surfaces was described where there was more than one type of preparation for the same surface.

Principal changes from CESMM 2

None.

Measurement Rules

M3 An isolated surface is one which is detached from the main body of the paintwork itself. The reference to 'groups' of surfaces refers to the surfaces of varying angles or widths contained within that isolated surface. Separate items should be given for each isolated group of surfaces unless they are identical in which case they can be collected together.

M4 This rule means that only the areas of the main members would be measured. It is recommended that no deduction is made for notches or splayed ends. This rule applies to items which are incidental to the steelwork involved but it could be argued that items such as stanchion bases are not incidental and should be included in the measurement. It is recommended that it be made clear in the Preamble that the areas included are those of the primary members only.

M5 In CESMM 1 fittings were measured 'extra over' the pipework and the measurement of painting the pipework was a simple matter of abstracting the various lengths of pipework measured under Class I. Fittings not in trenches are now measured full value and the painting will involve either adding together the effective lengths of all the fittings or actually measuring the pipework again through all the fittings. It should also be remembered that pipes and fittings are described by their nominal bore whereas painting will measured to their outside diameter. The rates for painting pipework are deemed to include for painting all fittings. However, items such as extension spindles to penstocks can hardly be taken as being part of the pipe and it is recommended that these are dealt with separately and a suitable preamble incorporated stating the deviation from CESMM 3.

CLASS V: PAINTING

Definition Rules

D1 For irregular shaped surfaces, it may be difficult at times to calculate exactly whether or not an isolated group of surfaces exceeds six square metres or not. Therefore an estimation would have to suffice, but if at tender stage an item was measured as an isolated group of surfaces, then it would be incorrect to try and change the method of measurement in the remeasurement if it was subsequently found to exceed the stated limit.

The criterion of six square metres should not be strictly adhered to if a group of surfaces lends itself to this method of measurement even if its area exceeds the stated amount.

Coverage Rules

C1 It follows that it must be clear from the item descriptions for the painting exactly what the preparation items are.

C2 See Measurement Rule M4.

C3 See Measurement Rule M5.

Additional Description Rules

A1 More often than not it would be the number of coats stated rather than the coat thickness. It is recommended that any constraints on the method of application should also be stated (i.e. brush, spray, etc.).

ITEM MEASUREMENT

V. * * * Painting generally

Applicable to all items

Divisions   State the material to be used. The First Division should be taken as being indicative only and further types of paint added as required. The actual description for several coats of paint could either be as separate items i.e. primer, undercoat, topcoat or gloss, or as a composite item i.e. one coat primer, plus one undercoat plus one coat gloss paint.

Rules   M1 Do not deduct areas of holes or openings unless they exceed 0.5m2.

A1 State the number of coats or film thickness.

A2 State the preparation required where more than one type of preparation is specified for the same surface.

Generally   Painting is only measured for work carried out on the site. Any painting which is done prior to components being delivered to the site is deemed to be included with the items.

See also Rule C1.

V. * 1-6 1-7 Painting to general surfaces

Divisions   Painting to general surfaces is measured in square metres for surfaces
exceeding one metre wide and linear metres for surfaces not exceeding one
metre wide. State the nature of the base. For surfaces which exceed one metre
wide state angle of inclination in accordance with the Third Division, (see
Figure V1). For surfaces less than one metre wide distinguish between surfaces
exceeding and not exceeding 300mm. The Second Division should be taken as
being indicative only and added to as required.

Rules   M2 Do not distinguish surfaces less than one metre wide by inclination
Problems arise when trying to classify painting to curved or domed structures,
see Figure V2.

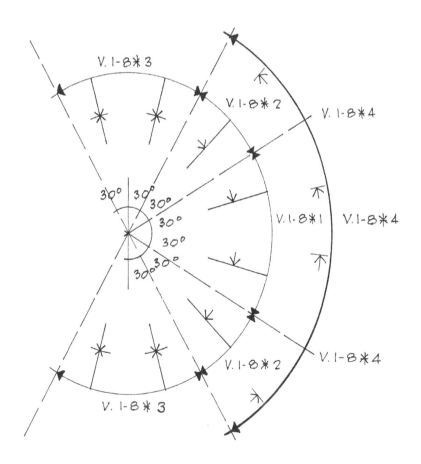

Figure V1

It is not possible to classify the painted surface according to any of the classifications listed in the Third Division because it cannot be given an angle to the horizontal. Therefore an additional classification needs to be created. In this case it could be described as 'soffit surfaces curved' or 'soffit surfaces of varying angles'. In either case a suitable preamble would have to be incorporated.

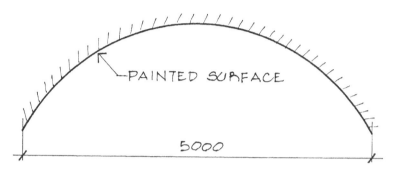

Figure V2

Generally   -

V. * 1-6 8   Painting to isolated groups of surfaces

Divisions   Painting to isolated groups of surfaces is enumerated. State the nature of the base. The Second Division should be taken as being indicative only and added to as required.

Rules   M3 Measure different shapes or dimensions of isolated groups of surfaces as different items.

A3 Identify the work to be painted and its location.

Generally   See also Rule D1.

V. * 7&8 0   Painting metal sections and pipework

Divisions   Painting to metal sections and pipework is measured in square metres.

Rules   M4 Ignore the presence of fittings and bolts to metalwork.

M5 Ignore the presence of fittings to pipework.

Generally   Where there are several types of pipe on a scheme it may be prudent to specify the principal type of joints.

See also Rules C2, C3.

STANDARD DESCRIPTION LIBRARY

One coat lead based primer as specification clause 26.1

Metal other than metal sections and pipework

V. 1 1 1	upper surfaces inclined at an angle not exceeding 30 degrees to the horizontal	m2
V. 1 1 2	upper surfaces inclined at 30-60 degrees to the horizontal	m2
V. 1 1 3	surfaces inclined at an angle exceeding 60 degrees to the horizontal	m2
V. 1 1 4	soffit surfaces and lower surfaces inclined at an angle not exceeding 60 degrees to the horizontal	m2
V. 1 1 6	surfaces of width not exceeding 300mm	m
V. 1 1 8	isolated groups of surfaces; open grid flooring; chamber 2 cover	nr

Metal sections

V. 1 7 0	generally	m2

Pipework

V. 1 8 0	generally	m2

Two undercoats emulsion paint as specification clause 26.2

Smooth concrete

V. 5 3 1	upper surfaces inclined at an angle not exceeding 30 degrees to the horizontal	m2
V. 5 3 2	upper surfaces inclined at 30-60 degrees to the horizontal	m2
V. 5 3 4	soffit surfaces and lower surfaces inclined at an angle not exceeding 60 degrees to the horizontal	m2
V. 5 3 7	surfaces of width 300mm-1m	m

One coat primer; two undercoats, one coat gloss paint as specification clauses 26.3 26.4 and 26.5

Timber

V. 1&4 2 3	surfaces inclined at an angle exceeding 60 degrees to the horizontal	m2
V. 1&4 2 6	surfaces of width not exceeding 300mm	m
V. 1&4 2 7	surfaces of width 300mm-1m	m

Chapter 25

CLASS W: WATERPROOFING

GENERAL TEXT

Principal changes from CESMM 1

There were no significant changes to this class.

Principal changes from CESMM 2

An item for 'tiling' has been included in the 2nd Division at 3 7*.

Measurement Rules

M1 Note that the actual measurement of waterproofing is the area of the 'surfaces covered'. The measurement would involve the calculation of the area of the base actually in contact with the waterproofing including the area in contact with aprons, flashings, skirtings and sides of gutters etc. The additional area of material in joints, laps, angles, fillets, etc. would not be measured.

M2 For surfaces which are less than one metre wide it is not necessary to state the angle of inclination or whether the surface is curved or domed.

M3 See Definition Rule D1.

M4 See Measurement Rule M3 of Class V.

Definition Rules

D1 This rule requires further explanation with regard to the classification of surfaces with multiple radii. The situation is clear concerning waterproofing to one radius in one plane (cylindrical), and one radius in two planes (spherical), but is not clear for varying radii (conical, elliptical, etc.).

    For conical work the measurement could be divided into two classifications as shown in Figure W1.

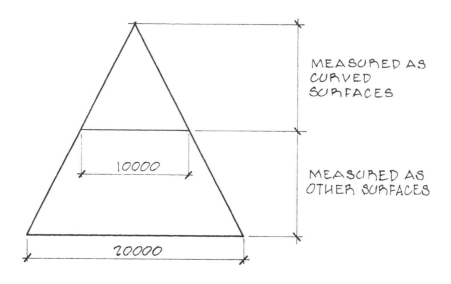

MEASURED AS CURVED SURFACES

MEASURED AS OTHER SURFACES

10000

20000

Figure W1

For elliptical work or other work with varying radii in different planes it is recommended that the work is classed to domed surfaces where the minimum radius does not exceed 10m.

It is difficult to classify the work to the lower half of the cone i.e. where the radius exceeds 10 m. As in Class V there is no classification in the Third Division to cover this situation. The problem could be overcome by writing out this rule altogether or by amending it to state that all curved and domed surfaces are to be so described, distinguishing between those which do and do not exceed 10m radius of curvature. In either case a suitable preamble would be needed stating the deviation from CESMM 3.

D2   See Definition Rule D1 of Class V.

Coverage Rules

C1   It follows that the Contractor must be supplied with complete drawings with details of all the items included in this rule. Details of any underlays and reinforcement should also be given in item descriptions. Although not specifically stated, all labours such as preparatory work, fair edges, turning into grooves, dressing to outlets and gutters and similar items would also be included.

Additional Description Rules

A1   Details should also be given of methods of fixing sheet metals and details of preparatory work to the base or subsequent layers where this would be otherwise unclear from the drawings or specification.

ITEM MEASUREMENT

W. 1-5 * * Waterproofing generally

Applicable to all items

Divisions   -

Rules   M1 Measure the area of the base actually in contact with the waterproofing. Do not deduct holes or openings unless they exceed 0.5m2.

A1 State the materials used and the number and thickness of coatings or layers.

Generally   See also Rule C1.

W.1-3 * 1-7 W.4 * 1-7 Damp proofing, tanking, roofing and protective layers to general surfaces

Divisions   Damp proofing, tanking, roofing and protective layers are measured in square metres for widths exceeding one metre and in linear metres for widths not exceeding one metre. State the classification of the surface in accordance with the Third Division. The Second Division should be taken as being indicative only and added to as required.

Rules   M2 Do not distinguish surfaces less than one metre wide by inclination or curvature.

M3. Do not distinguish curved or domed surfaces by inclination.

D1 Class curved or domed surfaces as such only when their radius or curvature is less than 10m.

Generally   Unlike Class V this Class has no provision in the Third Division for soffit surfaces not exceeding 60 degrees to the horizontal. If in practice the situation arises where this classification is required then it should be incorporated by way of a suitable preamble.

W. 1-3 * 8 W. 4 * 8 Damp proofing, tanking, roofing and protective layers to isolated groups of surfaces

Divisions   Damp proofing, tanking, roofing and protective layers to isolated groups of surfaces are enumerated. The Second Division should be taken as being indicative only and added to as required.

Rules M4 Measure different shapes or dimensions of isolated groups of surfaces as different items.

A2 Identify the work to be measured and its location.

Generally See also Rule D2.

W. 5 0 0 Sprayed or brushed waterproofing

Divisions Sprayed or brushed waterproofing is measured in square metres.

Rules -

Generally It is thought that this classification is not necessary. Any items which are considered as being suitable for measurement under this item would be more suitably measured under W. 1-3 * * or even under Class V (Painting). It is recommended that this classification is deleted by the inclusion of a suitable preamble.

STANDARD DESCRIPTION LIBRARY

Damp proofing

One layer 1200 gauge polythene sheet; fixed with adhesive as specification clause 27.1

W. 1 3 1	upper surfaces inclined at an angle not exceeding 30 degrees to the horizontal	m2
W. 1 3 3	surfaces inclined at an angle exceeding 60 degrees to the horizontal	m2
W. 1 3 6	surfaces of width not exceeding 300mm	m
W. 1 3 7	surfaces of width 300mm-1m	m

Tanking

Two coats asphalt; BS6577; 20mm first coat to classification T.1418; 15mm second coat to appendix A grade 3

W. 2 1 2	upper surfaces inclined at 30-60 degrees to the horizontal	m2
W. 2 1 4	curved surfaces	m2
W. 2 1 5	domed surfaces	m2

W. 2 1 8	isolated groups of surfaces; Valve Chamber 2; external walls	nr

Roofing

Profiled aluminium to BS4868 type A; as specification clause 27.2

W. 3 2 2	upper surfaces inclined at 30-60 degrees to the horizontal	m2
W. 3 2 3	surfaces inclined at an angle exceeding 60 degrees to the horizontal	m2

Protective layers

One layer self adhesive 1200 gauge polythene sheet

W. 4 2 1	upper surfaces inclined at an angle not exceeding 30 degrees to the horizontal	m2
W. 4 2 3	surfaces inclined at an angle exceeding 60 degrees to the horizontal	m2
W. 4 2 6	surfaces of width not exceeding 300mm	m
W. 4 2 7	surfaces of width 300mm-1m	m

Sand; 50mm thick

W. 4 3 1	upper surfaces inclined at an angle not exceeding 30 degrees to the horizontal	m2

Tiles; precast concrete flags to BS368; 450 x 450 x 50mm thick; as specification clause 27.3

W. 4 5 1	upper surfaces inclined at an angle not exceeding 30 degrees to the horizontal	m2

Chapter 26

CLASS X: MISCELLANEOUS WORK

GENERAL TEXT

Principal changes from CESMM 1

The only change to the Divisions in the Class was that rock filled gabions were added as an additional classification (X. 4 * 0). Coverage Rules were created for fences, gates and drainage above ground.

Principal changes from CESMM 2

Two new items have been included in the 2nd Division at 1 7* and 1 8* to cover metal guardrails and crash barriers.

A new Additional Description Rule A4 has been inserted to cover the case of gates where there is more than leaf. Existing Rules 4 and 5 are re-numbered.

Measurement Rules

M1 This means that gates and stiles are full value. The lengths measured should include lengths occupied by gate posts.

Definition Rules

D1 In certain instances such as stepped fencing on sloping ground it is possible that the height may cross from one Third Division classification to another. In this case it would be prudent to state the maximum height.

D2 The posts in this case are the posts on which the gate is supported and not the stiles or post which are part of the gate construction itself.

D3 There are other types of fittings from the ones listed such as branches, rainwater heads, access pipes etc., and these should all be enumerated separately.

Coverage Rules

C1 This rule does not deal the situation where hard materials are met with in post holes. If an inordinate amount of hard material were encountered then it is likely that the Contractor would claim for extra payment. This can be obviated by measuring the breaking out of the hard materials separately. This could be done by either enumeration per post hole or by measurement in cubic metres. Either way a suitable preamble would have to be incorporated.

There is no mention of on-site surface treatments such as preservative coatings and the like. The coverage rule could either be extended to include this or it could be measured separately under Class V.

The inclusion of concrete in the coverage rule would only apply where all the post holes were backfilled with concrete. In schemes where just the occasional post is in unsuitable ground, the Specification may say that these holes only have to be backfilled with concrete. Obviously the taker-off and the Contractor will have no way of knowing where or how many holes will have to be dealt with in this way. It is recommended that the items of concrete backfilling are measured separately. A suitable preamble would have to be incorporated stating the deviation from CESMM 3.

C2   This rule is adequate providing that the layout of the fence is shown and the Contractor can ascertain from the drawings and Specification exactly how many end, angle, straining, and gate posts there are, and also providing there is little or no likelihood of them being varied. If this is not the case it is recommended that these are all enumerated separately as 'extra over' the fence.

It is thought that because gate posts can often be substantially longer or different to the posts in the normal running length that it would be more appropriate to include them with the item for the gate or stile. If this is the case then Rules M1 and C2 would need to be amended and a further coverage rule created for the item of the gate. Alternatively the posts could be included with the item descriptions of the gates themselves.

C3   Supports include gutter brackets, gutter joint brackets, holderbats, hangers, etc.

Additional Description Rules

A1   Wherever possible reference should be made to the relevant drawing or A2 specification clause rather than relying on detailed description which could make A3 become unwieldy. Details of ironmongery should also be given in items for gates and stiles.

A4   State the number of leaves in the item description where there is more than one.

A5   It would also be necessary to state by what method the items are to be fixed to enable the Contractor to price the fixings.

A6   It is recommended that where gabions are placed below water, item descriptions should so state.

ITEM MEASUREMENT

X. 1 * * Fences

Divisions   Fences are measured in linear metres stating the type in accordance with the

Second Division and the height in accordance with the Third Division. The Second Division should be taken as being indicative only and should be added to as required.

Rules   M1 Do not include the length of gates and stiles.

D1 Measure the height from the Commencing Surface.

C1 Measure excavation of hard materials and backfilling with concrete if thought to be appropriate.

C2 Measure these items separately if thought to be appropriate.

A1 State when fences are erected to a curve not exceeding 100m or on surfaces exceeding 10 degrees to the horizontal.

A2 State the types and principal dimensions of fences or refer to the drawing or specification.

Generally   -

X. 2 * *   Gates and Stiles

Divisions   Gates and stiles are enumerated stating the type in accordance with the Second Division and width as classified in the Third Division. The Second Division should be taken as being indicative only and should be added to as required.

Rules   D2 The width should be measured between the inside faces of the posts.

A3 State the type and principal dimensions or refer to the drawing or specification.

A4 State the number of leaves in the item description where there is more than one.

Generally   Include the posts with the item if considered to be more appropriate.

X. 3 * *   Drainage to structures above ground

Divisions   Gutters and pipes are measured in linear metres and fittings are enumerated.

Rules   A5 State the type, principal dimensions and materials of the components.

Generally   It is recommended that the type of fitting is also stated and the type of fixing.

See also Rules D3, C3.

X. 4 * 0   Rock filled gabions

Divisions   Box gabions are enumerated stating the size. Mattress gabions are measured in square metres stating the thickness.

Rules   A6 State the type and grading of filling, size and diameter of mesh. Give details of protective coatings. State if the filling materials are other than imported.

Generally   It is also recommended that gabions placed below water are so described.

See also Rule D4.

STANDARD DESCRIPTION LIBRARY

Fences

Timber post and rail; treated sawn softwood; 100 x 100mm driven posts with 3nr 75 x 25mm rails

X. 1 1 2	height 1-1.25m	m
X. 1 1 2.1	height 1-1.25m; inclined at an angle exceeding 10 degrees	m
X. 1 1 2.2	height 1-1.25m; curved to radius not exceeding 100m	m
X. 1 1 2.3	height 1-1.25m; inclined at an angle exceeding 10 degrees, curved to radius not exceeding 100mm	m

Concrete post and wire; 100 x 100mm concrete posts in 400 x 400 x 400mm deep post holes back filled with C10P concrete; 2nr 4mm diameter galvanised line wires

X. 1 3 1	height not exceeding 1m	m

Gates and stiles

Timber field gate; BS3470; cast iron fittings; as drawing number MW/DWG/1; including posts

X. 2 1 4	gate width 2.7m	nr

CLASS X: MISCELLANEOUS WORK

Drainage to structures above ground

uPVC; BS4576; rubber seal joints; fixed with brackets to timber

X. 3 3 1	gutters 100mm	m
X. 3 3 2	stop ends	nr
X. 3 3 2.1	outlets 75mm	nr
X. 3 3 2.2	stop ends with 75mm outlets	nr
X. 3 3 2.3	angles	nr

uPVC; BS4576; rubber seal ring joints; fixed with eared fixings to brickwork

X. 3 3 3	pipes 75mm	m
X. 3 3 4	shoes	nr
X. 3 3 4.1	heads	nr
X. 3 3 4.2	swan necks	nr
X. 3 3 4.3	bends	nr

Rock filled gabions

Box type as specification clause 28.1; filled with imported graded material as specification clause 28.2

X. 4 1 0	2000 x 2000 x 2000mm	nr
X. 4 1 0.1	2500 x 2500 x 2500mm	nr

Mattress type as specification clause 28.3 filled with imported graded material as specification clause 28.4

X. 4 2 0	500mm thick	m2
X. 4 2 0.1	750mm thick	m2

Chapter 27

CLASS Y: SEWER AND WATER MAIN RENOVATION AND ANCILLARY WORKS

GENERAL TEXT

Principal changes from CESMM 1

Class Y was new to CESMM 2 and there were therefore no changes.

Principal changes from CESMM 2

A new division item (Y 1 3 0) has been inserted for close-circuit television surveys and this has caused a re-numbering of the previous Y. 1 3-5 * items.

A complete new range of items are included at Y.5 * * to deal with water main renovation and ancillary works. Associated with this work are new rules M6, D5, C7 and A11 and all previous rules beyond these have been re-numbered.

Coverage Rule C1 now refers to dead services instead of existing services and the words 'or water main' have been added to the end of Addituional Description Rule A5.

In CESMM 2, new manholes in new locations or replacing existing manholes were separately stated in the first division Y. 5 * * and Y. 6 * * respectively. The first division now only refers to 'new manholes' and new Additional Description Rule A14 says that where a new manhole replaces an existing one, it shall state it in the description.

The old Coverage Rule C8 has been reworded and now becomes C9 and C10.

Measurement Rules

M1　The length should also include the length of fittings encountered in the normal running length.

M2　There are basically three methods of sewer renovation. The first involves working manually from the inside of the sewer. The second is also working from the inside but using remote controlled machines. The third is where the sewer is exposed from above by excavating down from the surface.

Where work is expressly required to be carried out by excavation, then any crossings, reinstatement and other pipework ancillaries (K. 6-8 * *) must also be measured and any work encountered which is included in Class L must be measured. This rule may suggest that the work should be billed with other items measured in Classes K and L as opposed to in accordance with these classes. It is recommended that any work measured in Classes K and L for work in connection with excavation carried out under Class Y is kept apart from other items under a separate heading.

Unlike Class I there is no requirement to state the Commencing Surface and this is because this will almost always be the Original Surface. Class Y does not define from which point the depth is measured for work which is expressly required to be carried out by excavation. This should be measured from the Commencing Surface and it should also be remembered that Paragraph 5.21 must also be complied with for work which is expressly required to be carried out by excavation.

M3 Openings would be incoming laterals, access shafts and the like.

M4 The whole question of grouting in Class Y requires careful consideration. There
M5 are two types of grouting; annulus and external grouting. Annulus grouting is the
D3 operation of filling the void formed between a new lining and the existing sewer.
and External grouting is the operation in stabilising the ground around the outside
D4 of the existing sewer, (see Figure Y1).

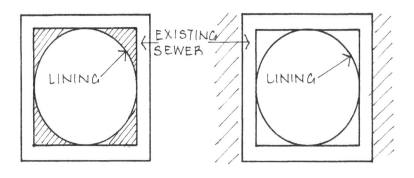

Annulus grout   External grout

Figure Y1

The first problem occurs in the unit of measurement for grouting measured in Class Y. Unlike Classes C and T which measure grouting in tonnes, this class measures it in cubic metres. Whilst this may be satisfactory for the tender documentation because the quantities given will, at best, only be an approximation, it is almost impossible to remeasure the actual grouting on this basis. It is possible to convert the total tonnage injected, by a conversion factor to obtain the cubic content, but this would not make sense. It may be possible to calculate the volume of the annulus grouting by deducting the volume of the lining from the internal volume of the sewer, but this would only be

an approximation because it would not take into account the filling of any other voids
such as minor depressions and joints in the sewer walls. It is almost impossible to
calculate the volume of external grouting by any physical means.

It would seem that the logical way to measure grouting is by the tonne, and it is
recommended that a suitable preamble be incorporated to this effect.

The second problem is that for admeasurement it may be difficult to completely
separate annulus grouting from external grouting. Whilst Measurement Rule M5
states that the volume of external grouting must not be included in the annulus
grouting it may be difficult if not impossible to comply with this rule and some
flexibility must be allowed.

The third problem is the different measurement classifications between this class,
Class C and Class T. Whilst it is appreciated that there is no reason why they should
all be identical, (after all the classes represent three different construction
techniques), it may have been preferable if there had been more standardisation
between them. After all, the same specialist subcontractor will probably grout all three
types of work, but is faced with the problem of having to deal with three quite separate
modes of measurement within the same document.

Classes C and T have separate classifications for drilling of holes and for packers
whereas Class Y does not. There is nothing necessarily wrong with this but the matter
could have been dealt with by a coverage rule.

M7   Measurement under this rule could only occur where the Engineer has specifically
stated that a minimum pumping capacity must be installed in the sewer. Where the
pumping capacity is left to the Contractor's discretion the situation is at the
Contractor's risk.

A potential argument arises in Rule M7 where the Contractor installs a larger
capacity pump than that specified (it may be that he owns a larger pump which is lying
idle and chooses to install it rather than hire the specified size at greater expense). If the
minimum pumping capacity is exceeded, but there is no interruption because the larger
pump than that specified can deal with the increased flow, is the Contractor entitled to
be paid for the theoretical interruption? The answer is not clear and would be
dependent upon the individual circumstances of the case.

It is thought that the pricing of items measured under this classification could prove
to be extremely difficult. For instance, the rate inserted for interruptions on manholes
would depend upon several factors. These would include:

a) how many manholes were being worked upon at any one time and how many
   operatives were involved

b) what type of work was involved

c) whether the labour force could be redeployed elsewhere

d) what type of plant is involved.

The rates for this classifications will be, at best, a rough approximation. It is unclear
whether the rates should include the cost of additional General Items or head

office overheads. It is considered that they should not, although the Preamble
should make this clear.

It may be necessary to define precisely in the Preamble or Specification exactly what
working hours constitute because these can vary greatly and have a bearing on
the final value.

Definition Rules

D1 Laterals are incoming branch sewers or drains. Artificial intrusions include dead
services, odd projecting bricks and any other features which project beyond the general
sewer face.

D2 Local internal repairs involve the cutting out and replacement of isolated areas of the
fabric of the sewer lining. This could be with new materials or utilizing all or part of
the existing materials and it is recommended that item descriptions should
differentiate.

D5 It may be necessary to carry out some work before a pipe sample inspection or closed-
circuit television survey can take place and this work is deemed to be coverd by the
item description.

D6 See Definition D2 in Class K.

Coverage Rules

C1 This rule is the same as Coverage Rule C2 in Class I and now refers specifically to
'dead rather than 'existing' services.

C2 This rule makes the Contractor responsible for any damage he may cause to the sewer
fabric due to his cleaning operations but it may be a very difficult rule to enforce.
Some old sewers nowadays are in such a serious state of disrepair that the
operation of cleaning them would inevitably lead to damage. The degree of damage
would be difficult to forecast and it would be very harsh to expect the Contractor to
absorb all the resultant costs. There may be an overlap between the repair work
involved and the permanent works which the Contractor has contracted to do.
Some degree of flexibility must be exercised when applying this rule.

C3 Although not specifically stated, the disposal of materials would also be included.

C4 See Definition Rule D2.

C6 There is far more to the reconnecting of laterals than is at first envisaged from a
cursory glance at the classification.
Take the example of an existing rectangular brick lined sewer with a incoming
300mm nominal bore clay lateral, which is to be lined with a preformed liner and
annulus grouting. The first operation would be to remove or cut the existing lateral
flush. The sewer would then be lined and grouted. In order to reconnect the lateral

the liner would have to be cut at the correct place (bearing in mind that the precise location would have to be determined first), the annulus grout would have to be removed (this may extend some way up the lateral unless it had been previously blanked off), a short connection of pipe may have to be inserted and the whole made good.

This rule limits the Contractor's liability to the first metre beyond the face of the liner in either direction.

C7   This rule should be taken mean that the Contractor will be paid for replacing the length of pipe necessary to make good the gap created by removing the sample and not just the length of sample removed.

C8   See Rule A1 on page 165. See also Coverage Rules C1 and C3 of Class K.

C9   See Rule C3 on page 164.

C10 Disposal would normally be off the site unless specifically stated otherwise. Item descriptions should make it clear whether any of the materials arising are to remain the property of the Employer and exactly how they are to the treated (for instance manhole covers and frames).

Additional Description Rules

A1   It is recommended that the work to preparation, stabilization, renovation and
to   laterals is billed under a general heading which collectively includes all these
A5   Additional Description Rules as applicable.

A6   It may also be prudent to state the size of artificial intrusions.

A9   The offset stated in this rule is demonstrated in Figure Y2. This Rule applies equally to bends as well as curved sewers.

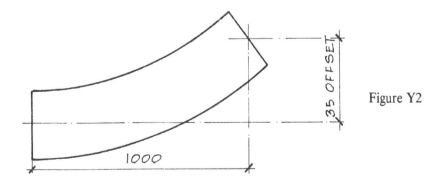

Figure Y2

A12 See Additional Description Rule A1 in Class K.

A15 This rule leaves the scope for the description of the work involved in existing manholes open. However, the taker-off should not become involved in lengthy item descriptions. It would be easier to refer to the drawings and specification which should contain details of the work involved.

ITEM MEASUREMENT

Y. * * *   Sewer renovation generally

Applicable to all items

Divisions   -

Rules   M1 Measure sewers along centre lines excluding manholes and backdrops.

            M2 Where applicable measure crossings, reinstatement and other pipework ancillaries (K. 6-8 * *), and extras to excavation and backfilling (Class L) where work is expressly required to be carried out by excavation.

            A1 State the location of the work.

            A2 State the principal dimensions and profiles of sewers.

            A3 State when work is expressly required to be carried out by remote control or manually.

            A4 State when work is to be carried out by excavation. State the maximum depth of excavation in stages of one metre measured to the invert of the sewer.

            A5 State the material of the existing sewer for items for preparation, stabilization, renovation and laterals.

Paragraph 5.21

            Where work is expressly required to be carried out by excavation state the Commencing Surface if it is not the Original Surface. Also state the Excavated Surface if it is not the Final Surface.

Generally   Use Additional Description Rules A1-A5 inclusive as a basis for a general heading under which to measure the work.
            The measurement of sewer renovation is heavily reliant on the specification and/or drawings. The descriptions given are brief and it is the responsibility of the Contractor to ascertain the precise nature of the work elsewhere.
            There is also a large responsibility placed on the taker-off with regard to temporary works which can be a major cost factor in sewer renovation work.
            Pumps, temporary stanks or cofferdams, safety equipment breathing

apparatus and lighting are all items which may require measurement under specified requirements in Class A. In addition any testing will have to be measured in Class A because there is no provision for it in this Class.

See also Rule C1.

Y. 1 1 0   Cleaning existing sewers

Divisions   Cleaning existing sewers is measured in linear metres.

Rules   -

Generally   Item descriptions should also state the mode of cleaning where this would otherwise be unclear.

See also Rule C2.

Y. 1 2 *   Removing intrusions

Divisions   Removing intrusions are enumerated. Laterals not exceeding 150mm are grouped together. Laterals over 150mm in one or more dimension must have their size and profile stated. State the nature of other artificial intrusions.

Rules   A6 State the materials of which the laterals are comprised.

Generally   Although not specifically stated it may be prudent to state the size of other artificial intrusions.

See also Rules D1, C3.

Y. 1 3 0 Closed circuit television surveys

Divisions   This work is measured in linear metres.

Rules   -

Generally   The bulk of the information the Contractor will need to price this item will be contained in Specification. The third division should be used to identify any difference between the surveys to be carried out, e.g. the bores of the pipes being surveyed, the location of the pipes (usually expressed as being between referenced manholes or chambers) etc.
   It is suggested that bores of the pipe are classified in the same ranges as those for television surveys of water mains in Y. 5 3 *.

Y. 1 3&4 *   Plugging and filling

Divisions   Plugging laterals is enumerated. Filling laterals and other pipes is measured in cubic metres. State the materials used. Items not exceeding 300mm nominal bore are grouped together. Items over 300mm in one or more dimension must have their size and profile stated.

Rules   -

Generally   It is important that items for filling pipes of any great length under 300mm nominal bore are measured separately and readily identified from the drawings so that the Contractor is allowed to price them separately.

Y. 1 5 * Local internal repairs

Divisions   Local internal repairs are enumerated stating the areas in the ranges given in the Third Division. The actual area must be stated where this exceeds 0.25m2.

Rules   D2 Areas shall be the finished surface areas.

Generally   It is recommended that item descriptions should state whether any existing materials are to be re-used.

    See also Rule C4.

Y. 2 1&2 0 Pointing and pipe joint seating

Divisions   Pointing is measured in square metres. Pipe joint sealing is enumerated. State the materials used in both instances.

Rules   M3 Do not deduct the area of voids from pointing unless their area exceeds 0.5m2.

Generally See also Rule C5.

Y. 2 3 1&2 External grouting

Divisions   External grouting is measured by the number of holes through which grout is injected and by the amount of grout injected given in cubic metres. State the type of grout.

Rules   M4 Only measure when expressly required to be carried out as a separate operation to annulus grouting.

    A7 State when required to be done through pipe joints in the item for number of holes.

Generally   It is recommended that the unit of measurement for grouting is changed from cubic metres to tonnes. A suitable preamble would have to be included.

See also Rule D3.

Y. 3 1-4 * Linings

Divisions   Linings are measured in linear metres stating the classification in accordance with the Second Division and the type in accordance with the Third Division.

Rules   A8 State the type, minimum finished internal size and thickness or grade.

A9 State the offset where this exceeds 35mm in any one metre for in situ jointed pipe lining and segmental lining.

Generally   -

Y. 3 5 0 Gunite coating

Divisions   Gunite coating is measured in linear metres stating the thickness.

Rules   -

Generally   The type, mixes and application requirements will normally be covered by the Specification to which reference should be made.

Y. 3 6 0 Annulus grouting

Divisions   Annulus grouting is measured in cubic metres.

Rules   M5 Do not include any volumes measured for external grouting.

Generally   The measurement of annulus grouting is even simpler than external grouting in that the number of holes or injections do not have to be stated.

See also Generally under Y. 2 3 *.

See also Rule D4.

Y. 4 1 * Jointing laterals to renovated sewers

Divisions   Jointing laterals to renovated sewers is enumerated stating the bore as classified. The profile and size must be stated when this exceeds 300mm in one or more dimension.

Rules   A10 State the type of lining to which the laterals are to be connected. Identify those laterals which are to be regraded.

Generally   Item descriptions should also make it clear where the responsibility lies for the supply of materials other than the actual jointing materials for which the

Contractor is responsible.

See also Rule C6.

Y. 4 2 * Flap valves

Divisions   Flap valves are enumerated stating the size. Separate items must be given for
removal and replacement of existing and new flap valves of stated type.

Rules   -

Generally   The Specification should normally detail exactly what treatment (if any) is to be
given to existing flap valves to be reused.

Y.5 * * Water main renovation and ancillary works

Divisions   The cleaning, television surveys and linings are given in linear metres and
removing intrusions and pipe sample inspections are enumerated.

Rules   M6   Measure lenghts of pipe over fitttings and valves.

C7   If the sample taken is less than the length to be replaced (the Engineer
may insist on the replacement being taken back to the nearest joints) measure
the length to be replaced, not the sample length.

A11   State the lining material, the nominal bore of the pipe being lined and the
lining thickness.

Y. 5&6 * * New manholes

Divisions   New manholes are enumerated stating the construction in accordance with the
Second Division and the depth as classified in the Third Division. State the
actual depth where this exceeds 4m. State whether built in new location or on
the site of an existing manhole.

Rules   D5   Measure the depth from the top of the cover to channel invert or top of
base slab whichever is lower.

A11   State type or mark number. Identify different configurations.

A12   State type and loading duties.

Generally   See also Rules C8, C9, C10, C11.

Y. 7 * * Existing manholes

Divisions   The abandonment of existing manholes is enumerated stating the depth as classified. The actual depth has to be stated where this exceeds 4m. The alteration of existing manholes is enumerated but it is not necessary to state the depth at all.

Rules   D5   Measure the depth from the top of the covers to channel invert or top of base slab whichever is lower.

A11   State type or mark number. Identify different configurations.

A13   Give details of the work required.

Generally   Whilst it is not strictly necessary to state the depth for the alteration of existing manholes this should be stated if it is considered to be cost significant.

See also Rule C8.

Y. 8 * * Interruptions

Divisions   Interruptions are measured in hours stating the nature of the work interrupted as classified in the Second Division. The work involving renovation of existing sewers must be further subdivided into the various types of lining, gunite coating, and annulus grouting.

Rules   M6 Only measure interruptions during normal working hours when the sewer flow exceeds the installed pumping capacity and a minimum pumping capacity is expressly required.

Generally   -

STANDARD DESCRIPTION LIBRARY

**Renovation of existing sewer between manholes MHSW1 and
MHSW5; square sewer internal size 3 x 3m; work to be
done manually; principal material brick**

Preparation of existing sewers

Cleaning

Y. 1 1 0   generally   m

Preparation of existing sewers (cont'd)

Removing intrusions

Y. 1 2 1	laterals; clay; bore not exceeding 150mm	nr
Y. 1 2 3	projecting single bricks	nr

'Clearview' closed-circuit television survey

Y. 1 3 0	bore 350mm	m

Plugging laterals with concrete grade C15P

Y. 1 3 1	bore not exceeding 300mm	nr

Filling laterals and other pipes with PFA

Y. 1 4 2	circular; 400mm nominal bore	m3

Local internal repairs; as specification clause 29.1

Y. 1 5 1	area not exceeding 0.1m2	nr
Y. 1 5 2	area 0.1-0.25m2	nr
Y. 1 5 3	area 0.4m2	nr

Stabilization of existing sewers

Y. 2 1 0	Pointing with waterproof cement mortar (1:3)	m2
Y. 2 3 1	External grouting; number of holes	nr
Y. 2 3 2	External grouting; injection; grout as specification clause 29.2	m3

Renovation of existing sewers

Sliplining

Y. 3 1 2	polypropylene; as specification clause 29.3	m

Annulus grouting

Y. 3 6 0	grout as specification clause 29.4	m3

CLASS Y: SEWER AND WATER MAIN RENOVATION AND ANCILLARY WORKS

Laterals to renovated sewers

Jointing; polypropylene lining

Y. 4 1 1	bore not exceeding 150mm	nr
Y. 4 1 3	circular; 400mm nominal bore	nr

Water main renovation and ancillary works; length between chambers MAG 1 to SAFC 9; steel pipe with welded joints

Cleaning

Y. 5 1 3	nominal bore 304.9mm	m

'Clearview' closed-circuit television survey

Y.5 3 1	nominal bore 154.1mm	m
Y. 5 3 3	nominal bore 304.9mm	m

New manholes

Brick; cover and frame to BS497 reference MB1-55

Y. 5 1 3	depth 2-2.5m; type SW1	nr
Y. 5 1 4	depth 2.5-3m; type SW1	nr

New manholes replacing existing manholes

Brick; cover and frame to BS497 reference MB1-55

Y. 6 1 3	depth 2-2.5m; type SSWR	nr
Y. 6 1 4	depth 2.5-3m; type SW2	nr

Existing manholes

Abandonment

Y. 7 1 4	depth 2.5-3m; breaking out complete; backfilling with lean concrete as specification clause 29.5	nr

Existing manhole (cont'd)

Alteration

Y. 7 2 0	depth 2.5-3m; breaking out and replacing benching; pointing walls internally; new cover and frame reference MB1-55; as specification clause 29.6	nr

Interruptions

Preparation of existing sewers

Y. 8 1 0	generally	h

Stabilization of existing sewers

Y. 8 2 0	generally	h

Renovation of existing sewers

Y. 8 3 1	sliplining	h
Y. 8 3 2	in situ jointed pipe lining	h
Y. 8 3 3	segmental lining	h
Y. 8 3 5	gunite coating	h
Y. 8 3 6	annulus grouting	h

Work on laterals to renovated sewers

Y. 8 4 0	generally	h

Work on manholes

Y. 8 5 0	generally	h

Chapter 28

CLASS Z: SIMPLE BUILDING WORKS INCIDENTAL TO CIVIL ENGINEERING WORKS

GENERAL TEXT

Measurement Rules

M1 This rule somewhat unnecessarily re-states the basic philosophy of CESMM3 that measurements taken shall be net with no allowance made for joints or laps. In this case the statement refers to carpentry and joinery work.

M2 The minimum size opening of 0.5m2 for boarding and the like, below which deductions need not be made, is in line with the measurement of areas of other work in CESMM3.

M3 This common sense rule requires that strutting is measured on plan. If the joints being strutted were particularly deep and the struts were herringbone or less than the full depth of the joists, it would be wise to enlarge the description to give the estimator a clear picture of what is needed.

M4 This rule stating that the measurement of patent glazing should be taken over glazing bars, should also be assumed to refer to similar members which are an integral part of the system.

M6 The effect of this rule makes the value of all fittings 'extra over' the pipe they are part of. Although this may seem at odds with the rule governing the provision of pipes not in trenches in Class I Measurement Rule M4, it should be remembered that Class I is usually dealing with pipes of large diameter with a high preponderance of fittings, whereas this Class is concerned with pipes of diameter ranging between 13 and 50mm.

M7 See M6

M8 It is unlikely that the information regarding sags and tails will be available to the taker off when he prepares the bill of quantities. In any case, they are not to be included in the measurement for cables.

Definition Rules

D1 The normal size of timber refers to the unplaned or 'ex' dimensions of the component. If a wrought or planed finish is required the actual size of the member could be stated (Additional Description Rule A1).

D2   The wording of this rule is less than clear but what is intened is that ceiling joists
      should come under the classification of pitched roofs where appropriate.

D4   This rule states that the surfaces of columns should be included with walls whilst
      surfaces of beams shall be included in soffits.

D   The classification of boarding and sheeting according to the angle of inclination is set
      out below in Figure Z1.

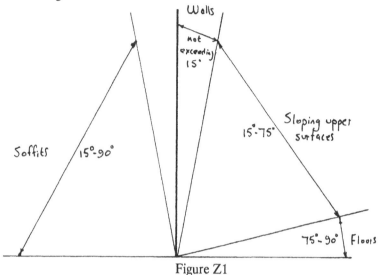

Figure Z1

D7   See figure Z1

Coverage Rules

C1   This rule also covers any labours on timber such as scarfing, rebating, grooving etc.

C2   The specification should describe the fixing requirements of windows and doors e.g.
      cramps, lugs etc. Specialist patent glazing systems would probably include these
      sundry items in the overall package. In the case of glass however it is probably better to
      conform to convention of describing the beads, putty, sprigs etc used for fixing the glass
      in the item descriptions.

C3   The taker off should ensure that the method of fixing, commissioning and other
to   sundry requirements are covered in the Specification or in the item description.
C14

Additional Description Rules

A1   See D1

A2   This rule seems to preclude stating the finished thickness of boarding. In the case of
      strip flooring, however, it is the wrought or planed size which is generally referred to by
      the suppliers. The taker off should follow the provisions of A1 in this case and note

it in the preamble.

A6   Although this rule does not mention all the items in 1 4 1-4 and 1 6 1-4, it should be
     taken as though it does.

A7   The thickness of loose fill refers to the overall thickness of the bed not the
     individual granule size.

A9   Not only should the materials ironmongery be stated in the item descriptions, but
     any other relevant information including possibly the manufacturer's name and the
     product's brand name.

A10 It could be thought that there is a slight ambiguity between C2 and this item but it
     can easily be removed by stating the method of glazing and of securing the glass in the
     item descriptions.

A11 Detailing the construction of hermetically sealed glazing units is the item
     descriptions may be an arduous and repetitive task in some cases. Where this occurs,
     the construction details should be transferred to the Specification with an appropriate
     note in the preamble stating the change.

A13 Care should be taken to distinguish between glass which is curved (to which this
     rule refers) and flat glass in a curved window, e.g. a bay bow window, which should be
     measured in the normal way.

A16 See A13.

A19 If the bulkhead is in a stairwell with difficult access problems or at an excessive
     height above a safe base, this information should be included in the item description.

A22 It should be sufficient in most cases to measure the pipework under headings such
     as 'Cold water Supply', 'Overflow', 'Wastes' etc rather than attempt to give locational
     information as inferred by this rule.

A23 If the information on joint types, BS reference and specified qualities is too large
     to be included in the item description, it should be placed in the Specification and a
     note to this effect given in the preamble.
     Where a fitting has multiple bores, the largest bore should be used for classification
     and coding purposes but *all* the bores should be stated.

A24 It will usually be necessary to give the manufacturer's name and the brand name
     of each individual fitting in addition to the information requested by this rule.

A26 See A22

A28 See A24

A27 See A23

A29 See A22

A31 See A24

ITEM MEASUREMENT

2.11 * Carpentry and joinery

Divisions   All structural and carcassing timbers are measured in linear metres. Trussed rafters, roof trusses are enumerated together with small components such as cleats.

Rules   M1 Do not make allowances for joints and laps.

M2 Do not deduct areas of boarding unless they exceed 0.5m2.

M3 Strutting between joists to be measured in linear metres on plan over joists.

C1 All labours on timber are deemed to be included.

A1 State whether timber is sawn or wrought together with any treatment or protection required.

A2 State thickness of boarding and where it is in narrow widths, measure it in linear metres in stages of 100mm up to 300mm wide.

A4 State cross-section dimensions of structural and carcassing timbers.

A5 State overall size and shape of stairs, walkways and the like.

Z.2**Insulation

A7 State overall nominal thickness of insulation.

Z.3 1-3 * Windows, doors and glazing

Divisions   All windows, doors, frames and the like are enumerated

Rules   C2 All materials needed to fix the components are deemed to be included together with all necessary cutting and drilling.

A8 State overall size, shape and limit of components.

Z.3 4 * Ironmongery

Divisions		All ironmongery to be enumerated.
Rules	A9	State the material in the item description together with the manufacturer's name and the product brand-name if further identification is necessary.

Z.3 5 * Glazing

Divisions		Glass is measured in square metres. Panes over 4m2 in size are measured separately. All special glass as defined in D5 is enumerated.
Rules	D4	Measure panes exceeding 4m2 as large panes and keep separate.
	C2	All materials for fixing are deemed to be included together with all necessary cutting and drilling.
	A10	State type of glass, thickness and method of glazing in item descriptions.
	A11	State construction of thermetically sealed units in item descriptions. If impracticable, state it in specification and make note in Preamble.
	A12	State shape and size of panes for special glass sealed units and mirrors.
	A13	State if glass is curved in item description.

Z.3 6 * Patent glazing

Divisions		Patent glazing measured in square metres.
Rules	M4	Measurements to be taken over glazing bars.
	M2	Do not deduct areas of patent glazing unless they exceed 0.5m2.
	C2	All materials for fixing are deemed to be included together with all necessary cutting and drilling.
	A14	State shape, size and limits of patent glazing.
	A15	Include any incidental metalwork in item descriptions.
	A16	Measure curved glazing separately.

Z.4**Surface finishes, linings and partitions

Divisions   All in situ finishes, beds backings, tiles, flexible sheet coverings, partitions and linings are measured in square metres except for surfaces not exceeding 300mm to 1m wide which are both measured in linear metres.

Rules

M2 Do not deduct areas unless they exceed 0.5m2.

M5 Do not measure proprietary system partitions over voids which extend the full height of the partition.

A17 Include the materials, surface finish and finished thickness in the item description for surface finishes, linings and partitions.

A18 Include lathing and baseboarding in item descriptions for in situ finishes beds and backings.

A19 State the girth of bulkheads in item description together with any other information such as height above floor level if it assists estimator.

A21 If the depth of suspension exceeds 500mm in suspended ceilings it shall be stated separately in stages of 500mm.

Z.5 ** Piped building services

Divisions   All pipework is measured linear metres. Fittings, equipment and appliances are enumerated.

Rules

M6 Measure pipes along centre line and over all fittings.

A22 State type of piped building service (e.g. 'cold water supply') or its location (e.g. 'pipework in boiler room') in the item description or heading.

A23 State material, joint type, nominal bore and BS references of pipework (if not transferred to Specification). Classify fittings with multiple bores by largest bore but state *all* bore sizes.

A24 Give the type of sanitary fittings together with size or capacity and methods of fixing. State the name of the manufacturer and the model if known.

A25 Include traps in the item description for sanitary fittings.

Z. 6 * * Ducted building services

Divisions		Ductwork is measured in linear metres and fittings are enumerated.
Rules	M7	Measure ducts along centre line and over fittings.
	A26	State type of ducted building service or its location in item description.
	A27	State material, joint type, size and BS reference of ducting. Classify fittings with multiple sizes by the largest but state *all* sizes.
	A28	Give the type of ducting together with size, capacity and method of fixing. State the name of the manufacturer and the model if known.

Z. 7 * * Cabled building services

Divisions		All cables and conduits are measured in linear metres. Fittings and equipment are enumerated
Rules	M8	Exclude sags and tails from cable lengths.
	M9	Measure conduits, trunking, busbar trunking, and cable trays over all fittings.
	M10	Measure the length of cables in conduits, trunking in ducts or on trays as the same length as the conduit, trunking, duct or tray.
	A28	State type of cable building service or its location in item description.
	A29	State material, size, capacity and BS references of cables.
	A30	Give the type, size, capacity and BS references of equipment and fittings. State the manufacturer's name and model if known.

STANDARD DESCRIPTION LIBRARY

CARPENTRY AND JOINERY

Structural and carcassing timber, sawn untreated softwood

Floor members

Z.1 1 1	50 x 150mm	m
Z.1 1 1.1	75 x 125mm	m

Structural and carcassing timber, sawn untreated softwood (cont'd)

Wall or partition members

Z.1 1 2	50 x 75mm	m
Z.1 1 2.1	50 x 100mm	m

Flat roof members

	50 x 200mm	m

Trussed rafters, pressed impregnated softwood, set in position 8m above ground level

Z.1 1 8	Type W truss, 22.5 degrees. span 6m	nr
Z.1 1 8.1	Type W truss, 22.5 degrees, span 8m	nr

Sheet boarding external quality plywood; 18mm thick

Z.1 3 2	Sloping upper surfaces	m2

Stairs and walkways

Z.1 4 1	Standard stairway, wrought softwood, 25mm treads with rounded nosings, 12mm plywood risers, 32mm strings once rounded, bottom tread bullnosed, 50 x 75mm hardwood handrail, two 32 x 140mm balustrade knee rails, 32 x 50mm stiffeners and 100 x 100mm newel posts with hardwood newel caps, overall rise 2600mm, width 838mm, going 2688mm	nr

Miscellaneous joinery; wrought softwood

Z.1 5 1	Skirting, moulded 19 x 75mm	m
Z.1 5 2	Architrave, splayed, 19 x 75mm	m

Units and fittings, Hygena QA range

Z.1 6 1	Base unit 600 x 900 x 600mm	nr
Z.1 6 2	Wall unit 1200 x 900 x 300mm	nr

INSULATION

Glass fibre quilt; Gypglass 1000

Z.2 2 1	Laid to floors, 80mm thick	m2
Z.2 2 2	Fixed vertically to stud partitions, 100mm thick	m2

Expanded polystyrene granules

Z.2 4 3	Injected in cavity wall, 65mm thick	m2

WINDOWS, DOORS AND GLAZING

Standard window, treated softwood, side hung casement window without glazing bars, 140mm wide softwood cill

Z.3 1 1	600 x 1050, reference 110 V	nr
Z.3 1 1.1	1200 x 750, reference 207 C	nr

Standard flush door; softwood faced both sides, lipped on two vertical edges

Z.3 1 3	610 x 1981 x 35mm	nr
Z.3 1 3.1	626 x 2040 x 40mm	nr

Standard joinery set; wrought softwood; for doors size 686 x 1981mm

Z.3 1 4	35 x 107mm rebated lining	nr
Z.3 1 4.1	44 x 94mm rebated frame with fanlight over	nr

Ironmongery to softwood

Medium pattern butt hinges

Z.3 4 1	50mm	nr
Z.3 4 1.1	75mm	nr
Z.3 4 2	Overhead door closers, 'Briton 2000', liquid check and spring, light door	nr
Z.3 4 3	Rim dead lock, japanned case, 150 x 100mm	nr

Ironmongery to softwood (cont'd)

Z.3 4 3.1	Mortice latch, stamped steel case, 75mm	
	Barrel bolt, extruded brass, round brass shoot, 25mm wide	
Z.3 4 4	100mm	nr
Z.3 4 4.1	150mm	nr

Glazing standard clear float glass, glazed to softwood with putty and sprigs

Z.3 5 1	4mm thick	m2
Z. 3 5 1.1	6mm thick	m2
Z.3 5 2	6mm thick in large panes	m2

Patent glazing, aluminium alloy members 2.4m long at 600mm centres, glazed with 7mm thick Georgian wired cast glass

Z.3 6 3	Vertical surface	m2

SURFACE FINISHES, LININGS AND PARTITIONS

In situ bed, cement and sand (1:3) screed; steel trowelled to floors

Z.4 1 1	32mm thick	m2
Z.4 1 1.1	50mm thick	m2

Red quarry tiles to BS6431 Part 1; laid on cement and sand screed bed 10mm thick; butt jointed and flush pointed with grout

Z.4 2 1	150 x 150 x 12.5mm	m2
Z.4 2 1.1	200 x 200 x 19mm	m2

Flexible sheet covering; linoleum to BS810, fixed with adhesive

Z.4 3 1	3.2mm thick, floors	m2

Flexible sheet covering; linoleum to BS810, fixed with adhesive (cont'd)

Z.4 3 6	3.2mm thick, surfaces width 300mm-1m	m

**Dry partitions; Paramount comprising panels secured
to softwood battens at vertical joints; walls**

| Z.4 4 3 | Square edge panels, joints filled with plaster and
jute scrim cloth, 57mm thick | m2 |
|---|---|---|

**Suspended ceilings; Gyproc M/F system hangers fixed
to softwood joists, 900 x 1800 x 12.7mm tapered edge
wallboard panels, joints filled with joint filler and taped to
receive decoration**

Z.4 5 2	Depth of suspension 150 - 500mm	m2

PIPED BUILDING SERVICES

**Pipework, copper pipes to BS2871 table X capillary fittings,
fixing with pipe clips, plugged and screwed**

Z.5 1 1	15mm nominal bore	m
	Extra over for	
Z.5 1 2	backplate elbow	nr
Z.5 1 2.1	equal tee	nr
Z.5 1 1 2.2	straight tap connector	nr
Z.5 1 1.1	22mm nominal bore	m
	Extra over for	
Z.5 1 2.3	reducing coupling, 22 to 15mm	nr
Z.5 1 2.4	elbow	nr
Z.5 1 2.5	equal cross	nr

Equipment, water tank; glass reinforced plastic

Z.5 2 4	Reference 899.25, 68 litres	nr
Z.5 2 4.1	Reference 899.70, 227 litres	nr

Sanitary fittings; lavatory basin; vitreous china; BS1188; 32mm chromium plated waste; chain, stay and plug; pair 13mm chromium plated easy clean pillar taps to BS 1010; painted cantilever brackets; plugged and screwed.

Z.5 3 0	560 x 405mm, white	nr
Z.5 3 0.1	635 x 455mm, coloured	nr

DUCTED BUILDING SERVICES

Galvanised sheet metal spirally wound circular ductwork to BS 2989; straight

Z.6 1 1	200mm nominal bore, 0.6mm sheet thickness	m
Z.6 1 1.1	450mm nominal bore, 0.8mm sheet thickness	m
	Extra over for	
Z.6 1 3	stopped end	nr
Z.6 1 3.1	90 degrees medium bend	nr
	Extra over for	
Z.6 2 3	taper, 450 to 300mm	nr
Z.6 3 3.1	transformation piece	nr

Galvanised sheet metal high pressure rectangular ductwork to BS2989; 0.8mm sheet thickness; curved

Z.6 2 2	Sum of two sides 400mm	m
	Extra over for	
Z.6 2 3	45 degrees medium bend	nr
Z.6 2 3.1	90 degrees square bend	nr
Z.6 2 2.1	Sum of two sides	m

Extra over for

Z.6 2 3.2	stopped end	nr
Z.6 2 3.3	transformation piece	nr
Z.6 3 3	Roof extract propellor fan with glass fibre reinforced weather cap and base, automatic shutters, curb mounted on flat roof	nr

CABLED BUILDING SERVICES

Z.7 1 3	Single core cable, 50mm2, fixed to brickwork	m
Z.7 2 1	Black enamelled heavy gauge plain conduit, 32mm diameter	m
Z.7 8 4	Equipment and fittings, 13 amp white moulded plastic socket outlet, box and cover plate, fixed to plastered wall	nr

PART THREE

APPENDICES

APPENDIX A

PUMPING STATION 1

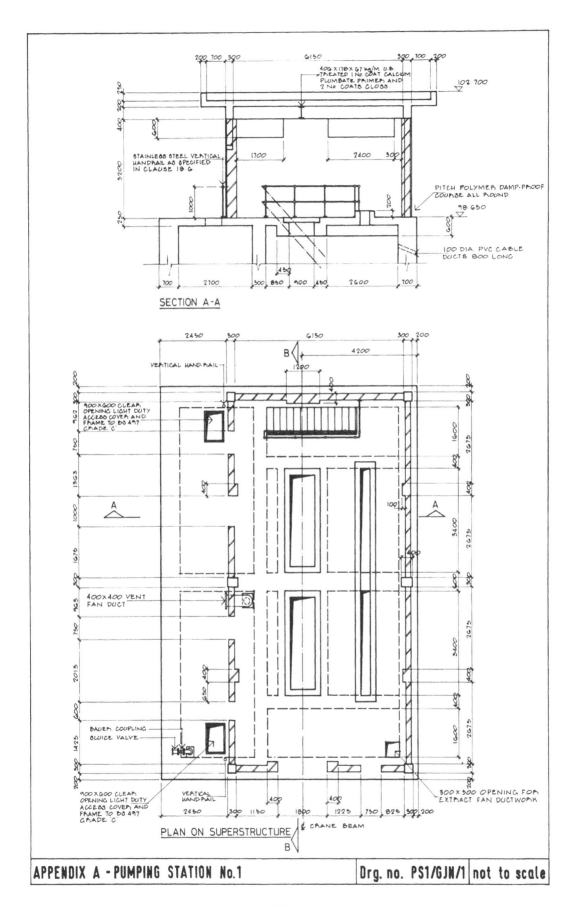

SECTION A-A

PLAN ON SUPERSTRUCTURE

| APPENDIX A - PUMPING STATION No.1 | Drg. no. PS1/GJN/1 | not to scale |

335

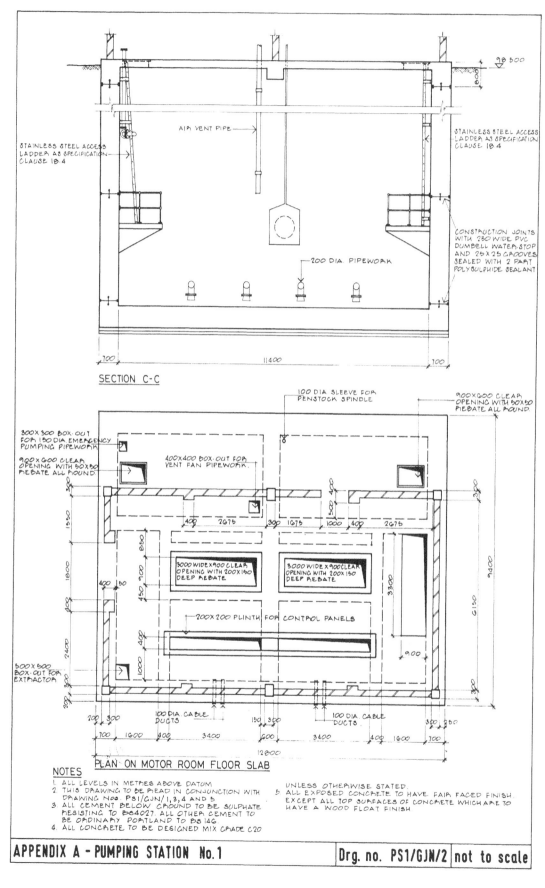

SECTION C-C

PLAN ON MOTOR ROOM FLOOR SLAB

NOTES
1. ALL LEVELS IN METRES ABOVE DATUM UNLESS OTHERWISE STATED.
2. THIS DRAWING TO BE READ IN CONJUNCTION WITH DRAWING Nos. PS1/GJN/1,3,4 AND 5
3. ALL CEMENT BELOW GROUND TO BE SULPHATE RESISTING TO BS4027. ALL OTHER CEMENT TO BE ORDINARY PORTLAND TO BS146.
4. ALL CONCRETE TO BE DESIGNED MIX GRADE C20
5. ALL EXPOSED CONCRETE TO HAVE FAIR FACED FINISH EXCEPT ALL TOP SURFACES OF CONCRETE WHICH ARE TO HAVE A WOOD FLOAT FINISH.

APPENDIX A - PUMPING STATION No. 1   Drg. no. PS1/GJN/2   not to scale

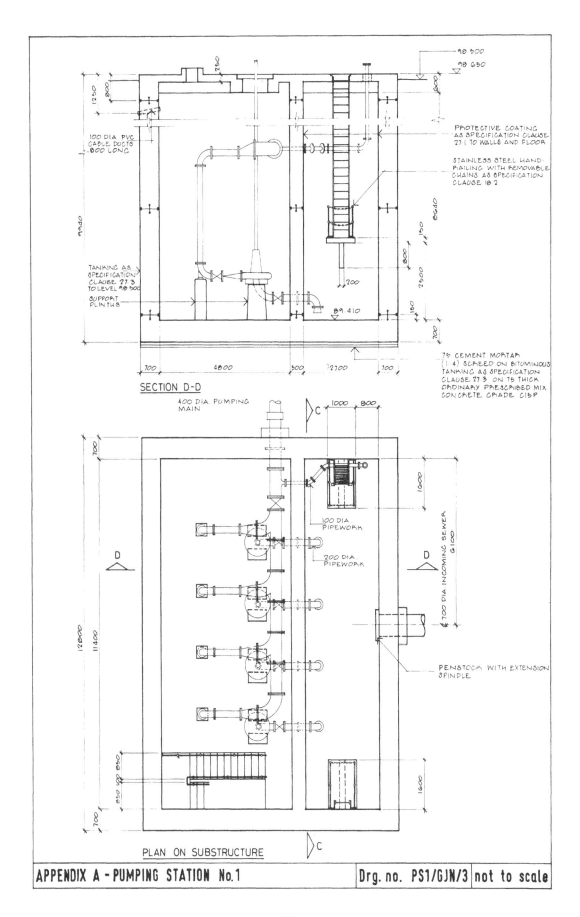

SECTION D-D

400 DIA. PUMPING MAIN

PLAN ON SUBSTRUCTURE

| APPENDIX A - PUMPING STATION No.1 | Drg. no. PS1/GJN/3 | not to scale |

337

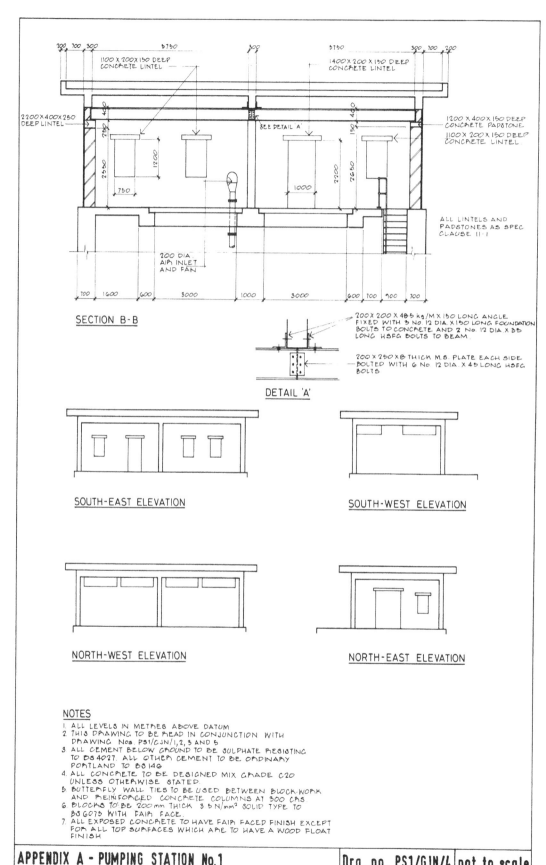

SECTION B-B

DETAIL 'A'

SOUTH-EAST ELEVATION

SOUTH-WEST ELEVATION

NORTH-WEST ELEVATION

NORTH-EAST ELEVATION

NOTES
1. ALL LEVELS IN METRES ABOVE DATUM
2. THIS DRAWING TO BE READ IN CONJUNCTION WITH
   DRAWING Nos. PS1/GJN/1,2,3 AND 5
3. ALL CEMENT BELOW GROUND TO BE SULPHATE RESISTING
   TO BS 4027. ALL OTHER CEMENT TO BE ORDINARY
   PORTLAND TO BS 146
4. ALL CONCRETE TO BE DESIGNED MIX GRADE C20
   UNLESS OTHERWISE STATED.
5. BUTTERFLY WALL TIES TO BE USED BETWEEN BLOCK-WORK
   AND REINFORCED CONCRETE COLUMNS AT 300 CRS.
6. BLOCKS TO BE 200 mm THICK 3.5 N/mm² SOLID TYPE TO
   BS 6073 WITH FAIR FACE.
7. ALL EXPOSED CONCRETE TO HAVE FAIR FACED FINISH EXCEPT
   FOR ALL TOP SURFACES WHICH ARE TO HAVE A WOOD FLOAT
   FINISH

| APPENDIX A – PUMPING STATION No.1 | Drg. no. PS1/GJN/4 | not to scale |

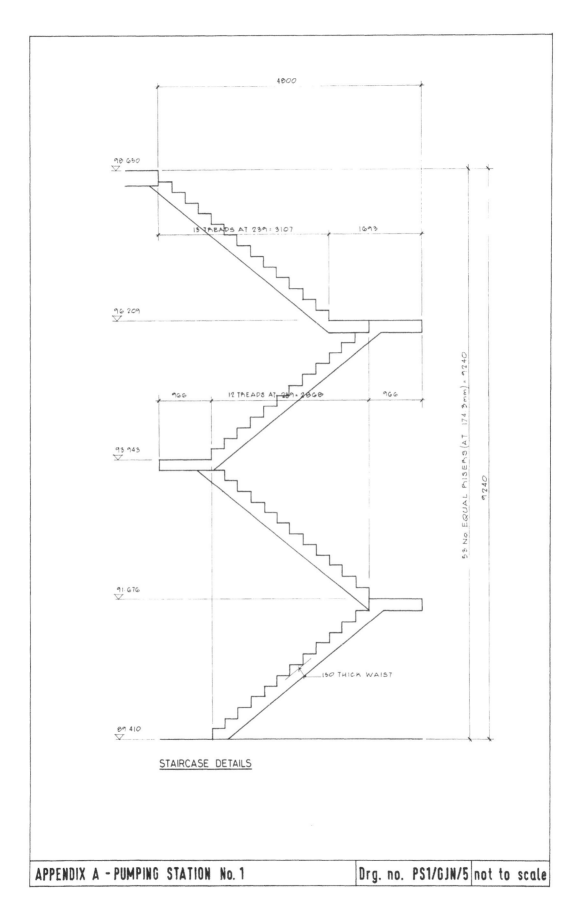

STAIRCASE DETAILS

| APPENDIX A - PUMPING STATION No. 1 | Drg. no. PS1/GJN/5 | not to scale |

Pumping Station 1

Drawing Numbers PS1/GJN/1

PS1/GJN/2

PS1/GJN/3

PS1/GJN/4

PS1/GJN/5

The following example shows a typical reinforced concrete pumping station with a concrete and blockwork superstructure.

It is assumed that the existing site is level with 200 mm of topsoil, with a rockhead at level 94.600. The surrounding finished ground level is also taken as being the existing level.

This example demonstrates measurement in accordance with Classes E, F, G, H, N, U, V and W only. The pipework has not been measured, nor has the reinforcement to concrete. A separate example of reinforcement measurement is given elsewhere.

The structure has been measured in two elements, substructure and superstructure.

Pumping Station 1

					SUBSTRUCTURE	Explanatory Notes

SUBSTRUCTURE

ETHWKS

(depth for clasofctn)

					98 500
			89 410		
Slab	700				
Screed	75				
blindg	75	850	88 560		
			9940		

Explanatory Notes

This is the depth classification for all the materials in the excavation. This is because there is no express requirement to excavate the void in stages (M5), and the Commencing and Excavated Surface for all materials would be the same i.e. the top and bottom of the void.

Excavn for foundns
Pumping Station 1

9·40		Topsoil; max
12·80		depth 5-10m
0·20		(E 316)
	24·06	
		$
		Disposal
		(E 531)

Although drawing PSI/GJN/2 shows the overall dimensions, the prudent taker off would check these by adding the individual dimensions. The volume of topsoil is based on the net plan area, as is all excavation. It is not necessary to state the Commencing Surface as it is also the Original Surface. The Excavated Surface is not stated because it is also the Final Surface (Paragraphs 1.12 and 1.13 and Rule A4)

None of the excavated material is required for re-use and there are no particular requirements for its disposal. It is therefore deemed to be taken off site (D4).

Pumping Station 1

<table>
<tr><td colspan="3"><u>Excaon for foundns</u>
(cont)</td><td><u>Explanatory Notes</u></td></tr>
</table>

		<u>Dep.</u>
		98500
topsoil		200
		98300
rock level		94600
		3700

Matl other than topsoil, rock or artfl hard matl, max depth 5-10m.

(E 326)

The depth of material is calculated by deducting the rock head level from the ground level after topsoil has been stripped.

9·40
12·80
3·70

445·18

&

Disposal

(E532)

		<u>Dep</u>	
rock level		94600	
	89410		
base	700		
blinding	75		
screed	75	850	88560
			6040

Rock; max depth 5 –10m

(E 336)

The depth of rock is calculated from the top of the rock head to the underside of the blinding.

9·40
12·80
6·04

726·73

&

Disposal

(E 533)

Pumping Station 1

			Excavn. Anc.	Explanatory Notes

Excavn. Anc.

Explanatory Notes

Prepn of excavated surfs; rock (E523)

Preparation of surfaces is measured only to the underside of the blinding as this is the only surface to receive Permanent Works. The sides of the excavation will have formwork measured and no preparation is measured (M11). As no angle of inclination is stated, it is assumed that the preparation is less than 10° to the horizontal (A7)

9·40
12·80
120·32

IN-SITU CONCRETE

Provn. of conc.

Standard mix ST3; sulphate resistg; cement to B.S. 4027; 20 agg. to B.S. 882 (F137)

The quantities for provision of concrete are calculated last, and are abstracted from the squared dimensions for the placing of concrete items.

9·63
9·63m³ (blinding)

Designed mix grade C20; sulphate resistg cement to B.S. 4027; 20 agg. to B.S.882 (F247) (structure)

Because of the rounding up and down of dimensions, it is possible to have provision of concrete items which are slightly different to the total billed quantities for the placing items.

84·22
0·74
Ddt 26·69
3·62
52·67
Ddt 263·89
0·27
0·65
1·91
1·85
428·73 m³

343

			Placg of mass conc.	Explanatory Notes

Placg of mass
conc.

Blinding; thness
n. e. 150mm
(F511)

9.40		
12.80		
0.08		
	9.63	

It is not necessary to state that blinding is placed against the excavated surface (A2). The top surface of the blinding would habe a general levelling and no further items would be measured under G81*

Placg of r/f'd
conc

Bases, footgs, pile caps and grnd slabs; thness ex. 500mm
(F624)

9.40		
12.80		
0.70		
	84.22	

Note that up to this point the annotation or 'signposting' of dimensions has been minimal. This is because it has been quite obvious where they apply i.e there is only one blinding layer and only one base.

Susp. slabs;
thness n. e. 150mm
(F631)
(platform at bottom of ladders)

2/	1.00		
	1.60		
	0.15	0.48	
2/½/	1.60		
	0.20		
	0.80	0.26	
		0.74	

The supporting concrete to the platform is an integral part of the slab and therefore classed and measured as part of it.

				Placg of r/f'd conc (cont)	Explanatory Notes

Susp. slabs;
thness 150–300mm
(F632)

The suspended slabs are measured from the inside face of the walls. The classification of the thickness ignores the presence of the attached beams and upstands (D7)

		4·80		(pump (roof
		11·40		chamber),
		0·25	13·68	
		2·70		(penstock
		11·40	7·70	chamber)
		0·25		(beams
		2·70		(penstock
		0·60	0·57	chamber)
		0·35		
2/	4·80		1·34	
	0·40			
	0·35			(pumpchamber
	4·80			across width)
	0·60		1·01	
	0·35			
2/2	0·90		0·25	
	0·20			
	0·35			
2/2	3·40		2·14	(pump chamber
	0·45			spanning between
	0·35			previous)
			26·69	

Beams integral with suspended slabs are classed as suspended slabs (M4). In this case the upstand to the control panels is measured separately. Beams are measured to the inside face of the walls from which they span.

Placg of r/f'd conc (cont'd)

Explanatory Notes

Ddt
Susp. slabs; thness 150-300mm abd   (F632)

2/0.90	
3.00	
0.10	0.54

(rebated opening)

$$\begin{array}{r} dep \\ 600 \\ \underline{150} \\ 350 \quad 500 \\ \underline{100} \end{array}$$

2/1.30	
3.40	
0.15	1.33

0.40	
7.40	
0.25	0.74

(control panel opening)

0.90	
3.30	
0.25	0.74

(staircase opening)

2/0.90	
0.60	
0.25	0.27

(access covers above penstock chamber)

	3.62

$$\begin{array}{r} \text{height} \\ 98650 \\ \underline{89410} \\ \underline{9240} \end{array}$$

Add
Walls; thness 300-500mm   (F643)

11.40	
0.50	
9.24	

(int wall)

	52.67

Walls; thness ex 500mm   (F644)

2/9.00	
0.70	
9.24	116.42

(exte walls)

2/11.40	
0.70	
9.24	147.47

	263.89

These openings must be deducted from the concrete volume as their area exceeds 0.5 m². All other openings in the slab are measured as voids and are therefore not deducted (see D3 of Class G)

The walls are measured from the top of the base slab to the top of the suspended slab. No deductions are made for joint components or cast in components unless they exceed 0.1 m³ (M1).

Pumping Station 1

		Placg. of r/f'd conc (cont)	**Explanatory Notes**

		Ddt	The incoming sewer is the
		Walls; thness	only cast in component
		Ex. 500mm abd	large enough to warrant
Π/	0.35	(F644)	a deduction from the volume
	0.35		
	0.70		
		0.27	

		Add	Although these have been
		Upstands; size	classified under other concrete
		200 × 200 mm	forms, they could have been
2/	7.40	(F680)	measured under the
	0.20	0.59	classification for walls.
	0.20	(control panel)	
2/	0.80		
	0.20		
	0.20	0.06	
		0.65	

		Stairs and landings (F680.1)	Plan of top landing
	1.69		1690
	0.85	0.26	850 1800
	0.18	(top landing)	966
	0.97		
	0.95	0.17	
	0.18		
2/	0.97	(intermediate landings)	The top step and associated
	1.80	0.63	waist of the lower three flights
	0.18		will have been measured
3/12/1/2/	0.85	(bottom flights steps)	with the landings. This is
	0.24	0.62	acceptable because both items
	0.17		are combined under the
13/1/2/	0.85	(top flight steps)	same description. (See next
	0.24	0.23	item for calculation of
	0.17		step and waist volumes).
		1.91	

			Placg + 1f'd conc (cont)	**Explanatory Notes**

Starts & landings (cont)

Len

$$x^2 = 239^2 + 174.34^2$$
$$x = 295.83$$

	$\times 13$	
	3846	
	295.83	
	$\times 12$	
	3550	

(waists)

The calculation of the length of the waist is done by calculating the length for each individual step and multiplying it by the number of steps in the flight. The calculation of the volume of the steps is done by calculating the cross-sectional area of the step and multiplying by its length.

	3.85		(top flight)
	0.85		
	0.15	0.49	
3/	3.55		(bottom
	0.85	1.36	flights)
	0.15		
		1.85	

CONCRETE ANCILLARIES

Fmwk. Rough Fin

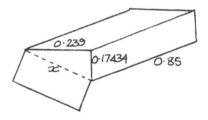

dep.

		98650
Top joint		800
		97850
	89410	
bottom joint	150	89560
		8290

Plane vert. Ex.
1.22m

(G145)

Formwork to the top and bottom of the wall is classed as width 0.4-1.22 because of the position of the joints and is therefore excluded from this measurement.

2/	9.40		(extl wall
	8.29	155.85	outside face)
2/	12.80		
	8.29	212.22	
		368.07	

			Fmwk. Rough. Fin (cont')	Explanatory Notes

Fmwk. Rough. Fin (cont')   **Explanatory Notes**

		dep
top joint	800
fair face	150
	650

2/	9.40		Plane vert' 0.4–1.22m
	0.65	12.22	(G 144)
2/	12.80		(top joint)
	0.65	16.64	(extl wall outside face)
2/	9.40		
	0.85	15.98	(bottom joint)
2/	12.80		
	0.85	21.76	
		66.60	

The formwork to the areas above and below the top and bottom joints are classed as 0.4–1.22m. The upper 150mm of the top joint formwork is exposed to view and must be fair faced. It is deducted from the total area and measured separately, classed as 0.1–0.2m wide.

Fmwk. Fair Fin

2/	9.40	18.80	Plane vert' 0.1–0.2m
2/	12.80	25.60	(G 242)
2/	4.80	9.60	(top 150mm of outside face of extl wall)
2/	2.70	5.40	(internally at kicker level to all walls)
2/2/	11.40	45.60	
		105.00	

The formwork to the inside face of the walls is classed as 0.1–0.2m at the base because of the presence of the kicker joint.

2/	4.80		Plane vert' ex. 1.22m
	8.29	79.58	(G 245)
2/	2.70		(internally between
	8.29	44.77	bottom and top
2/2/	11.40		joint to all walls)
	8.29	378.02	
		502.37	

			Fmwk . Fair Fin (cont')	Explanatory Notes
			dep	
			800	
			256	
			550	
			Plane vert. 0.4-1.22m	It is recommended that no
			(G 244)	deduction is made from the
2/	4.80			area of formwork
	0.55	5.28		between the top joint and the
2/	2.70		(internally between	underside of the slab for
	0.55	2.97	top joint and	intersections with beams because
2/2/	11.40		U/s of slab)	their cross-sectional area does
	0.55	25.08		not exceed $0.5 m^2$
		33.33		
			Len	
			1600	
			400	
			3400	
			5400	
			Plane horiz. ex 1.22m	Narrow width formwork caused
2/	2.70		(G 215)	by openings is only classed
	5.40	29.16	(penstock chamber	as such where formwork to the
			roof)	openings is measured under the
Ddt			(for access	general formwork classification
2/	0.90		covers)	and not as voids as defined in
	0.60	1.08		D.3. The only areas to the
				soffit of the pump chamber
	1.60		(pump chamber	roof which are not below 1.22m
	4.80	7.68	roof)	wide are the area at the opposite
				end of the stairs and the area
	1.60			between the staircase opening
	1.50	2.40		and the long external wall
		38.16		(see diag.)

350

			FMwk. Fair. Fin (cont')	Explanatory Notes

FMwk. Fair. Fin (cont')

Dott
Plane horiz.
0.4 - 1.22m

	0.70			
	3.30	2.31	(next stairs) (G214)	
2/	3.40		(either side of plinth panel opening)	(soffit of pump chamber)
	1.00	6.80		
2/	3.40			
	1.20	8.16		
2/	3.40		(between intl wall & rebated opening)	
	0.45	3.06		
		20.33		

Add
Plane. horiz.
0.4 - 1.22m

Because of the smaller nature of the work, formwork to beams is measured in accordance with G1-4 1-4* and not G1-4 8*

	4.80		(central beam across (beam width of pump chamber) (G214) (beam soffits)	
	0.60	2.88		
2/	0.90			
	0.20	0.36		
2/	0.90		(outside beams across width of pump chamber)	
	0.66	1.08		
2/	3.40		(long beams to rebated opening)	
	0.45	3.06		
		7.38		

The beams which span the width of the pump chamber widen out where they cross the rebated beams. In the case of the central beam this simply means an additional area of formwork in the 0.4-1.22 classification. The outside beams however are classified as 0.2-0.4m and this widening means that the beam is classified as 0.4-1.22m for the length of the additional width

Plane horiz.
0.2 - 0.4m
(G213)

| 2/ | 4.80 | | (outside beams across width of pump chamber) | |
| | 0.40 | 3.84 | | |

Dott

2/	0.90			
	0.40	0.72		
		3.12		

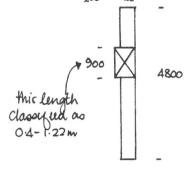

200 / 400

900

4800

this length classified as 0.4 - 1.22m

Fmwk. Fair Fin (cont)

2/	2.70		
	0.35	1.89	(penstock chamber) (beam sides
2/	4.80		
	0.35	3.36	(outside beams across
2/	2.60		width of pump
	0.35	1.82	chamber)
2/	0.40		
	0.35	0.28	
2/2/	3.40		(long beams to
	0.35	4.76	rebated opening
			- outside face only)
		12.11	

Plane vert. 0.2-0.4m (G243)

2/2/	3.00		(opngs in
	0.45	5.40	(rebated roof slab
2/	0.90		openings)
	0.45	0.81	
2/	7.40		(control panel
	0.45	6.66	openings - long side)
		12.87	

Plane vert. 0.4-1.22m (G244)

Plane. vert. 0.2-0.4m (G243)

2/	0.40		(control panel (opngs in
	0.45	0.36	opng - short roof slab
			sides)
2/2/	0.90		(access covers
	0.25	0.90	in penstock
2/2/	0.60		chamber)
	0.25	0.60	
	3.30		
	0.25	0.83	(staircase)
	0.90		
	0.25	0.23	
		2.92	

Explanatory Notes

The sides of the long beams on either side of the rebated openings are in two different width classifications

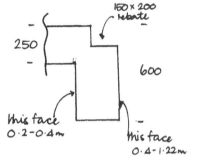

These holes are above the minimum size to enable measurement under the voids classification and have to be measured under the general formwork classification. Their area must therefore be deducted

352

Pumping Station 1

Fmwk. Fair Fin (cont')

Plane vert. 0.1-0.2m (G242)

2/2	3.40	13.60
2/2	1.30	5.20
2/	8.00	16.00
2/	0.80	1.60
		36.40

(rebates (opngp i to rebated roof slab opngp)
(control panel opng- side g plinth)

Plane horiz; 0.2-0.4m (G213)

2/2	0.40	
	1.60	
		2.56

(landing to ladders)

Plane vert. 0.1-0.2m (G242)

2/2	1.60	6.40
2/	1.00	2.00
		8.40

(landing to ladders)

Plane vert; average width 0.4m (G246)

2/2/½	0.80	
	1.60	
		2.56

(landing to ladders)

Plane slopg; 0.1-0.2m (G222)

$$x^2 = 1.6^2 + 0.8^2$$
$$x = 1.78$$

2/	1.78	
		3.56

(landing to ladders.

353

Explanatory Notes

1	1.60	1

0.1 0.2 0.4

0.80

n.e. 0.1	0.1 -0.2	0.2-0.4	0.4- 1.22

Strictly speaking the formwork to the sides of the support to the landing should be classified as above. With such small quantities this would be unhelpful and therefore an additional description in accordance with Paragraph 5.10 is incorporated into the description and the average width is stated.

Pumping Station 1

			Fmwk. Fair Fin (cont')	Explanatory Notes
			Plane vert. 0.1-0.2m (G 242)	Because of the 100mm gap between the stairs, the top riser in the bottom three flights will effectively be 100mm longer
14/	0.85	11.90	(risers-top flight) (Stairs	
3/	0.95	2.85	(risers-bottom flight)	
3/12/	0.85	30.60		Len.
	0.73	} 1.69	(top landing)	1693
	0.96			966
				727
2/	0.96	1.92	(intermediate landing)	
		48.96		
			Plane vert. 0.2-0.4m (G 243)	The classification for the width of formwork to the waist is taken as the maximum width and the area is taken as the gross area. The top flight and flight second from bottom have formwork measured to one side only as the other side is cast against the external wall. The calculation of the area is done by adding the area of the waist to twice the area of the side of the step.
	3.85		(top flight) (sides of stairs	
	0.15	0.58		
13/1/2/	0.24			
	0.17	0.27		
2/2/	3.55			
	0.15	2.13		
2/12/1/2/	0.24			
	0.17	0.49	(bottom flight)	
	3.55			
	0.15	0.53		
12/1/2/	0.24			
	0.17	0.24		
		4.24		
			Plane. slopg 0.4-1.22m (G 224)	
	3.85	3.27	(soffits) (stairs	
	0.85			
3/	3.55	9.05		
	0.85			
		12.32		

area of step x 2 to account for / width

354

			Fmwk. Rough Fin.	Explanatory Notes

Fmwk. Rough Fin.   **Explanatory Notes**

For voids

small void depth
n.l. 0.5
(emergency   (G171)
pumping   (in roof
pipework)

It is assumed that the formwork to the sides of the voids will have a rough finish because they will be eventually filled in around the insert.

| 1 | | | | |
| | 1 | | | |

large void depth
n.e. 0.5
(vent fan)   (G175)
(extractor) (in roof.

1				
	1			
	2			

Joints

Open surf plain
0.5 - 1m
(G612)

(jnts at all
three levels
extl walls)

The average width of the joint is measured from outside face to outside face with no deduction for the width occupied by the rebate (M11) There is no surface treatment to the joints so they are classed as plain.

2/3	9.00			
	0.70	37.80		
2/3	11.40			
	0.70	47.88		
		85.68		

Open surf plain
n.e. 0.5m
(G611)
(int
wall)

3	11.40			
	0.50			
		17.10		

			Joints (cont)		Explanatory Notes

PVC waterstop, dumbell type, 230 wide

As the width of the waterstop is stated it is not necessary to also state the average width in the ranges given in the Third Division. Note that there is no separate measurement for angles as these are deemed to be included (C4)

(G 653)

	3/	36.00	108.00	(extl walls)	guth
	3/	12.10	36.30	(int walls)	11.400
			144.30		8000

```
            11.400
             8000
            19400
             ×2
            38800
4/2/½/700   2800
            36000

            lin
            11.400
             .700
            12.100
Ends 2/½/700
```

Sealed rebate or groove; 25×25mm with two part polysulphide sealant (G670)

Although not specifically mentioned, the measurement of the groove is taken at its exposed face. No formwork is measurable to any work classed as joints (C3)

2/3/	2/3/	36.00	216.00	(extl walls)
	Dett			
0.70		4.20		
	2/3/	11.40	68.40	(intl walls)
			280.20	

Conc. Accessories

Fin top surfs; wood float

(G 811)

No deduction is made for the areas of the support plinths as they do not exceed 0.5m² (M14)

	11.40			(penstock chamber floor)
	2.70	30.78		
	11.40			(pump chamber floor)
	4.80	54.72		
2/	1.00			(ladder landings)
	1.60	3.20		
		88.70		

			Conc. Accessories (Cont)	Explanatory Notes

Fin top surfs; wood float (G811)

13/	0·24		
	0·85	2·65	(treads) (stairs
3/12/	0·24		
	0·85	7·34	
	1·69		(top
	0·85	1·44	landing)
	0·97		
	0·95	0·92	
2/	0·97		(intermediate
	1·80	3·49	landings)
		15·84	

Fin. top surfs; Wood float (G811) (roof slab

12·80		
9·40		
	120·32	

Ddt
Fin. top surfs; Wood float (G811)

	0·90		
	3·30	2·97	(stairs opng)
2/	3·00		
	0·90	5·40	(rebated opngs)
	7·40		(control panel
	0·40	2·96	opng)
2/	0·90		(access
	0·60	1·08	covers)
		12·41	

Although not clear from CESMM no deduction is made from the area for that which is subsequently covered by the brick walls. No deduction is made for openings less than 0.5m² (M14)

357

Pumping Station 1

			Inserts	Explanatory Notes
4/	1		100mm dia. PVC cable ducts, 800 long; proj. two sides (pump chamber (G832) extl wall at high level)	This item would also include for the supply of the duct as it has not been otherwise stated (C7)
		4		
	1		700mm dia. pipe proj. one side (supply incl. elsewhere) (sewer thro extl wall) (G832·1)	This is classified as projecting from one surface because the end of the pipe is flush with the internal face of the wall.
		1		
	1		400mm dia. pipe proj. two sides (supply incl. elsewhere) (pumpg main (G832·2) thro extl wall)	
		1		
4/	1		200mm dia. pipe proj. two sides (supply incl. elsewhere (pumpg main (G832·3) thro extl wall)	
		4		
2/	1		100mm dia. pipe proj. two sides (supply incl. elsewhere) (G832·4) (emergency pumpg pipe thro extl wall) (sleeve for penstock spindle)	
		2		

Pumping Station 1

			Inserts (cont.)	Explanatory Notes
			150mm dia pipe in 300 × 300 × 250mm deep preformed boxout proj. two sides as Spec Clause 12.1 (G 832.5) (emergency pumping pipe thro' roof)	Inserts grouted into preformed openings must be described. The work in grouting into preformed openings can be quite involved so reference is made to the relevant Specification Clause rather than give a detailed description
1				
	1			
			200mm dia pipe in 400 × 400 × 250mm deep preformed boxout proj. two sides as Spec Clause 12.1 (G832.6) (Air inlet thro' roof)	
1				
	1			
			300 × 300mm duct in 500 × 500 × 250mm deep preformed box-out proj two sides as Spec Clause 12.1 (G832.7) (Extractor fan thro' roof)	
1				
	1			

			Miscellaneous Metalwork	Explanatory Notes

Miscellaneous Metalwork | **Explanatory Notes**

Ladders, stainless steel as Spec. Clause 18.4 and drwg no. PS1/GSN/2; incl. all fixing to conc.

(N 130)

$2/6.59$ 13.18

Len
98650
89410
9240

2500
150   2650
      6590

Alternatively the ladders could have been enumerated stating the length. The item includes for all fixing to concrete (C1). Alternatively the pockets and bolts could have been measured in Class G.

Handrails; level; stainless steel as Spec. Clause 18.2; incl. all fixing

(N 140)

$2/2/ 1.60$ 6.40 (ladder landings)
8.53 8.53 (stair landings)
14.93

Len
topof stairs   1200
3300
200
727
1st inter down   200
966
2&3rd inters   966
966
8525

Although not a specific requirement, level handrails and raking handrails are kept separate because it is considered that they have different cost considerations. Alternate flights will have a handrail either one or both sides depending on whether it is next to the external wall or not.

Ditto sloping

(N 140.1)

3.95
$2/2/ 3.66$
3.66
22.25

$x^2 = 3.1^2 + 2.45^2$
$x = 3.95$

$x^2 = 2.87 + 2.28$
$x = 3.66$

x is the length of the sloping handrail which is the hypotenuse of a right angled triangle (Pythagoras' theorem)

360

Pumping Station 1

			Miscellaneous Metalwork (cont)	Explanatory Notes

Miscellaneous
Metalwork (cont)

Explanatory Notes

Handrail; level;
stainless steel
as spec. Clause
18·6, incl. all
fixing
(extl)   (N140·2)

11·40

11·40

Although this handrail and
part of the handrail before
is strictly speaking a
superstructure item, they
have been included here
in order to keep all similar
items of work together.

Access covers &
fr. 900 × 600 mm
light duty to
B.S. 497 grade C
(N190·1)
(penstock chamber
roof)

2/ 1

2

There is no separate
classification for access
covers so this is a rogue
item.

Safety chains
900 mm long;
stainless steel
as spec. Clause
18·2 incl. all
fixing.
(N190)

2/ 1

2

There is no separate
classification for safety
chains so this is a rogue
item.

Pumping Station 1

			Waterproofing	Explanatory Notes

Waterproofing

Bituminous Tanking as Spec Clause 27.3

Because the description contains reference to the relevant specification clause, it is not necessary to go into detail about the number of coats and thickness of material.

Upper surfs.
n.l. 30° to horiz
(W241)
(u/s of structure)

12·80		
9·40		
	120·32	

Surfs. ex. 60°
(W243)

(outside of Extl. walls.)

2/	12·80		
	9·87	252·67	
2/	9·40		
	9·87	185·56	
		438·23	

```
          hk
        98500
        89410
base     9090
screed    700
           75
         9865
```

Damp proofing

Protective coating as Spec. Clause 27.1 n.l 30° to horiz
(W141)

(floor penstock chamber)

2·70		
11·40		
	30·78	

Pumping Station 1

Damp proofing (cont')

Explanatory Notes

Protective coating as Spec. Clause 27.1 ex. 60° to horiz.

No deduction is made where the beams meet the walls or the sewer as the cross-sectional area does not exceed 0.5m²

(W143)
(penstock chamber

2/2.70		
8.99	48.55	
2/11.40		
8.99	204.97	
	253.52	

Protective layers

Sand and cement (1:4) screed 75mm thick in one coat n.e. 30° to horiz.

(W441)
(u/s of structure)

12.80	
9.40	
	120.32

			SUPER STRUCTURE	EXPLANATORY NOTES

SUPER STRUCTURE

Provn. of conc.

Designed mix grade C20; OPC to B.S. 146; 20agg to B.S. 882

(F243)

		1.30		
		0.86		
		4.24		
		24.28		
		2.24		
			32.92m³	

Placg of r/f'd conc.

Columns and piers; X-sect. area 0.03 – 1m²

(FC52)

The columns are measured from the top surface of the pumphouse floor to the underside of the roof slab. The beams are then measured between the columns.

```
  3200
   400
  3600
```

4/	0.30			
	0.30			
	3.60			
		1.30		

Ditto 0.1 – 0.25 m²

(F653)

2/	0.30			
	0.40			
	3.60			
		0.86		

Beams; X-sect area 0.1 – 0.25m²

(F663)

Again note that there is no annotation or 'signposting' of the dimensions as it is quite apparent on the drawing where the descriptions are derived from

2/	6.15			
	0.30	1.48		
	0.40			
2/2	5.75			
	0.30	2.76		
	0.40			
		4.24		

Pumping Station 1

				Placg. of r/f'd conc. (cont)	Explanatory Notes

Placg. of r/f'd conc. (cont) | Explanatory Notes

				Len
			2/200	400
			2/700	1400
			2/300	600
			2/5750	11500
				300
				14200
				Wid
			2/200	400
			2/700	1400
			2/300	600
				6150
				8550

Susp. slabs thness — The suspended slab is
150 - 300 mm   measured over the beams
            (F632)   and under the perimeter kerb

(roof)

14.20			
8.55			
0.20			
	24.28		

Upstands; size — Again the perimeter kerb to
200 × 250 mm (F680)   the roof has been classified
        gurth   as an upstand rather than a
(roof)   wall as it is felt that this
                                more accurately reflects the
                                work involved.

44.70	
0.20	
0.25	
	2.24

	4200
	8550
	22750
	×2
	45500
4/2/½/200	800
	44700

365

Pumping Station 1

			CONCRETE ANCILLARIES	Explanatory Notes

<u>CONCRETE ANCILLARIES</u>

<u>Explanatory Notes</u>

<u>Fmwk. Fair Fin.</u>

To components of const'ant X-sect; columns size 300 × 300mm; extl walls of superst.

(G282)

4/	3·60		
		14·40	

Ditto 300×400mm
(G282·1)

2/	3·60		
		7·20	

To components of const'ant X sect; beams size 300 × 400mm; roof slab
(G281)

2/	6·15	12·30	
2/2	5·75	23·00	
		35·30	

Plane horiz; 0·4 − 1·22m
(G214)
(perimeter o'hang of roof)

2/	14·20		
	0·90	25·56	
2/	6·75		
	0·90	12·15	
		37·71	

The measurement of the columns and beams to the superstructure is done as components of const'ant cross-section in order to demonstrate the classification. In practice on a contract of this size it is not of much benefit to the contractor in that the number of uses of the formwork will be very small.

To be strictly correct, the area of the columns in cont'act with the brickwork should be deducted from the area of the fair finish and added back as rough because it is not exposed to view. However, it is felt that this is totally impracticable and the fair finish is carried through.

Forwk. Fair Fin.   Explanatory Notes

(Cont)

Plane. horiz;
ex. 1.22m

2/ 5.75
6.15   (soff. of   (G215)
       roof intl.)

| 70.73 |

Plane vert.
0.1 - 0.2m

2/ 13.80   27.60   (intl. face of   (G242)
2/ 8.15   16.30   upstand)

| 43.90 |

Plane vert.
0.2 - 0.4m

2/ 14.20   (G243)
0.45   12.78

      (exle. face

2/ 8.55   of upstand)
0.45   7.70

| 20.48 |

Concrete Accessories

Fin. top surp;
wood float

14.20   (G.811)
8.55   (top of roof
      slab)

| 121.41 |

			Concrete Accessories	Explanatory Notes
			(cont)	
			Butterfly wall ties; 300mm centres; proj from one surf (G831) (betw. column & block wall)	Alternatively the measurement of the wall ties could have been included in Class U, but it would still have been necessary to measure an item in Class G for the building in.
2/6/	3·20			
		38·40		Although the ties have been measured under the linear inserts classification stating the spacing; they could have been enumerated.
			fndn. bolts 12 dia × 150mm long; proj from one surf (G832) (fixing crane rail to central conc. beam)	
2/3/	1			
		6		
			PRECAST CONCRETE	
			Lintels	There is no separate classification for lintels in Class C and this is therefore a rogue item. Because the actual size is stated it is considered that it is not necessary to also state the weight. The specification clause would contain details of reinforcement etc., or alternatively this may be contained on a drawing
			1100 × 200 × 150mm deep as Spec. Clause 11.1 (window) (H900)	
3/	1			
		3		
			1400 × 200 × 150mm deep as Spec. Clause 11.1 (personnel door) (H900.1)	
	1			
		1		

Lintels (cont')

Explanatory Notes

2200 × 400 × 150 mm
deep as Spec. Clause
11.1

(door under
crane beam)   (H900.2)

| | 1 | | |
| | | 1 | |

Padstones

1200 × 400 × 150 mm
deep as Spec.
Clause 11.1

(H9003)

| | 1 | | |
| | | 1 | |

STRUCTURAL METALWORK

Fabrication of
crane beams;
straight on plan,
406 × 178 × 67 kg/m
U.B.

The weight of the plate and angle must also be included in the mass of steel measured, but not the bolts (M3). The section size and weight is also given to assist the contractor although this is not strictly necessary.

11.80			(M421)
0.067	0.79	(UB.)	
2/0.15		(angles)	
0.049	0.02		
2/0.25		(plate)	
0.25	0.01		
0.06			
	0.82 kg	&	
	÷1000		
	0.01 t	Permanent	
		Erection	
		(M720)	

Len
5750
300
5750
11800

			Site bolts	Explanatory Notes

HSFG general grade

Although not a specific requirement the actual diameter and length of bolts is also stated. The bolts which fix the angles to the concrete are measured in Class G.

6/	1		12 dia × 45mm long (plate) (M741)
		6	

2/2/	1		12 dia × 35mm long
		4	(flange/angle) (M741·1)

BRICKWORK AND BLOCKWORK

<u>Dense concrete blocks to BS 6073 part 1 3·5 N/mm² solid; 450 × 225 × 200mm; jointing in cement mortar (1:3); stretcher bond.</u>

			Vert. straight walls; 200mm th.	The measurement of the external walls is taken over all openings and holes for which the deductions are made separately.
2/2	5·75		(U521)	
	3·20	73·60	(extl. walls)	
2/	6·15			
	3·20	39·36		
		112·96		

			Vert. straight walls; 400mm th.
	1·20		(U531)
	2·65		(pilaster to S.W. wall.)
		3·18	

			Dense conc. blocks (cont)	Explanatory Notes

Ddt

Vert, straight walls 200mm th.

The length of the attached pilaster to the N.W wall exceeds 4 times its projecting width so that the area of wall is measured separately as 300 thick. The area of the padstone is less than $0.25 m^2$ in area and is not deducted (M2)

1.70		1.02	(US2)
0.60			(SW Elev. (for opng)
2.40		1.44	
0.60			
1.20		3.18	(for attached pilaster)
2.65			(NE elev.
0.75		0.90	(window)
1.20			
1.80		4.59	(door)
2.55			

Although the smaller lintels are less than $0.25 m^2$ they are deducted from the overall area because they form part of the window openings and their combined area exceeds the minimum.

2.20		0.88	(lintels to same
0.40			
1.10		0.22	
0.20			
1.00		2.20	(door) (SE Elev
2.20			
2/ 0.75		1.80	
1.20			(windows)
0.60		0.45	
0.75			
1.40		0.28	(lintels to same)
0.20			
2/ 1.10		0.44	
0.20			(NW elev.
4/ 2.40		5.76	(opgp)
0.60			
		23.16	

			Surface features	Explanatory Notes

Add
Pilasters size
400 × 100 mm

2/2/	3.20	12.80	(to N.W. & S.W walls) (U576)	Although the cross-sectional area of the pilasters is less than that specified in A7, the size is still stated because it is considered that this gives the contractor the information necessary to identify any cutting or special blocks in accordance with A.6.
2/	2.55	5.10	(either side of main door)	
		17.90		

Fair facing;
flush pointg as
work proceeds

(U 578)

2/2/2	5.75			Fair facing is measured separately to all items of brickwork and blockwork. In this case it is measured by doubling the area of the wall, deducting the openings and adding the area of reveals etc.
	3.20	147.20	(both sides of gross brickwk measurement)	
2/2	6.15			
	3.20	78.72		
2/	0.20			
	2.20	0.88	(personnel door) (reveals	
2/	0.20			
	2.55	1.02	(main door)	
2/3/	0.20			
	1.20	1.44	(windows)	
2/	0.20			
	0.75	0.30		
2/5	0.20			
	0.60	1.20	(opengs)	
	0.20			
	0.60	0.12		
2/2/	0.10			
	3.20	2.56	(pilasters)	
2/	0.10			
	2.55	1.02		
2/	0.20		(sides of wall thickening under padstone)	
	2.65	1.06		
		235.52		

Pumping Station 1

Surface features
(cont')

Explanatory Notes

<u>Ddt</u>
Ditto

2/ 1·70			(SW Elev
0·60	2·04	(opngo)	
2/ 2·40			
0·60	2·88		
2/ 1·20		(attached pilasters)	
2·65	6·36		
			(NE Elev
2/ 0·75		(window)	
1·20	1·80		
2/ 1·80		(door)	
2·55	9·18		
2/ 2·20		(lintels	
0·40	1·76	to same)	
2/ 1·10			
0·20	0·44		(SE Elev
2/ 1·00		door	
2·20	4·40		
2/2 0·75			
1·20	3·60	(windows)	
2/ 0·60			
0·75	0·90		
			(NW Elev
2/ 1·40		(lintels	
1·20	3·36	to same)	
2/2 1·10			
0·20	0·88		
2/4 2·40		(opngs)	
0·60	11·52		
	49·12		

373

			Ancillaries	Explanatory Notes
			Add	
			Damp proof course; pitch polymer; 200 wide	It is necessary to state the nature and width of damp proof courses (A8). Consequently
2/2/	5.75	23.00	(U582)	the length of damp proof
2/	6.15	12.30		course over the wall
Ddt		35.30		thickening and pilaster has to be classified
1.20	1.20		(for door opngs)	separately. The damp proof
1.80	1.80			course does not carry
6/ 0.40	2.40		(for pilasters)	through the door openings
1.20	1.20	6.60	(for wall thickening to SW elevation)	and is deducted.
		28.70		
			Ditto 300 wide	
6/	0.40		(pilasters) (U582.1)	
		2.40		
			Ditto 400 wide (U582.2)	
	1.20		(wall thickening to SW elevation)	
		1.20		

			Ancillaries (cont)	Explanatory Notes

Building in
pipes and ducts
X-Sect area
0.16m²
(U588)
(vent fan duct.

| | | | Painting | |

1 coat calcium
plumbate primer
2 ts gloss paint
to metal sections
(V1$470)

(crane
rail)

girth
4/178 712
2/406 812
1524

11.80				
1.52				
	17.94			

The girth calculated for the
crane rail will be slightly
overmeasured as it does not
deduct the area where the
web meets the flange or allow
for the curves etc. However,
it is considered to be
sufficiently accurate for
measurement purposes

APPENDIX B

PUMPING STATION 2

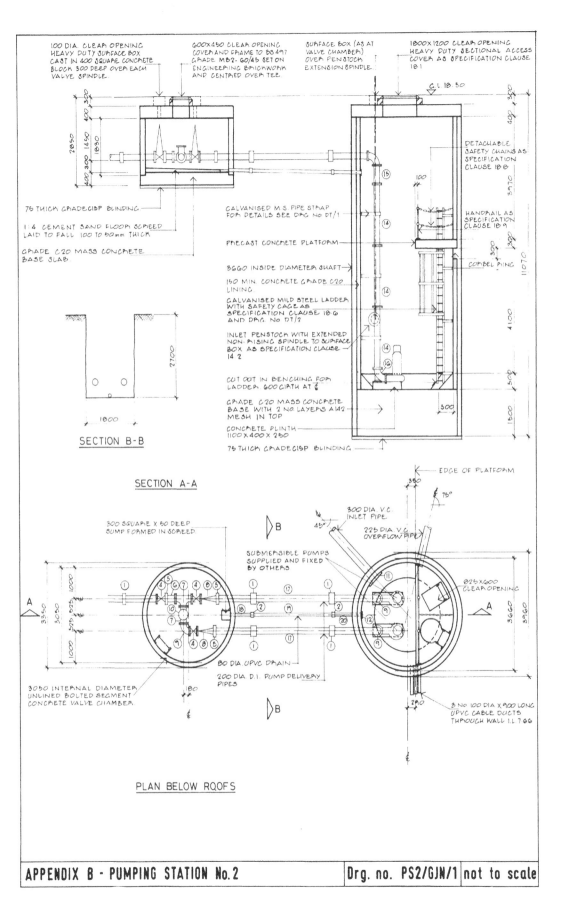

100 DIA. CLEAR OPENING HEAVY DUTY SURFACE BOX CAST IN 400 SQUARE CONCRETE BLOCK 300 DEEP OVER EACH VALVE SPINDLE.

600X450 CLEAR OPENING COVER AND FRAME TO BS 497 GRADE. MB2- 60/45 SET ON ENGINEERING BRICKWORK AND CENTRED OVER TEE.

SURFACE BOX (AS AT VALVE CHAMBER) OVER PENSTOCK EXTENSION SPINDLE.

1800X1200 CLEAR OPENING HEAVY DUTY SECTIONAL ACCESS COVER AS SPECIFICATION CLAUSE 18·1

G.L. 18·50

DETACHABLE SAFETY CHAINS AS SPECIFICATION CLAUSE 18·8

HANDRAIL AS SPECIFICATION CLAUSE 18·9

CORBEL RING

75 THICK GRADE C15P BLINDING

1:4 CEMENT SAND FLOOR SCREED LAID TO FALL 100 TO 50mm THICK

GRADE C20 MASS CONCRETE BASE SLAB.

GALVANISED M.S. PIPE STRAP FOR DETAILS SEE DRG No DT/1

PRECAST CONCRETE PLATFORM

3660 INSIDE DIAMETER SHAFT

150 MIN CONCRETE GRADE C20 LINING.

GALVANISED MILD STEEL LADDER WITH SAFETY CAGE AS SPECIFICATION CLAUSE 18·6 AND DRG No DT/2

INLET PENSTOCK WITH EXTENDED NON-RISING SPINDLE TO SURFACE BOX AS SPECIFICATION CLAUSE 14·2

CUT OUT IN BENCHING FOR LADDER 600 GIRTH AT ₵

GRADE C20 MASS CONCRETE BASE WITH 2 No LAYERS A142 MESH IN TOP

CONCRETE PLINTH 1100 X 400 X 250

75 THICK GRADE C15P BLINDING

SECTION B-B

SECTION A-A

300 SQUARE X 50 DEEP SUMP FORMED IN SCREED.

EDGE OF PLATFORM

300 DIA. V.C. INLET PIPE.

225 DIA. V.C. OVERFLOW PIPE.

825 X 600 CLEAR OPENING

SUBMERSIBLE PUMPS SUPPLIED AND FIXED BY OTHERS

A

80 DIA. UPVC DRAIN

200 DIA. D.I. PUMP DELIVERY PIPES

3050 INTERNAL DIAMETER UNLINED BOLTED SEGMENT CONCRETE VALVE CHAMBER.

3 No 100 DIA X 900 LONG UPVC CABLE DUCTS THROUGH WALL I.L 7·66

PLAN BELOW ROOFS

APPENDIX B - PUMPING STATION No. 2 | Drg. no. PS2/GJN/1 | not to scale

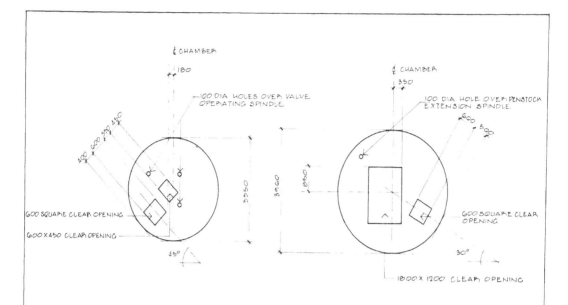

PLAN ON ROOF SLABS

FITTINGS SCHEDULE

ITEM No	No OFF	NOM DIA	DESCRIPTION
1	5	200	DETACHABLE FLEXIBLE COUPLING
2	2	80	DETACHABLE FLEXIBLE COUPLING
3	1	200	FLANGE SPIGOT 1200 LONG
4	3	200	DOUBLE FLANGE RISING SPINDLE SLUICE VALVE WITH GEARBOX AND EXTENSION SPINDLE TO GIVE NON-RISING VALVE OPERATING CAP APPROX. 100 BELOW SOFFIT OF ROOF SLAB
5	3	200	FLANGE ADAPTOR
6	1	200	FLANGE SPIGOT 250 LONG
7	2	200	ALL FLANGE EQUAL TEE
8	2	200	DOUBLE FLANGE CHECK VALVE
9	3	200	DOUBLE FLANGE 90° BEND
10	1	200	BLANK FLANGE
11	1	300	PENSTOCK AND EXTENSION SPINDLE SUITABLE FOR AN OFF SEATING PRESSURE OF 8.5M OF WATER FOR BOLTING TO THE STRUCTURE
12	1	80	FLAP VALVE FOR BOLTING TO THE STRUCTURE
13	2	200	FLANGE SPIGOT 1300 LONG
14	6	200	DOUBLE FLANGE STRAIGHT 2000 LONG
15	2	200	DOUBLE FLANGE STRAIGHT 605 LONG
16	2	200	DOUBLE FLANGE DUCKFOOT BEND
17	2	200	DOUBLE SPIGOT STRAIGHT 2900 LONG
18	1	80	DOUBLE SPIGOT UPVC STRAIGHT 750 LONG
19	1	80	DOUBLE SPIGOT UPVC STRAIGHT 3000 LONG
20	1	80	DOUBLE SPIGOT UPVC STRAIGHT 1000 LONG

COVERS SCHEDULE

No OFF	CLEAR OPENING	DESCRIPTION
1	1800 X 1200	HEAVY DUTY ACCESS COVER
2	600 X 600	HEAVY DUTY HINGED LIFT OUT ACCESS COVER
1	600 X 450	HEAVY DUTY COVER
4	100 DIA	HEAVY DUTY SURFACE BOX

NOTES

1. ALL DIMENSIONS ARE IN MILLIMETRES AND LEVELS IN METRES
2. ROCK HEAD LEVEL AT APPROXIMATELY 10.35M
3. SHAFT TO BE PRECAST CONCRETE BOLTED SEGMENTAL LININGS
4. ALL EXPOSED CONCRETE TO HAVE A TYPE 'B' FINISH
   ALL UNEXPOSED CONCRETE TO HAVE A TYPE 'A' FINISH

APPENDIX B - PUMPING STATION No.2	Drg. no. PS2/GJN/2	not to scale

Pumping Station 2

Drawing Numbers PS2/GJN/1

PS2/GJN/2

Explanatory Notes

The following worked example shows a second type of pumping station constructed from bolted precast concrete shaft rings.

It is assumed that the existing site is level with no topsoil. Rockhead is at level 10·350 and the existing site levels are to be maintained.

This example demonstrates measurement in accordance with Classes E, F, G, H, J, N and T. As all the structure is below ground it is not necessary to divide it into substructure and superstructure. The work measured includes only for pipework which runs between and in the shafts. Inlet and overflow pipework are not included.

TUNNELS	Explanatory Notes

TUNNELS

Exc. straight shaft in mat. other than topsoil, rock or artificial hard mat.; dia 3·96m

(T143)

	dep
g.l.	18500
rock head	10350
	8150

π/1·98
1·98
8·15

106·39

Exc. straight shaft in rock; dia. 3·96m

(T133)

	dep
total dep	11070
blinding	75
	11145
Less rock	8150
	2995

π/1·98
1·98
3·00

36·95

Exc. straight shafts in mat. other than topsoil, rock or artificial hard mat.; dia. 3·35m

(T143.1)

	dep
	2850
blinding	76
	2926

π/1·68
1·68
2·93

25·98

Explanatory Notes

As there are no payment lines shown on the drawings the excavation is measured to the net dimensions of the volumes to be excavated (M2). Due to the small size of the work, it is considered that the classification for 'other stated material' should be kept as simple as possible and the description of the material is taken from Class E. Detailed annotation of the dimensions is not necessary as it is quite apparent from the drawings where they apply.

Although the smaller of the two shafts falls in the same Third division classification as the larger shaft, it is necessary to keep it separate because the actual external diameter must be stated.

Pumping Station 2

π/	1.98			Exc. straight shaft in mat. other than topsoil, rock, or artificial hard mat.; dia. 3.96m used as filling above cover slab.
	1.98			
	0.30	3.70		(T143.2)
Ddt				
1.80				
1.80				
0.30	0.97			
0.60				(ddts for opening)
0.60				
0.30	0.11			
π/	0.05			
	0.05			$
	0.30	0.01	1.09	
			2.61	Ddt
				Exc. straight shaft in mat. other than topsoil, rock or artificial hard mat.; dia 3.96m a.b.d
				(T143)

Explanatory Notes

This is the approximate volume of material required for filling above the shaft cover slabs. A4 requires item descriptions for excavated material used as filling to be so stated and consequently it has to be measured separately.

The volume measured for filling has to be deducted from the overall volume for excavation. The balance of material is deemed to be disposed of off site. (C1)

Explanatory Notes

			Add Exc. straight shaft in mat. other than topsoil, rock or artificial hard mat.; dia. 3·35m used as filling above cover slab. (T143·3)
π/	1·68		
	1·68		
Ddt	0·30	2·66	
0·60			
0·60			
0·30	0·11		
0·60			(ddts for opngs)
0·45			
0·30	0·08		
3/π/ 0·05			&
0·05			
0·30	0·01	0·20	
		2·46	Ddt Exc. straight shaft in mat. other than topsoil, rock or artificial hard mat.; dia 3·35m a.b.d. (T143.1)

				Explanatory Notes
			Exc. surfs. in mat. other than topsoil, rock or artificial hard mat.; filling voids with grout. c.s. Spec. Clause 24.10	It is not necessary to classify the measurement of excavated surfaces according to the external diameter and therefore the quantities for both shafts can be combined. It is necessary to state the nature of the material to be used as filling for voids (A5) and this is best done by reference to the relevant specification clause.
π/	3.96			
	8.15	101.40	(large shaft sides) (T180)	
π/	3.35		(small shaft sides)	
	2.93	30.84		
		132.24		
			Exc. surfs. in rock; filling voids with grout. c.s. Spec. Clause 24.10	It is important that the base of the shaft is not forgotten, but as this will not be treated in the same way as the sides, then it must be measured separately.
π/	3.96			
	3.00		(large shaft sides) (T170)	
		37.33		
			Exc. surfs. in mat. other than topsoil, rock or artificial hardmat. to receive blinding conc.	
π/	1.68			
	1.68		(T180.1)	
		8.87		
			Exc. surfs in rock to receive blinding conc.	
π/	1.96		(T170.1)	
	1.96			
		12.07		

Explanatory Notes

Preformed
segmental linings
to shafts

Precast conc.
bolted rings;
flanged; nom.
ring width 610mm
max. piece weight
0.27t.; 4No. O
Segments, 2No. T
segments, 1No key
Segment; 34No.
7" x 7/8" circle and
cross bolts; 2No
10½" x 7/8" key bolts;
3mm bituminous
packing to
longitudinal
joints; intl. dia. 3.60m

All the details of the
constituent components of
the shaft rings would be
contained in the relevant
manufacturers catalogue or
in the specification. The
number of rings is
calculated by dividing
the total shaft depth
by the nominal ring
width. One of the rings
has a corbel and this
is measured separately.
The diameter given is the
internal diameter.

(T613)

$$\begin{array}{r} \underline{No} \\ 1500 \\ 500 \\ 4100 \\ 300 \\ 3970 \\ \hline 610)\overline{10370} \\ \hline 17No \\ corbel\ ring\ \ 1No \\ \hline 16No. \end{array}$$

16/ 1

16

Ditto but with
corbel ring for
half the dia.

1

1

Preformed
segment'al linings (cont)

Precast conc.
bolted rings;
flanged; nom.
ring width 610mm;
max. piece weight
0·25t; 4 No O
segments, 2No T
segments, 1No key
segment; 28No
6½" × ¾" circle
and cross bolts;
2No 10½" × ¾" key
bolts; 3mm
bituminous
packing to
longitudinal
joints; intl. dia. 3.05m

(T 613.1)

$\dfrac{No}{}$

610) 1830

3No

3/ 1

3

Pumping Station 2

Lining ancillaries

Caulking with mat. do Spec. Clause 24.8

(T674)

6/	10·37	62·22	(longitudinal caulking)
6/	1·83	10·98	
π/16/	3·66	184·00	circumferential caulking)
π/2/	3·05	19·17	
		276·37	

There is no requirement to distinguish between longitudinal and circumferential caulking and they are therefore measured together providing the materials are the same, and they are the same size. It is not a requirement to state the size, but this should not be automatically discounted, particularly if there are various sized caulking grooves.

Support and stabilization

Pressure grouting

No permanent support is required. It is considered undesirable to measure temporary support and M8 would have to be suitably amended in the Preamble. It is further considered that the number of sets of drilling and grouting plant should be left up to the Contractor's discretion and therefore item T831 would have to be taken out. The specification requires the grouting to be done in two stages. The first stage is to drill to a depth of 4m and then grout. The second stage is to drill a further 2m and grout.

388

Pumping Station 2

Explanatory Notes

Support and stabilization (cont)

Pressure grouting (cont)

			face packers	
			(T832)	
2/17/3/	1	102	(large shaft)	
2/3/3/	1	18	(small shaft)	
		120		

The specification calls for 3 injection points per ring. As each injection point is injected twice, the number of face packers measured is double the number of injection points (M10)

			Drilling and flushing 100mm dia. in holes of dep. n.e. 5m	
			(T834)	
17/3/	4.00	204.00	(large shaft)	
3/3/	4.00	36.00	(small shaft)	
		240.00		

The length of the holes has to be stated in stages of 5m (A17). As the initial stage of drilling is 4m this is classified as not exceeding 5m.

			Re-drilling and flushing 100mm dia. in holes of dep. 5-10m	
			(T835)	
17/3/	4.00	204.00	(large shaft)	
3/3/	4.00	36.00	(small shaft)	
		240.00		

For the second stage of grouting, the total hole depth is 6m, ie. stage 5-10m, and this is used for the classification of the re-drilling item, even though the re-drilling depth is only 4m. The depth of the re-drilling is the same as the initial drilling, with the second 2m stage being classed as 'initial' drilling in depth of hole 5-10m

Explanatory Notes

Support and
stabilization (cont)

Pressure grouting (cont)

Drilling and
flushing 100mm
dia. in holes
of depth 5-10m

(T834.1)

17/3	2·00	102·00	(large shaft)
3/3	2·00	18·00	(small shaft)
		120·00	

Injection of grout
to Spec. Clause
24.9

The total tonnage of grout
measured will be an
estimate of the likely
quantity based on the
Engineer's knowledge of the
prevailing ground conditions.
The quantity inserted here
is based on 1 tonne of dry
material per ring.

No of rings
large shaft 17
small shaft 3
———
20
× 1E/ring
———
20E

	20E	
		20E

390

Explanatory Notes

<u>ALL THE FOLLOWING WORK</u>
<u>IN SHAFTS</u>

<u>IN-SITU CONCRETE</u>

<u>Provn of conc.</u>

Standard mix, ST3,
OPC to B.S.146;
20agg. to B.S.882

(F133)

| | 1·70 | | |
| | | 1·70m³ | (binding) |

Designed mix
grade C20; OPC
to B.S.146; 20agg
to B.S.882.

(F243)

(structure)

	18·48		
	3·43		
Dd't	30·60		
10·77			
	1·04		
		42·78m³	

Due to the fact that there are
several different classifications
of concrete involved, it is
considered that it is more
appropriate to measure the
in-situ concrete and
ancillaries in accordance
with Classes F and G.

Because the concrete is
measured in Class F instead
of Class T it would be
prudent to state in a
general heading that all
the work is in shafts

IN-SITU CONCRETE (CONT)

Placg. of mass conc.

Blinding; thness n.e. 150mm (F611)

π/	1.98		
	1.98		
	0.08	0.99	(large shaft)
π/	1.68		
	1.68		
	0.08	0.71	(small shaft)
		1.70	

Bases, footgs, pile caps & ground slabs thness ex. 500mm (F524)

Although the base to the large shaft has two layers of mesh in the top it is not considered 'reinforced' for the purposes of the concrete classification.

π/	1.98		
	1.98		(bottom of
	1.50		large shaft)
		18.48	

ditto 300-500mm (F523)

A deduction is made from the volume where the bottom ring 'sits' in the top of the base because the cross sectional area exceeds $0.01\,m^2$ (M1(d))

π/	1.68		(bottom of
	1.68		small shaft)
	0.40	3.55	
Ddt			
π/	3.20		
	0.15		
	0.08	0.12	
		3.43	

```
           dep
300       2850
400        700
          2150
          1830
           320

           400
           320
            80

           dia
          3350
          3050
       2)6400
          3200
```

392

			IN-SITU CONCRETE (CONT)	Explanatory Notes

Placg. of mass conc. (cont)

Lining to shaft walls; intl dia. 3360 mm; min. thcness 150mm (F580)

The minimum thickness of the lining is 150mm. This thickness is deducted from the internal diameter of the shafts to give the internal diameter of the lining

				Len
				500
				4100
	30·60			300
				3970
				8870

				Dia.
			extl.	3960
			intl	3360
			2)	7320
				3660

π	3.66			
	8.87			
	0.30			

				Wid
			Hrip	150
			lming	150
				300

The volume of the lining is derived by calculating the total volume of the lining including the space occupied by the rings, and then deducting the volume of the rings which is stated in the manufacturer's literature. This is the easiest way of taking into account the flanges in the rings. Further adjustments are made for the volume occupied by the corbel and the platform.

Ddt
ditto

14.54	0.71		(rings)	No.
	1·00	10·32		rings
	1·00			610) 8870
				14.54

	5·05		(corbel	Vol.
	0·15	0·45	platform)	0.71m³
	0·60			

	10·77			Len
			½/π/3·660	5749
			2/350	700
				5049.

			IN-SITU CONCRETE (CONT)	Explanatory Notes

Placg. of mass
conc. (cont')

Benching; 500 ×
500 o/all; bott.
of shaft

(F580·1)

The volume of the benching
is calculated by multiplying
its cross-sectional area
by the mean diameter.
The volume of the cut out
for the ladder is deducted,
but the overlap with the
concrete plinths is not.

½/π/	2·86		
	0·50		
	0·50	1·12	
Dott			
½/	0·60		
	0·50		
	0·50	0·08	(for cut out for ladder)
		1·04	

dia
3360
2/½/500   500
2860

Explanatory Notes

CONCRETE ANCILLARIES

<u>Fmwk. type A fin</u>

π/3.35		
0.32		
	3.37	

Plane curved to 3.35m rad. in one plane 0.2-0.4m

(extl face of base to small shaft) (G153)

The radius of the curved formwork ~~formwork~~ must be stated in item descriptions (A4). As the ~~formwork~~ to the outside face of the base to the small shaft is exposed, it has a type A finish.

<u>Fmwk. type B fin.</u>

Plane curved to 3.36m rad. in one plane ex 1.22m (G255)

π/3.36		
8.37	88.36	(lining to large shaft)
0.60		(ladder benching
0.50	0.30	cut out)

Len
8870
500
8370

Ddt
4.58
0.60   2.75

(for corbel platform) ½π/3360
2/350

Len.
5278
700
4578

85.91

Note that the measurements of the lining ~~formwork~~ does not include the area behind the benching, but does include for the area to the ladder cut out.

Plane curved to varying radii; min 2360; max 3360 (G260)

π/2.86		
0.71	6.38	(benching)
Ddt		(for ladder
0.60		cut out)
0.70	0.42	
	5.96	

Wid
x: 500²+500²
x= 707

It is not clear whether curved ~~formwork~~ which is also the upper surfaces of concrete should be so described. It is considered that it is not necessary in this particular case, although it should be made clear in the Preamble.

Explanatory Notes

CONCRETE ANCILLARIES (CONT)

Fmwk. type B fin
(cont)

2/ 1/3	0.50		Plane. vert. average Width 0.25m
	0.50		(G243)
		0.25	(sides of cut out to benching)

This is not strictly the correct method of measuring this item, but the quantity is so small that it is considered to be more applicable than dividing the item into the various width classifications which could give inordinately high quantities.

Reinforcement

2/π/	1.98		Fabric to B.S. 4483 ref. A142.
	1.98		(G562)
		24.64	(base to large shaft)

The area of fabric is the net area with no allowance for fabric in laps (M9).

Pumping Station 2

			Explanatory Notes

CONCRETE ANCILLARIES (CONT)

Concrete Accessories

			Fin. of top surf; Steel trowel fin.	The specification calls for a steel trowel finish to the base
π/	1.18			
	1.18	4.37	(base to large shaft) (G812)	of the large shaft. No deduction is made for the
	0.60		radius	support plinths as their
	0.50	0.30	(ladder cut'out) 2)2360	plan area does not exceed
			1180	0.5 m² (M14)
		4.67		

The specification calls for a steel trowel finish to the base of the large shaft. No deduction is made for the support plinths as their plan area does not exceed 0.5 m² (M14)

			C. & S. (1:4) screed; av. 75mm th. with steel trowel fin.	A13 requires the materials, thickness and surface finish of
π/	1.53		(G815)	applied finishes to be stated. No deduction is made for the
	1.53		radius	sump as this is less than
		7.36	2)3050	0.5m². The forming of the sump
			1525	could be measured separately

A13 requires the materials, thickness and surface finish of applied finishes to be stated. No deduction is made for the sump as this is less than 0.5m². The forming of the sump could be measured separately or as in this case Coverage Rule C6 could be extended to include it.

Inserts

		80mm dia. pipe through precast conc. shaft linings; proj. both sides (supply incl. elsewhere) (G832)
1		(drain i small shaft)
	1	

Explanatory Notes

CONCRETE ANCILLARIES (CONT)

Inserts (cont)

			200mm dia. pipe through precast conc. shaft lining; proj. both sides (supply incl. elsewhere) (G 832·1)
3/ 1			(pump pipes - small shaft)
		3	

One area not covered by Class T is the work involved in breaking through linings for incoming pipes and similar inserts. It is therefore necessary to create new items which inform the Contractor of the work involved.

All the following inserts projecting both sides of precast conc. shaft segments with in-situ conc. lining

			100mm dia × 900mm long U.P.V.C cable ducts (G 832·2)
3/ 1			(large shaft)
		3	

The supply and fixing of the ducts is included in this item (C7)

			80mm dia. pipe; (supply included elsewhere) (G 832.3)
1			(drain - large shaft)
		1	

<u>Pumping Station 2</u>

<u>Explanatory Notes</u>

<u>CONCRETE ANCILLARIES (CONT)</u>

<u>All the following
inserts projecting (cont)</u>

200 mm. dia. pipe
(supply incl.
elsewhere)

(pump delivery   (G 832·4)
pipes - large shaft)

2/ 1		
	2	

225 mm dia. pipe
(supply incl.
elsewhere)

(overflow pipe -   (G 832·5)
large shaft)

1		
	1	

300mm dia. pipe
(supply incl.
elsewhere)

(inlet pipe -   (G 832·6)
large shaft)

1		
	1	

Pumping Station 2

PRECAST CONCRETE

Slabs; conc. designed mix grade C30, OPC to B.S. 146; 10agg to BS. 882.

Areas
large cover slab

Ty 1·98
1·98
12·31 m²

small cover slab
π/1·68
1·68
8·86 m²

platform
½/π/1·83
1·83   5·24
less
0·35
1·83   0·64
4·62 m²

Weights
12·31
0·40
2·40t/m³
11·82 k

8·86
0·40
2·40t/m³
8·51 k

4·62
0·30
2·40t/m³
3·33 t.

A1 requires the Specification of concrete to be stated in items for precast concrete. The Second and Third divisions of the slab's classification require the area and weight of the slabs to be stated in bands as classified. These only have to be approximate, as the Contractor will ascertain from the drawings the exact nature of the work involved.

400

Explanatory Notes

PRECAST CONCRETE (CONT)

Slabs (cont)

Roof slab to
large shaft;
area 4-15 m²;
weight 10-20t;
thness 400mm

(H537)

A1 requires the position in
the works of each unit
to be stated. The units
do not have mark or type
numbers and these cannot
be stated in accordance
with A2.

Roof slab to
small shaft;
area 4-15 m²;
weight 5-10t;
thness 400mm

(H536)

Intermediate
platform to
large shaft;
area 4-15m²;
weight 2-5t;
thness 300mm

(H535)

Explanatory Notes

It was recommended in the text that the measurement of pipelines is considered on an individual basis for each contract. The majority of the pipes and fittings in this example are in the shafts but there are the lengths between the shafts to take into account. To try and measure these pipes as 'in trenches' would cause undue complication because the fittings into the chamber would be then part 'extra over' the pipe in the trench and part 'full value' for the length in the chamber. In this instance the recommended procedure is to measure the excavation for all three pipes separately and to enumerate all the pipes and fittings regardless of length as not in trenches. It would be necessary to state the measurement philosophy adopted in the Preamble.

Pumping Station 2

EARTHWORKS TO PIPEWORK

Pipe tr. exc.
as drwg PS2/GSN/1
mat. other than
topsoil, rock
or artificial hard
mat.; max dep.
2-5m

		(E425)
4·75		
1·80		750
2·70		3000
	23·09	1000
		4750

&

Disposal; backfilling
around pipes

(E532)

There is no classification for pipe trench excavation and backfilling so these are rogue items. This is a slight overmeasurement as the length is taken as the length of the UPVC drain which lies between the inside faces of the chamber.

403

Explanatory Notes

PIPEWORK - FITTINGS
AND VALVES; NOT
IN TRENCHES

<u>Ductile iron pipe
fitting</u>; B.S. 4772;
flanged joints;
nom. bore 200mm;
as Spec. Clause
13.1

	1		Flanged spigot pipe 1200mm long (ref 3) (J381)
		1	

Whilst all the fittings are scheduled, the prudent taker off would still check the fittings off the drawing as they were measured.

Although the fittings are described as having flanged joints, the first fitting has a spigot one end and this tells the estimator that this end does not have a flanged joint. This end has a flexible coupling which is measured separately and the cost of the joint to the spigot end is therefore included elsewhere.

	1		Flanged spigot pipe 250mm long (ref 6) (J381.1)
		1	

2/	1		Flanged spigot pipe 1300mm long (ref 13) (J381.2)
		2	

6/	1		Double flanged pipe 2000mm long. (ref 14) (J381.3)
		6	

Explanatory Notes

PIPEWORK (CONT)

<u>Ductile iron pipe fittings (cont)</u>

Double flanged pipe 605mm long

(J381.4)

2/ 1

(ref 15)

2

Although this fitting does not have flanged joints at either end it is still classed as such for convenience. The estimator will know that the flexible couplings are measured separately and that he does not have to allow for joints for this fitting.

Double spigot pipe 2900mm long

(J381.5)

2/ 1

(ref 17)

2

Double flanged 90° bend

(J311)

3/ 1

(ref 9)

3

As both these bends are less than 300mm it is not necessary to describe them as vertical.

Double flanged 90° duckfoot bend

(J311.1)

2/ 1

(ref 16)

2

Explanatory Notes

PIPEWORK (CONT)

Ductile iron pipe
fitting (cont)

All flanged Equal
tee
(J 321)

2/ 1   (ref 7)

2

Blank flange
(J 391)

1   (ref 10)

1

Flexible coupling
(J 391·1)

5/ 1   (ref 1)

5

Flange adaptor
(J 351)

3/ 1   (ref 5)

3

Pumping Station 2

PIPEWORK (CONT)

U.P.V.C pipe fittings
B.S. 4660; nom.
bore 80mm

double spigot
pipe 750mm long.
(Ref18)   (J481)

	1		
		1	

double spigot
pipe 3000mm long

(Ref19)   (J481·1)

	1		
		1	

double spigot
pipe 1000mm long

(Ref20)   (J481·2)

	1		
		1	

Flexible coupling

(Ref2)   (J491·2)

2/	1		
		2	

As the flexible couplings are measured separately it is not necessary to state the nature of the joints in the item descriptions

407

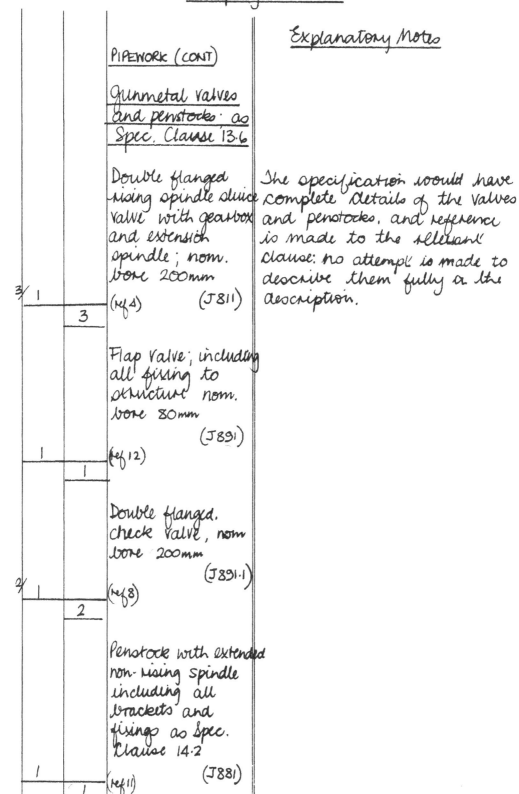

Pumping Station 2.

PIPEWORK (CONT)

Gunmetal valves and penstocks: as Spec. Clause 13.6

Double flanged rising spindle sluice valve with gearbox and extension spindle; nom. bore 200mm

3/1

(ref 4)   (J811)

3

Flap valve; including all fixing to structure nom. bore 80mm

(J891)

1

(ref 12)

1

Double flanged. check valve, nom bore 200mm

(J891.1)

2/1

(ref 8)

2

Penstock with extended non-rising spindle including all brackets and fixings as Spec. Clause 14.2

(J881)

1

(ref 11)

1

Explanatory Notes

The specification would have complete details of the valves and penstocks. and reference is made to the relevant clause: no attempt is made to describe them fully in the description.

Pumping Station 2.

Explanatory Notes

PIPEWORK - SUPPORTS
AND PROTECTION

Concrete stools and
thrust blocks;
concrete grade
C20, OPC to
B.S. 146, 20agg to
BS 882; unreinforced

	vol.
	1·1
	0·4
	0·25
	$\overline{0·11m^3}$

Vol. 0·1 - 0·2m³. nom
bore 200mm

(large shaft) (L721)

A5 requires the specification
of concrete to be stated
and whether it is reinforced.
If there were several
reinforced stools within the
same volume classification,
it would be necessary to
itemise them individually
if their reinforcement
requirements were different.

2/	1	
		2

Other isolated
pipe supports; as
drwg DT/1

height n.c. 1m.
nom. bore 200mm
(L811)
(large shaft)

Although A6 states that the
principal dimensions and
materials should be stated
in item descriptions, the
reference to the drawing would
in effect be giving the same
information

2/3/	1	
		6

Pumping Station 2

<table>
<tr><td>1</td><td></td><td></td><td colspan="2"><u>MISCELLANEOUS</u>
<u>METALWORK</u></td></tr>
</table>

			MISCELLANEOUS METALWORK	Explanatory Notes
1			Ladders; galvanised mild steel as Spec. Clause 18.6 & drwg DT/2 with safety cage; ladder 5700mm long incl. all fixing. (N130)	The unit of measurement for ladders in Class N is linear metres, but this ladder has a safety cage for part of its length which would mean measuring it as two items, one with and one without the cage. It is considered that enumerating the ladder stating the length is a fairer method of measurement.
	1			
3.20			Handrail as Spec. Clause 18.9 incl. all fixings (N140)	
	3.20			
2/ 1			Safety chain as Spec. Clause 18.8; nett. length 1300mm including all fixings (N190)	There is no separate classification for safety chains and this is a rogue item. They could have been measured in linear metres as an alternative to enumeration.
	2			

Pumping Station 2

<u>MISCELLANEOUS METALWORK (CONT)</u>

<u>Access covers and frames on two course Engineering brickwork</u>

There is no separate classification for access covers and frames, so these are rogue items.

B.S. 497 grade MB2 - 60/45

1			(small shaft)	(N190)
		1		

B.S. 497 grade MB2 - 60/60

2/	1		(small & large shafts)	(N190.1)
		1		

1800 x 1200 clear opening as Spec. Clause 18.1

1			(large shaft)	(N190.2)
		1		

411

Explanatory Notes

MISCELLANEOUS
METALWORK (CONT)

Heavy duty
surf. box incl
casting in 400×
400 × 300mm
deep concrete

The concrete surround is given in the item description, which is considered more appropriate than measuring it in detail in accordance with Classes F & G.

100mm dia. clear
opng.

(J130·3)

3/ | 1 | 3

| 1 | 1

| | 4

(for valve
spindles)

APPENDIX C

RETAINING WALL

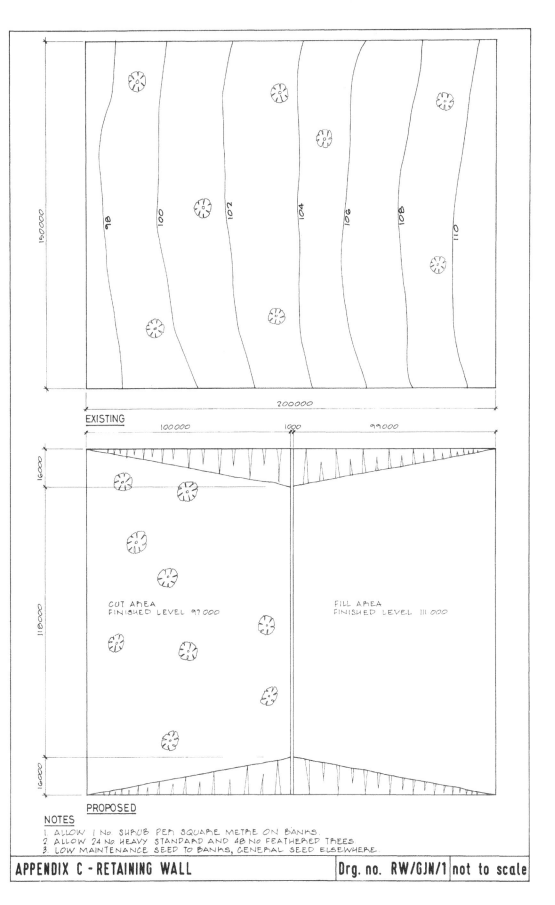

EXISTING

PROPOSED

NOTES
1. ALLOW 1 No SHRUB PER SQUARE METRE ON BANKS.
2. ALLOW 24 No HEAVY STANDARD AND 48 No FEATHERED TREES.
3. LOW MAINTENANCE SEED TO BANKS, GENERAL SEED ELSEWHERE.

| APPENDIX C - RETAINING WALL | Drg. no. RW/GJN/1 | not to scale |

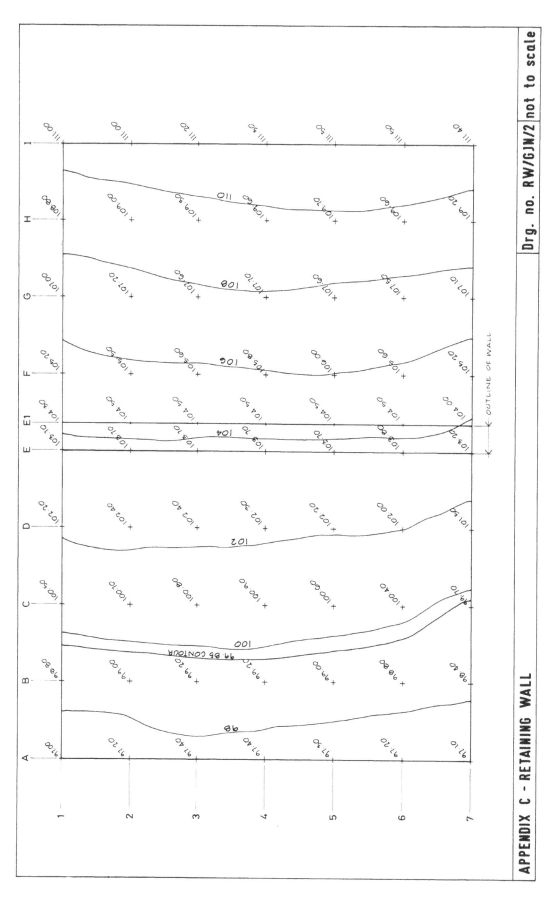

APPENDIX C - RETAINING WALL

Drg. no. RW/GJN/2 | not to scale

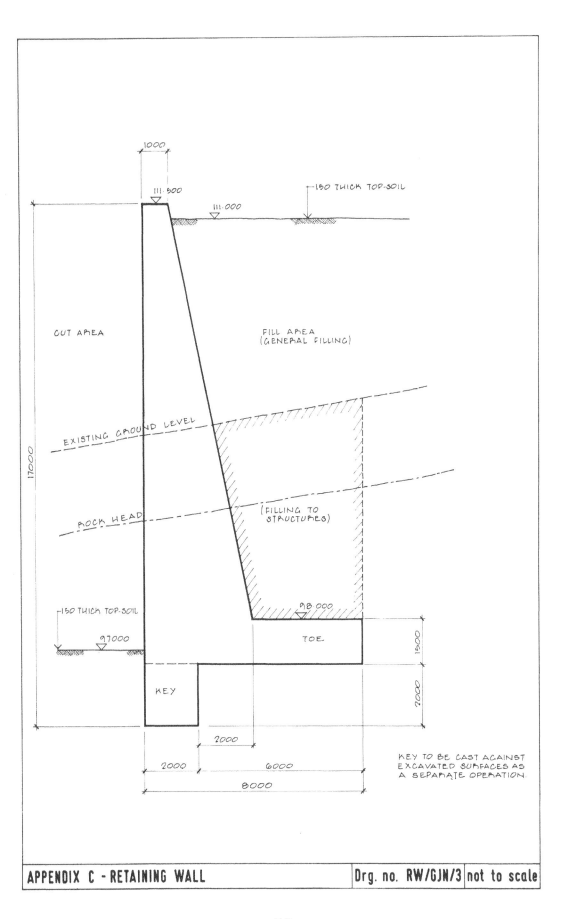

1000

111·500

111·000

‑150 THICK TOP-SOIL

CUT AREA

FILL AREA
(GENERAL FILLING)

17000

EXISTING GROUND LEVEL

ROCK HEAD

(FILLING TO
STRUCTURES)

‑150 THICK TOP-SOIL

98·000

97000

TOE

1500

KEY

2000

2000

2000

6000

8000

KEY TO BE CAST AGAINST
EXCAVATED SURFACES AS
A SEPARATE OPERATION.

APPENDIX C - RETAINING WALL	Drg. no. RW/GJN/3	not to scale

Retaining Wall

Drawing Numbers RW/GJN/1

RW/GJN/2

RW/GJN/3

The following example shows a rectangular virgin field with a reasonably steep gradient. The example involves the construction of a major concrete retaining wall across the middle of the site to enable one half of the area to be filled to a constant level whilst lowering the other half.

It is assumed that the site will have a constant 100mm of good quality topsoil together with various old trees which will have to be removed. Rock level is 3m below original ground level. This example deals only with Class E: Earthworks, the removal of the trees and the retaining wall itself would be measured under classes D, F and G.

The measurement of Earthworks invariably produces large amounts of waste calculations. If these are particular to any one item or group of items it is recommended that they are done adjacent to the item of work to which they apply as in this particular example.

418

Retaining Wall

Explanatory Notes

Alternatively, large amounts of general calculations may be done at the beginning of the take off with relevant annotation as to how and to which sections it applies.

For the purposes of measurement the site is divided into the following

1. Topsoil strip.

2. Excavation for the retaining wall.

3. Excavation over cut area.

4. Excavation ancillaries.

5. Filled area.

6. Topsoil and landscaping.

419

Retaining Wall

1. TOPSOIL STRIP

			gen. exc; topsoil
			max. dep. n.e
			0.25m; exc. surf
			u/s topsoil
			(E411)
200.00			
150.00			
0.10			
	3000.00		

It is not necessary to state the Commencing Surface as it is also the Original Surface. The Excavated Surface and hence the maximum depth is different depending upon whether the topsoil is in the cut or filled area. However due to the nature of the work it is felt that it would be more helpful to state the Excavated Surface as underside of topsoil and therefore give the maximum depth as the depth of the topsoil in accordance with paragraph 5.21.

The original site survey drawing RW/GJN/1 has level information by way of contours. This has been processed into a 25 metre grid of levels by interpolation as follows:-

420

Retaining Wall

1. TOPSOIL STRIP (CONT)

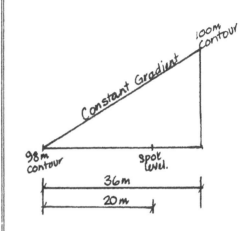

The height of the spot level is derived by calculating the vertical rise per metre of horizontal distance between the contours and multiplying it by the distance of the spot height from the lower contour. Add this value to the value of the lower contour to give the spot height e.g.

$$\text{spot height} = \left(\frac{2}{36} \times 20\right) + 98$$

$$= \underline{99 \cdot 111}$$

421

Retaining Wall

2. RETAINING WALL
EXCAVATION

Level		Avge G.L.
E1	1/103700	103700
E2	2/103700	207400
E3	2/103700	207400
E4	2/103700	207400
E5	2/103700	207400
E6	2/103800	207600
E7	1/103200	103200
E1.1	1/104300	104300
E12	2/104500	209000
E13	2/104500	209000
E14	2/104500	209000
E15	2/104500	209000
E1.6	2/104500	209000
E1.7	1/104000	104000
	24)	2497400
		104058
L1SS	topsoil	100
Average level	after topsoil strip	103958
	Form lw.	
top of toe		98000
toe thiness		1500
		96500

It is an express requirement of the Specification to carry out the bottom 'key' excavation as a separate operation and concrete immediately against the excavated surface and this is deemed to be a separate stage for excavation under M5. The first part of the excavation is from the stripped level to the underside of the toe.

This is the calculation of the average ground level after topsoil strip over the retaining wall area. Drawing RW/GJN/2 has the outermost outline of the foundations shown together with the interpolated levels. (see the calculation of the average ground level for the lower side of the site for an explanation of the weighting of the levels)

422

Retaining Wall

2 RETAINING WALL
EXCAVATION (CONT)

__Max dep of excav'n__

highest spot lev.	104500
topsoil	100
	104400
form lev.	96500
	7900

i.e. 5-10m

The maximum depth of the excavation must be stated and is calculated in accordance with paragraph 5.21

__Overall av. dep of excav__

av. lev after topsoil strip	103958
form lev.	96500
	7458

This is the calculation of the average overall depth of the excavation and will include excavation of natural material and rock. It is necessary to calculate the average depth of both in order to measure both separately.

__Av. dep of natural material__

overall av. dep. gl to top of rock	3000
Less topsoil	100
	2900

Exc. foundns;
max dep 5-10m;
comm. surf. u/s of topsoil; exc surf u/s of wall toe (lev 96500)

(E 326)

The Commencing Surface in this case is the underside of the topsoil which is not also the Original Surface and therefore must be stated. The Excavated Surface is bottom of the first stage of excavation i.e. underside of the toe. As this is not also the Final Surface it also must be stated (A4). It is not necessary to state the type of material as it is deemed to be natural material other than topsoil, rock or artificial hard material (D1).

150.00	
9.00	
2.90	
3915.00	

Retaining Wall

Explanatory Notes

2. RETAINING WALL
EXCAVATION (CONT)

Av. dep of
rock

Av. dep of total excav.		7458
less		
Av. dep of nat. mat.		2900
		4558

Exc. foundns; rock,
max. dep. 5-10m;
comm. surf. U/S of
topsoil; exc. surf
U/S of wall toe
(lev 96500)   (E 336)

150·00
9·00
4·56

6156·00

Although the rock is in a
different location in the
excavation and of a different
depth, the depth classification
is exactly the same as the
previous item because in
accordance with definitions of
1·12 and 1.13, the commencing
and excavated surfaces are the
same for both items, and
therefore the depth in
accordance with paragraph 5·21
is the same. It is the overall
depth of the void which is
critical, not the depth of
the individual materials
within it.

toe excav

Exc. foundns; rock,
max. dep. 1-2m;
comm. surf U/S
of wall toe (lev.
96500)   (E 334)

150·00
2·00
2·00

600·00

This is a separate stage of
excavation in accordance
with M5. The commencing
surface must be stated as it
is not also the Original surface.
It is not necessary to state
the excavated surface as it
is also the Final surface (A4)
The commencing surface may
be identified by stating
the level or by it's
relationship to any other
fixed point.

Retaining Wall

3 CUT AREA

Level		Av. gl after topsoil strip
A1	1/97000	97000
B1	2/98800	197600
C1	2/100500	201000
D1	2/102200	204400
E1	1/103700	103700
A2	2/97200	194400
B2	4/99000	396000
C2	4/100700	402800
D2	4/102400	409600
E2	2/103700	207400
A3	2/97400	194800
B3	4/99200	396800
C3	4/100800	403200
D3	4/102400	409600
E3	2/103700	207400
A4	2/97400	194800
B4	4/99200	396800
C4	4/100900	403600
D4	4/102300	409200
E4	2/103700	207400
A5	2/97300	194600
B5	4/99000	396000
C5	4/100600	402400
D5	4/102200	408800
E5	2/103700	207400
A6	2/97200	194400
B6	4/98800	395200
C6	4/100400	401600
D6	4/102000	408000
E6	2/103800	207600
		8853500

This is the calculation of the average ground level of the lower side of the site after topsoil strip.

The grid of levels extends from A-E / 1-7. The corner levels (A1) only have on sphere or 'square' of influence and are used only once. The side levels (B1, C1) have two spheres or 'squares' of influence and are used twice and the internal levels have four spheres or 'squares' of influence and are used four times. The total is then divided by the total number of spheres or 'squares' of influence to give the average level.

425

Retaining Wall

3. CUT AREA (CONT)

Level

		b/f 8 853500
A7	1/97100	97100
B7	2/98400	196800
C7	2/99700	199400
D7	2/101 500	203000
E7	1/103200	103200

$$96)\ 9653000$$

Av. Ex. g.l.	100552
topsoil	100
	100 452

final surf lev — The level to which excavation is to be taken to is the finished site level less the thickness of topsoil to be placed

Fin lev.	97000
Less topsoil thrness	150
	96850

Max dep of excav — The maximum depth of the excavation must be stated and is calculated in accordance with paragraph 5.21

highest spot lev	103800
topsoil	100
	103 700
final surf lev	96 850
	6850

ie. 5-10m

Overall av. dep. of excav.

av. g.l. after topsoil strip	100452
final surf lev	96850
	3602

Retaining Wall

3. CUT AREA (CONT)

			Gen. exc.; max dep. 5-10m; comm surf U/S of topsoil	
100.00				
150.00				
3.60				
	54000.00		(E426)	

It is not necessary to state the Excavated Surface as this is also the Final Surface. The Commencing Surface must be stated as it is not also the Original Surface (A4)

This quantity is overmeasured as it also contains the rock excavation which has to be measured separately and because it also assumes that the sides of the excavation are vertical when in fact they are sloping. Unlike the excavation for the foundations to the wall itself where the natural material and rock were measured separately, here it is easier if the total excavation is measured as if it were all natural material and the rock then measured and deducted from the gross quantity of natural material.

427

Retaining Wall

Explanatory Notes

3. CUT AREA (CONT)

Level		Av. rockhead lev.
		Av. rockhead lev.
C1	1½/100500	150750
C2	3/100700	302100
C3	3/100800	302400
C4	3/100900	302700
C5	3/100600	301800
C6	2½/100400	251000
C7	1/99700	99700
D1	2/102200	204400
D2	4/102400	409600
D3	4/102400	409600
D4	4/102300	409200
D5	4/102200	408800
D6	4/102000	408000
D7	2/101500	203000
E1	1/103700	103700
E2	2/103700	207400
E3	2/103700	207400
E4	2/103700	207400
E5	2/103700	207400
E6	2/103800	207600
E7	1/103200	103200

53) 5407150

102022

Less dep to rock head   3000

99022

As stated, the previous excavation item will include rock excavation for which adjustment will have to be made and it is therefore necessary to determine the area and depth over which the rock excavation will occur. In our example the rock head is 3 metres below original ground level. If we take a section through this area we

need to determine the line at which the final surface level intersects the rock head. This is most easily done by adding 3m onto the final surface level (96850) and plotting this contour on the original ground levels plan by interpolation (see RW/GSN/2). The area bounded by the 99850 contour and the retaining wall is the plan area of rock.

Retaining Wall

3 CUT AREA (CONT)

		Av. dep of rock		The average rock depth is

Av. dep of rock

	Av. rock lev	99022
	Fin. surf	96850
		2172

The average rock depth is calculated by deducting the Final Surface from the average rock level as shown opposite.

Gen exc; rock; max.dep. 5-10m; comm. surf u/s of topsoil

(E 436)

The easiest method of calculating the rock area is to divide it up into equal 25m strips as on the grid and measure the area of each.

63·00		
25·00		
2·17	3417·75	

&

65·00		
25·00		
2·17	3526·25	

Ddt
Gen exc; max. dep 5-10 m; comm surf u/s of topsoil a.b.d.

(E 426)

This is the deduction of the rock volume from the total quantity of the general excavation in natural materials a.b.d. a.b.d. stands for 'as before described' and shows that the item has already been measured.

67·00		
25·00	3634·75	
2·17		

66·00		
25·00	3580·50	
2·17		

62·00		
25·00	3363·50	
2·17		

55·00		
25·00	2983·75	
2·17		

20506·50

Retaining Wall

Adjustment must also be made to the general excavation in natural material quantities for the two sloping banks on either side which are not excavated. These are in the rough shape of a triangle based pyramid as follows :-

Total vol of both pyramids

$$2 \times \frac{1}{3} \times \frac{1}{2} \times 16 \times 6 \cdot 50 \times 100$$
$$= 3467 m^3$$

Total vol. of rock within both pyramids

$$2 \times \frac{1}{2} \times (8 \cdot 90 \times 3 \cdot 50 \times 54 \cdot 00) +$$
$$2 \times \frac{1}{3} \times (0 \cdot 5 \times 7 \cdot 1 \times 3 \cdot 50 \times 54 \cdot 00)$$

$$\cdot = 2130 m^3$$

$\therefore$ Vol. of nat. mat.

3467.00	
2130.00	
1337.00 m³	

Ddt
Gen exc; max. dep 5-10m; comm. surf U/S topsoil a.b.d.
(E426)

1337.00		
1.00		
1.00		
	1337.00	

Ddt
Gen. exc; rock; max dep. 5-10m; comm surf. U/S topsoil a.b.d.
(E436)

2130.00		
1.00		
1.00		
	2130.00	

With the volume of rock being enclosed by the dotted lines.

430

Retaining Wall

Explanatory Notes

4. EXCAVATION ANCILLARIES

	<u>Add</u>	
	Prep. of surfs; rock	Bottom of retaining wall
150.00	(E 523)	excavation. No preparation is
9.00		measurable over the general
	1350.00	cut area as it is to receive
		topsoil (M11)

	Prep. of surfs; rock;	Either side of key where the
	vert.	concrete is cast against
2/ 150.00	(E 523.1)	excavated surfaces.
2.00		
2/ 9.00		
2.00	(ends)	
	636.00	

	Disposal of exc. mat;	In our example, the rock
	rock	material is deemed not to be
150.00	(E 533)	suitable for the general filling
9.00		behind the retaining wall and
4.56	6156.00 (wall fndns)	therefore removed from site
150.00		
2.00		
2.00	600.00	
63.00		
25.00		
2.17	3417.75	
65.00		
25.00		
2.17	3526.25	
67.00		
25.00		
2.17	3634.75 (gen exc.)	
66.00		
25.00		
2.17	3580.50	
62.00		
25.00		
2.17	3363.50	
55.00		
25.00		
2.17	2983.75	
	27262.51	

Retaining Wall

				Explanatory Notes

4. EXCAVATION ANCILLARIES (CONT)

Ddt

Disposal of exc. mat;
rock

2130.00			(E 533)	Adjustment for the banks.
1.00				
1.00				
	2130.00			

There are no trimming or
preparation items measurable
over the general cut area
as all these surfaces are
to receive topsoil (M10 & M11)

432

Retaining Wall

Explanatory Notes

5. FILLED AREA

	mat. available	
	ddt	Add
		3915.00
		54000.00
	20506.50	
	1337.00	
	21843.50	57915.00
		21843.50
		36071.50

There are two types of filling involved. The first is backfilling to the structure above the toe level to bring the levels up to the stripped level, and then general filling to the required finished levels. The first calculation should be to compute the amount of excavated materials available for filling.

fill. to structure

dep.

av. lev. after topsoil strip	103958
top of toe	98000
	5958

wid

	5000
$\frac{6483}{13500} \times 3000$	1441
	6441

Fill to structures; non-selected exc. material

(E 613)

Backfilling above toe of retaining wall to stripped level.

150.00		
6.44		
5.96		
	5757.36	

433

Retaining Wall

5. FILLED AREA (CONT)

	Ddt	
	Fill to structures;	
	non-selected exc.	
2/½/ 5·96	material a.b.d.	
16·00	(E613)	
6·44		
	614.11	

Adjustment on last item for either end of the filling to the wall which has a sloping bank and is therefore not filled

N.B. in actual fact this is very slightly over size but the calculation to compute the actual size would be so complicated that it would never be done in practice.

434

Retaining Wall

5. FILLED AREA (CONT)

Level	Av. g.l after topsoil Strip	
E1	1/ 103700	103700
E2	2/ 103700	207400
E3	2/ 103700	207400
E4	2/ 103700	207400
E5	2/ 103700	207400
E6	2/ 103800	207600
E7	1/ 103200	103200
F1	2/ 105200	210400
F2	4/ 105500	422000
F3	4/ 105600	422400
F4	4/ 105800	423200
F5	4/ 106000	424000
F6	4/ 105600	422400
F7	2/ 105200	210400
G1	2/ 107000	214000
G2	4/ 107200	428800
G3	4/ 107600	430400
G4	4/ 107700	430800
G5	4/ 107600	430400
G6	4/ 107500	430000
G7	2/ 107100	214200
H1	2/ 108800	217600
H2	4/ 109000	436000
H3	4/ 109300	437200
H4	4/ 109600	438400
H5	4/ 109700	438800
H6	4/ 109600	438400
H7	2/ 109200	218400
I1	1/ 111000	111000
I2	2/ 111000	222000
I3	2/ 111200	222400
I4	2/ 111300	222600
I5	2/ 111500	223000
I6	2/ 111500	223000
I7	1/ 111400	111400
	96)	10 317 700
		107476
Less topsoil		100
		107 376 435

Retaining Wall

5. FILLED AREA (CONT)

Total quant. of Exc.
Mat. left for fill.

		36071·50
5757·56		
Less 614·11	5143·45	
	30928·05 m³	

This is the calculation for the total quantity of excavated material still left for filling after backfilling to the structure

Gen. fill

	Dep.
Fin. lev.	111000
topsoil	150
	110850
Lev. of extg grnd after topsoil strip	107376
	3474

Len.

Len. at lower lev. =

$$99000 - \left(\frac{111500 - 103700}{111500 - 98000} \times 3000\right)$$

= 97267

Av. len.	97267
	99000
2)	196267
	98134

Gen. fill; imported mat. as spec. Clause 9.1

(E 635)

The precise nature of the material would be given in the Specification. The item description should therefore refer to the clause or give the relevant details.

98·13			
100·00			
3·47			
	34051·11		

Retaining Wall

5. FILLED AREA (CONT)

Deduct

Gen fill; imported mat. Spec. Clause 9.1 abd. (E 635)

Adjustment on previous item for the banks which are not filled.

2/1/3/1/2	16.00	
	3.47	1816.06
	98.13	
	30928.05	
	1.00	
	1.00	30928.05
		32744.11

For filling with the excavated material still surplus which reduces the amount of import. (see next item)

Add

Gen. fill; non-selected exc. mat. other than topsoil or rock (E 633)

This is the total quantity of excavated material left for general filling.

30928.05		
1.00		
1.00		
	30928.05	

There are no filling or preparation items to be measured as all the filled surfaces are to receive topsoil (M 22 and 23)

6. TOPSOIL AND LANDSCAPING

	0.15) 3000 m³		
	= 20000 m²		

Fill to 150mm depth; Exc. topsoil

(E 641)

20000·00			
1·00			
	20000·00		

All the excavated topsoil will be re-used because the total quantity required exceeds the total available. The total volume of excavated topsoil is converted to m² of 150mm thick spread material and this is then deducted from the total amount required to give the volume of imported material.

Fill to 150mm depth; imp. topsoil

(E 642)

100·00			
150·00	15000·00	(cut area)	
99·00			
150·00	14850·00	(fill area)	
	29850·00		

Ddt
Ditto

(E 642)

2/1½/100·00			
16·00	1600·00	(for sloping banks)	
2/½/99·00			
16·00	1584·00		
20000·00		(for exc. mat)	
1·00	20000·00		
	23184·00		

The deduction for the sloping areas is because the CESMM requires that filling to a constant thickness be measured according to its angle of inclination to the horizontal

Explanatory Notes

6. TOPSOIL AND LANDSCAPING (CONT)

Cut side base length:
$$\sqrt{3.60^2 + 16.00^2}$$
$$= \underline{16.40}$$

Fill side base length:
$$\sqrt{3.47^2 + 16.00^2}$$
$$= \underline{16.37}$$

Add

Fill; 150mm dep;
imp. topsoil; 10-45°
to the horiz.

(E 642.1)

2/1½	101·27		
	16·40	1660·83	
2/1½	100·28		
	16·40	1644·59	
		3305·42	

&

Seeding with low maintenance grass seed as spec. Clause 2.13; over 10° to horiz.

(E 830)

Seeding as spec. Clause 2.12; n.e. 10° to the horiz.

(E 830·1)

100·00		
150·00	15000·00	
99·00		
150·00	14850·00	
	29850·00	

The angle of inclination is most easily calculated either by the use of logarithms or by scaling as shown.

101·27 or
100·28

3·47 or
3·60 m

14°
approx

16m

The sloping areas can be treated as triangles for the purposes of measurement. The area is slightly undermeasured, but is considered sufficiently accurate for this purpose

The precise method, types, and laying rates would be contained in the specification to which reference should be made; item coverage should define exactly what treatments are to be included in the rates

439

Retaining Wall

6. TOPSOIL AND LANDSCAPING (CONT)

			Ddt
			Seeding as spec.
			Clause 2.12; n.c.
			10° to the horiz.
2/1½	100.00		a.b.d. (E 830.1)
	16.00	1600.00	(for sloping banks
2/½	99.00		measured elsewhere)
	16.00	1584.00	
		3184.00	

		Plant shrubs;
		Acer griseum
3305		45-60 cms high
	3305	(to sloping bank (E 850) 1/m²)

Stating that trees are bare root is not a CESMM requirement but is an additional description under 5.10.

		Plant trees; Acer
		campestre ;
		heavy standard;
24		bare root
	24	(E 860)

The specification will normally detail the planting, treatment and techniques required, and it may be necessary to refer to these in the item description; item coverage rules should be extended or included to state exactly what is to be allowed for in the rates.

		Plant trees; Acer
		platanoides ;
		feathered; bare
48		root
	48	(E 860.1)

APPENDIX D

ROAD

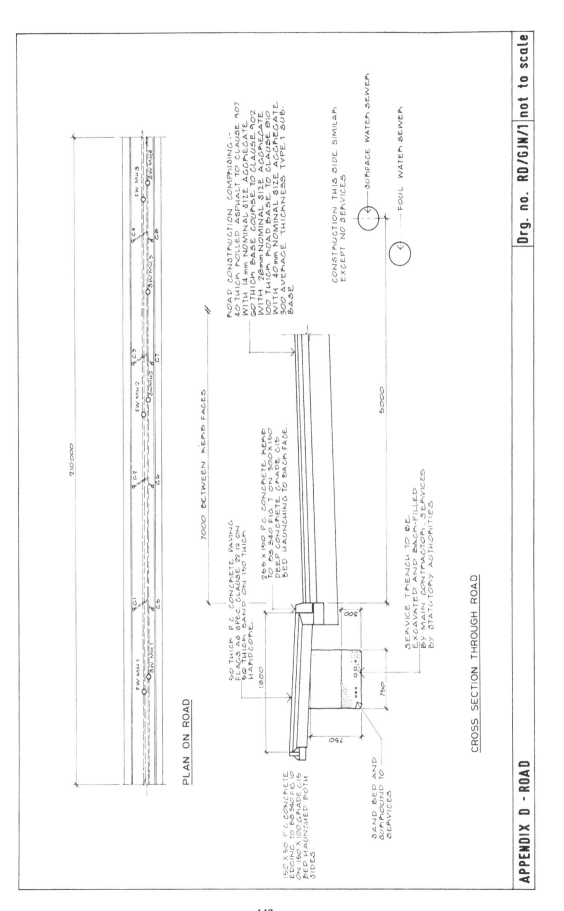

PLAN ON ROAD

CROSS SECTION THROUGH ROAD

ROAD CONSTRUCTION COMPRISING:-
40 THICK ROLLED ASPHALT TO CLAUSE 907
WITH 14mm NOMINAL SIZE AGGREGATE
60 THICK BASE COURSE TO CLAUSE 902
WITH 28mm NOMINAL SIZE AGGREGATE
100 THICK ROAD BASE TO CLAUSE 810
WITH 40mm NOMINAL SIZE AGGREGATE
300 AVERAGE THICKNESS TYPE 1 SUB-
BASE

CONSTRUCTION THIS SIDE SIMILAR
EXCEPT NO SERVICES

SURFACE WATER SEWER

FOUL WATER SEWER

7000 BETWEEN KERB FACES

50 THICK P.C. CONCRETE PAVING
FLAGS AS SPEC. CLAUSE 22.12 ON
50 THICK SAND ON 150 THICK
HARDCORE

255 × 150 P.C. CONCRETE KERB
TO BS 340 FIG. 7 ON 300 × 150
DEEP CONCRETE GRADE C15
BED HAUNCHING TO BACK FACE.

150 × 50 P.C. CONCRETE
EDGING TO BS 340 FIG 10
ON 150 × 100 GRADE C15
BED HAUNCHED BOTH
SIDES

SERVICE TRENCH TO BE
EXCAVATED AND BACK-FILLED
BY MAIN CONTRACTOR, SERVICES
BY STATUTORY AUTHORITIES

SAND BED AND
SURROUND TO
SERVICES

APPENDIX D - ROAD

Drg. no. RD/GJN/1 not to scale

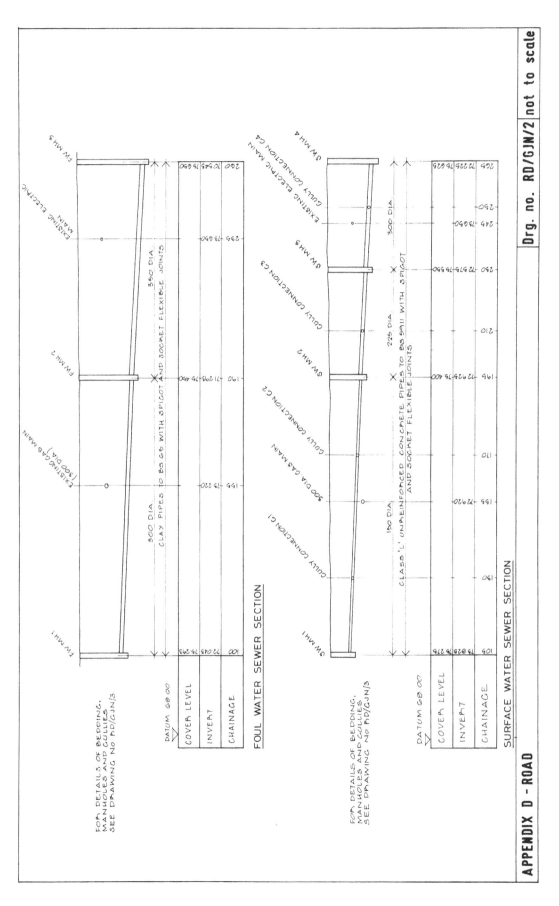

FOUL WATER SEWER SECTION

SURFACE WATER SEWER SECTION

APPENDIX D - ROAD

Drg. no. RD/GJN/2 not to scale

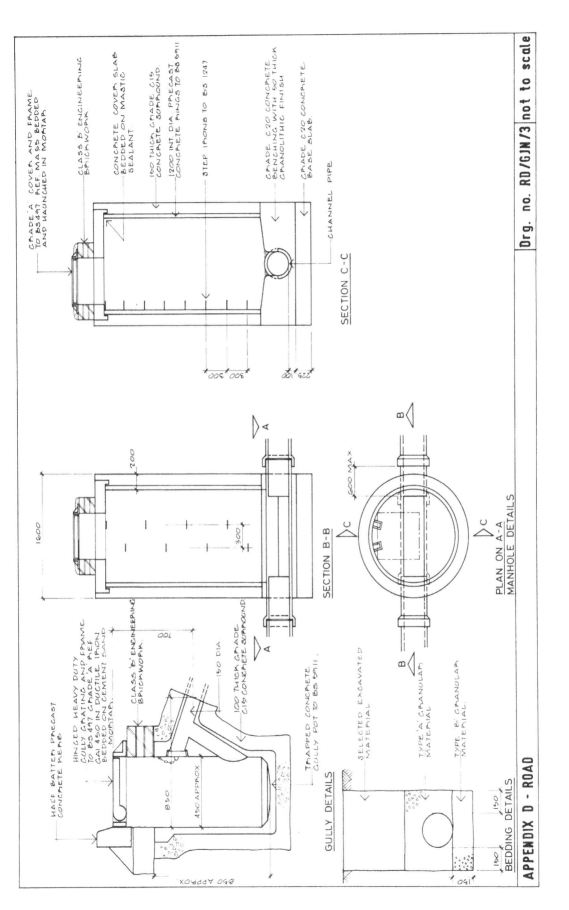

GRADE 'A' COVER AND FRAME TO BS497 PREF MASS BEDDED AND HAUNCHED IN MORTAR.

CLASS 'B' ENGINEERING BRICKWORK.

CONCRETE COVER SLAB BEDDED ON MASTIC SEALANT.

150 THICK GRADE C15 CONCRETE SURROUND.

1200 INT DIA. PRECAST CONCRETE RINGS TO BS 5911

STEP IRONS TO BS 1247

GRADE C20 CONCRETE BENCHING WITH 50 THICK GRANOLITHIC FINISH

GRADE C20 CONCRETE BASE SLAB.

CHANNEL PIPE

SECTION C-C

HALF BATTER PRECAST CONCRETE KERB

HINGED HEAVY DUTY GULLY GRATING AND FRAME TO BS497 GRADE 'A' PREF GALVANISED IN DUCTILE IRON BEDDED ON CEMENT SAND MORTAR.

CLASS 'B' ENGINEERING BRICKWORK

100 THICK GRADE C15 CONCRETE SURROUND

150 DIA

TRAPPED CONCRETE GULLY POT TO BS 5911.

GULLY DETAILS

SECTION B-B

PLAN ON A-A
MANHOLE DETAILS

SELECTED EXCAVATED MATERIAL

TYPE 'A' GRANULAR MATERIAL

TYPE 'B' GRANULAR MATERIAL.

BEDDING DETAILS

APPENDIX D - ROAD

Drg. no. RD/GJN/3 | not to scale

445

Drawing Numbers RD/GJN/1

RD/GJN/2

RD/GJN/3

The following example shows a length of road and footpath with associated drainage and services.

The example demonstrates measurement in accordance with Classes I, J, K, L and R only. Rock head is at level 71.50.

The foulwater pipework is measured between manholes FWMH1 and FWMH3 and the surface water pipework between SWMH1 and SHMH4. In practice the pipework would be scheduled in order to simplify the taking off process and this is shown in the example, demonstrating how each length of pipework is calculated within each of the depth classifications.

FOUL WATER PIPEWORK

RUN	TYPE	DIA	DEPTH CLASSIFICATION				
			2.5-3	3-3.5	3.5-4	4-4.5	4.5-5
FWMH1-MH2	CLAY	300	24.48	48.95	15.37		

CALCS

1st dep.
```
75293
72043
 3250
  500
 2750
```
Leg road

2.5-3 Length:
$$\frac{3-2.75}{0.907} \times 88.80$$
$$= 24.48$$

2ND dep.
```
75450
71293
 4157
  500
 3657
```
Leg road

3-3.5 Length:
$$\frac{3.5-3}{0.907} \times 88.80$$
$$= 48.95$$

3.5-4 Length:
$$\frac{3.657-3.5}{0.907} \times 88.80$$
$$= 15.37$$

total dep.
of fall
```
3657
2750
 907
```

Total pipe length
```
190000
100000
 90000
  1600
 88400
```

MH 2/4½/1600
```
  400
88800
```

Walls 2/200

The calculation of the pipe length within each depth classification is done by interpolation. The total pipe length is measured from the inside faces of manholes (M5)

FOUL WATER PIPEWORK

RUN	TYPE	DIA			DEPTH CLASSIFICATION			
			2.5-3	3-3.5	3.5-4	4-4.5	4.5-5
FWMH 2-MH3	CLAY	350			24.84	36.21	7.75

CALCS

1ST dep.

```
75450
71293
 4157
  500
 3657
```
Less road

2ND dep.
```
75650
70543
 5107
  500
 4607
```
Less road

total dep.
of fall
```
4607
3657
 950
```

Total pipe
length
```
260 000
190 000
 70 000
  1 600
 68 400
    400
 68 800
```
MH 2/44/1600

Nails 2/200

Length:
$$\frac{4-3.657}{0.95} \times 68.80$$
$$= 24.84$$

Length:
$$\frac{4.5-4}{0.95} \times 68.80$$
$$= 36.21$$

Length:
$$\frac{4.607-4.5}{0.95} \times 68.80$$
$$= 7.75$$

SURFACE WATER PIPEWORK

RUN	TYPE	DIA	DEPTH CLASSIFICATION				
			n.e. 1.5	1.5-2	2-2.5	2.5-3	3-3.5
SN/MH1-MH2	CONC	150	47.65	41.15			

CALCS

1st dep.
75275
73825
1450

Less road
500
950

2nd dep.
75400
72925
2475

Less road
500
1975

total dep. of fall
1975
950
1025

Total pipe length
195000
105000
90000
1600
88400

MH 2/4/1600
Walls 2/200
400
88800

Length =
$\dfrac{1.5 - 0.95}{1.026} \times 88.80$
= 47.65

Length =
$\dfrac{1.975 - 1.5}{1.025} \times 88.80$
= 41.15

449

SURFACE WATER PIPE WORK

RUN	TYPE	DIA	n.e. 1·5	DEPTH CLASSIFICATION 1·5-2	2-2·5	2·5-3	3-3·5
SWMH 2-MH3	CONC	225		1·69	32·11		

CALCS

1st dep.
$$\underline{\begin{array}{r} 75400 \\ 72925 \end{array}}$$
$$\underline{\begin{array}{r} 2475 \end{array}}$$
Less road
$$\underline{\begin{array}{r} 500 \end{array}}$$
$$\underline{1975}$$

2nd dep.
$$\underline{\begin{array}{r} 75550 \\ 72575 \end{array}}$$
$$\underline{\begin{array}{r} 2975 \end{array}}$$
Less road
$$\underline{\begin{array}{r} 500 \end{array}}$$
$$\underline{2475}$$

total dep. of fall
$$\begin{array}{r} 2475 \\ 1975 \\ \underline{500} \end{array}$$

Total pipe length
$$\underline{\begin{array}{r} 230000 \\ 195000 \end{array}}$$
$$\underline{35000}$$

MH 2½/1600
$$\underline{\begin{array}{r} 1600 \\ 33400 \end{array}}$$

MaHb 2/200
$$\underline{\begin{array}{r} 400 \\ 33800 \end{array}}$$

Length:
$$\frac{2-1.975}{0.50} \times 33.80 = 1.69$$

Length:
$$\frac{2.475-2}{0.50} \times 33.80 = 32.11$$

SURFACE WATER		PIPEWORK	DEPTH CLASSIFICATION				
RUN	TYPE	DIA	n.l. 1.5	1.5 - 2	2 - 2.5	2.5 - 3	3 - 3.5
SWMH 3 – MH4	CONC	300			1.99	31.81	

CALCS

1st dep.

$$\begin{array}{r} 75550 \\ \underline{72575} \\ 2975 \end{array}$$

Less road

$$\begin{array}{r} 500 \\ \underline{2475} \end{array}$$

2ND dep.

$$\begin{array}{r} 75625 \\ \underline{72225} \\ 3400 \end{array}$$

Less road

$$\begin{array}{r} 500 \\ \underline{2900} \end{array}$$

Total dep.
fall

$$\begin{array}{r} 2900 \\ \underline{2475} \\ 425 \end{array}$$

Total pipe length

$$\begin{array}{r} 265000 \\ \underline{230000} \\ 35000 \end{array}$$

MH 2/½/1600

$$\begin{array}{r} 1600 \\ \underline{33400} \end{array}$$

Walls 2/200

$$\begin{array}{r} 400 \\ \underline{33800} \end{array}$$

Length:
$$\frac{2.5 - 2.475}{0.425} \times 33.80 \quad = 1.99$$

Length:
$$\frac{2.9 - 2.5}{0.425} \times 33.80 \quad = 31.81$$

SURFACE WATER DRAINAGE – GULLY CONNECTIONS

– N.B. it is assumed that the inverts of adjacent gully connections are the same because their connections will be very little difference between them so they are very close together.

Depth at gully

Invert from wearing course
road construction

	40
	60
	100
	300
	400
	500
	200

Commencing surface levels at sewer

G1/G5 $75400 - 75275 \times \dfrac{244}{88.8} + 76275 = 75309$
less road $\quad\dfrac{500}{74809}$

G2/G6 $75400 - 75275 \times \dfrac{64.4}{88.8} + 75275 = 75366$
less road $\quad\dfrac{500}{74866}$

G3/G7 $75550 - 76400 \times \dfrac{14.4}{33.8} + 75400 = 75464$
less road $\quad\dfrac{500}{74964}$

G4/G8 $75625 - 75550 \times \dfrac{19.4}{33.8} + 75550 = 75693$
less road $\quad\dfrac{500}{75093}$

Inverts at sewer

G1/G5	$73825 - \left(\dfrac{244}{88.8} \times 1.025\right) =$	$\underline{73543}$
G2/G6	$73825 - \left(\dfrac{64.4}{88.8} \times 1.025\right) =$	$\underline{73082}$
G3/G7	$72925 - \left(\dfrac{144}{33.8} \times 500\right) =$	$\underline{72712}$
G4/G8	$72575 - \left(\dfrac{19.4}{33.8} \times 425\right) =$	72.331

Depths at sewer

G1/G5	$74809 - 73543 =$	$\underline{1266}$
G2/G6	$74866 - 73082 =$	$\underline{1784}$
G3/G7	$74964 - 72712 =$	$\underline{2252}$
G4/G8	$75093 - 72.331 =$	$\underline{2762}$

Pipe lengths

G1, G2, G3, G4. 5000
G5, G6, G7, G8.

gully $\dfrac{850}{4150}$

gully $\begin{array}{r} 4000 \\ 5000 \\ 2000 \\ 850 \\ \hline 11150 \end{array}$

SURFACE WATER DRAINAGE – GULLY CONNECTIONS

GULLY RUN	TYPE	DIA	n.l.1.5	1.5-2	2-2.5	2.5-3
			DEPTH CLASSIFICATION			
G1/G5	CONC	150	$\dfrac{1\cdot15}{4\cdot15}$			
G2/G6	CONC	150	$\dfrac{1\cdot5-0\cdot2}{1\cdot784-0\cdot2}\times4\cdot15$ $\times1\cdot15$ $=3\cdot41$ $=0\cdot94$	$\dfrac{1\cdot784-1\cdot5}{1\cdot784-0\cdot2}\times4\cdot15$ $\times1\cdot15$ $=0\cdot74$ $=0\cdot21$		
G3/G7	CONC	150	$\dfrac{1\cdot5-0\cdot2}{2\cdot252-0\cdot2}\times4\cdot15$ $\times1\cdot15$ $=2\cdot63$ $=0\cdot73$	$\dfrac{2\cdot0-1\cdot5}{2\cdot252-0\cdot2}\times4\cdot15$ $\times1\cdot15$ $=1\cdot01$ $=0\cdot28$	$\dfrac{2\cdot252-2\cdot0}{2\cdot262-0\cdot2}\times4\cdot15$ $\times1\cdot15$ $=0\cdot51$ $=0\cdot14$	
G4/G8	CONC	150	$\dfrac{1\cdot5-0\cdot2}{2\cdot762-0\cdot2}\times4\cdot15$ $\times1\cdot15$ $=2\cdot11$ $=0\cdot58$	$\dfrac{2\cdot0-1\cdot5}{2\cdot762-0\cdot2}\times4\cdot15$ $\times1\cdot15$ $=0\cdot81$ $=0\cdot22$	$\dfrac{2\cdot5-2\cdot0}{2\cdot762-0\cdot2}\times4\cdot15$ $\times1\cdot15$ $=0\cdot81$ $=0\cdot22$	$\dfrac{2\cdot762-2\cdot5}{2\cdot762-0\cdot2}\times4\cdot15$ $\times1\cdot16$ $=0\cdot42$ $=0\cdot12$

Explanatory Notes

FOUL WATER DRAINAGE

PIPES; FWMH 1 - FWMH3

Vitrified clay pipes; B.S. 65 normal quality; spigot & socket flexible joints; Comm. Surf underside of road construction

It is normal to measure and bill pipework separately according to its function. In this case this means measuring the foul and surface water separately

A1 requires the location of the pipework to be stated and in this case it forms part of the general heading. The Commencing Surface is stated as it is not also the Original Surface (A4.)

Nom. bore 300mm; in tr. dep. 2·5-3m
(I 125)
24·48 | 24·48 | (FWMH 1 - MH2)

Nom. bore 300 mm; in tr. dep 3 - 3·5m
(I 126)
48·95 | 48·95 | (FWMH 1 - MH2)

Nom. bore 300mm; in tr. dep 3·5-4m
(I 127)
15·37 | 15·37 | (FWMH 1 - MH2)

Road

FOUL WATER (CONT)
PIPES; FWMH1-MH3(CONT)

Vitrified clay
pipes (cont)

Nom. bore 350mm;
in tr. dep 3.5-4m
(I 137)
(FWMH 2-MH3)

| 24.84 | |
| 24.84 | |

Nom. bore 350mm;
in tr. dep 4-4.5m
(I 138)
(FWMH 2-MH3)

| 36.21 | |
| 36.21 | |

Nom. bore 350mm;
in tr. dep. 4.5-5m
(I 138.1)
(FWMH2-MH3)

| 7.75 | |
| 7.75 | |

Trenches over 4m deep must
be stated in increments of
0.5m (A6)

455

Explanatory Notes

FOUL WATER (CONT)

MANHOLES AND
PIPEWORK ANCILLARIES

Precast conc. man-
holes with in-situ
surr; as dwg
no. RD/GJN/3;
cover & fr. to
B.S. 497 grade A
ref. MA 55

			dep
			75293
			72043
			3250
			100
			3350

dep. 3-3.5m; FWMH1

(K155)

			dep
			75450
			71293
			4157
			100
			4257

dep. 4.26m; FWMH2

(K157)

			dep
			75650
			70543
			5107
			100
			5207

dep. 5.21m; FWMH3

(K157.1)

1

1

1

1

1

1

A1 requires that the type or
mark numbers be given in
the item descriptions. This
is done by giving their
actual reference. The type
and loading duty must
be stated (A2) and this
is given in the general
heading. As an alternative
to enumerating the manholes
they could have been
measured in detail in
accordance with other
classes of CESMM (see
note at bottom of page 55)

The actual depth of the
manhole must be stated
where this exceeds 4m.
It is recommended that
this be kept to 2 places
of decimals. The depth of the
manholes is measured to the
top of the base slab, which
in this case is 100mm below
the invert (D2)

FOUL WATER (CONT)

MHS. AND PIPEWK.
ANCILL. (CONT)

Crossings

300 dia. gas main;
pipe bore n.e.
300mm
                    (K681)
(FWMH1-MH2)

| 1 | | | |
| | 1 | | |

Although not a specific
requirement, the diameter
of the gas main is stated
in order to assist the
estimator.

Electric main;
pipe bore 300-900mm

(FWMH2-MH3) (K682)

| 1 | | | |

PIPEWORK-SUPPORTS

FNMH1   72043
FWMH2   71293
               ─────
                 750

Rockhead   71750
               71293
               ─────
                 457

∴ rock meets invert
at chainage

        130-(90x$\frac{457}{750}$)

        = 135.16

In order to calculate the
quantity of rock excavation, the
point at which the rock
head crosses the invert of the
pipe must be determined. This
will give the total length of
trench in rock. It must be
borne in mind that the
inverts on the section are
pipe inverts, and that there
is also a 150mm bed
underneath the pipe which
must be accounted for.

457

FOUL WATER (CONT)

PIPEWORK-SUPPORTS(CONT)

Total pipe length FWMH1
-MH2 in rock

	190000
	135160
	54840
Less ½ × FWMH2	800
	54040

Av. dep. FWMH1-MH2
in rock

1st.		0000
2nd.	71750	
	71293	457
	2) 457	
	229	

Nom. tr. wid.

	pipe	300
		500
		800

Total pipe length
FWMH 2-MH3 in rock

	260000
	190000
	70000
Less ½×FWMH2 800	
½ × FWMH3 800	1600
	68400

Av. dep. FWMH 2- MH3
in rock

1st.	71750	
	71293	457
2nd	71750	
	70543	1207
	2) 1664	
	832	

Nom. tr. wid.

	pipe	350
		500
		850

Although the pipes are measured to the inside face of the manhole walls, the volumes of extras to excavation are based on the trench length which is shorter by the width of the manhole walls. Extras to excavation must distinguish between those in trenches and those in manholes, and the length of the manhole is therefore deducted from the pipe length.

The nominal trench width is 500mm greater than the nominal bores of the pipes (D1).

FOUL WATER (CONT)

PIPEWORK SUPPORTS (CONT)

Explanatory Notes

		Extras to exc. in pipe tr.; exc. of rock (L111)	
54·04 0·80 0·23	9·94	(FWMH1-MH2)	
54·04 0·80 0·15	6·48	(ditto for bed)	
68·40 0·85 0·83	48·26	(FWMH2-MH3)	
68·40 0·86 0·15	8·72	(ditto for bed)	
	73·40		

All the foregoing calculations are based on the invert of the pipe. The volume occupied by the bed underneath the pipe is also in rock and is included in the dimensions separately.

```
                    dep
rock level       71750
invert           71293
                   457
base  100          
      200         300
                   757
rock level       71750
invert           70543
                  1207
base  100
      200          300
                  1507
```

The depth of rock excavation in manholes must take into account the depth of the base below the invert.

FOUL WATER (CONT)

PIPENORK SUPPORTS(CONT)

Extras to exc. in mhs ; exc. of rock   (L121)

π/	0·80		
	0·80		
	0·76	1·53	(FWMH2)
π/	0·80		
	0·80		
	1·51	3·04	(FWMH3)

Beds 150dp in type B gran. mat.; with surr. in type A gran. mat.

Because the surround is in a different material to the bed, the item description must so state. The actual nominal bore of the pipe is given because this is considered to be more helpful to the estimator than the ranges given in the Third Division. Note that the total length of the bed is slightly shorter than the pipe length because of the manhole walls.

	Len
	190000
	100000
	90000
2/½/manhole	1600
	88400

pipe nom. bore 300mm   (L332)

(FWMH 1–MH2)

| 88·40 | |
| | 88·40 |

	Len
	260000
	190000
	70000
2/½/manhole	1600
	68400

pipe nom. bore 350 mm   (L333)

(FWMH 2–MH3)

| 68·40 | |
| | 68·40 |

<u>Road</u>

SURFACE WATER
DRAINAGE

PIPES; SWMH 1 - SWMH4

<u>Precast conc. pipes
and fittings; B.S.
5911 class 2
unreinforced;
spigot & socket
flexible joints;
comm. surf.
underside of
road construction</u>

In order to prevent undue
repetition, the fittings to
the surface water pipework
are included with the
general measurement of
the pipes rather than being
classified under a separate
heading.

			Nom. bore 150mm; in. tr. dep. n.e. 1·5m
			(I 212)
47·65			(SWMH 1-MH2)
	47·65		
			Nom. bore 150mm; in. tr. dep. 1·5-2m
			(I 213)
41·15			(SWMH 1- MH2)
	41·15		
			Branches; 150×150 ×150mm
			(J 221)
2/ 1	2		(gully connections G.1 G.2
2/ 1	2		G.7 G.8)
	4		
			Nom. bore 225mm; in. tr. dep 1·5-2m
			(I 223)
1·69			SWMH 2-MH 3
	1·69		

Explanatory Notes

SURFACE WATER (CONT)

PIPES; SWMH1 - MH4 (CONT)

Precast conc. pipes & fittings (cont)

Nom. bore 225mm; in tr. dep. 2-2.5m

(I224)

32·11		(SWMH 2-MH3)
	32·11	

Branches; 225×225 ×150mm

(J222)

2/1	2	(gully connections G.3. G7)
	2	

Nom. bore 300mm; in tr. dep. 2-2.5m

(I224.1)

1·99		(SWMH3 - MH4)
	1.99	

Nom. bore 300mm; in tr. dep. 2.5-3m

(I225)

31·81		(SWMH3- MH4)
	31·81	

Branches; 300×300 ×150mm

(J222.1)

2/1	2	(gully connection G4. G8)
	2	

Road

SURFACE WATER (CONT)

PIPES; SWMH I–MH4(CONT)

Precast conc. pipes
& fittings (cont)

Nom. bore 150mm;
in tr. dep. n.e 1·5m
(I 212)

1·15		G5	(gully connections)
4·15		G1	
3·41		G6	
0·94		G2	
2·63		G7	
0·73		G3	
2·11		G8	
0·58		G4	
	15·70		

Nom. bore 150mm;
in tr. dep. 1·5–2m
(I 213)

0·74		G6	(gully connections)
0·21		G2	
1·01		G7	
0·28		G3	
0·81		G4	
0·22		G8	
	3·27		

Explanatory Notes

The length of the gully connections is measured from the centreline of the sewer (M3). Although not specifically stated the pipe is not measured through the gully itself as this is not a fitting (M3)

463

Explanatory Notes

SURFACE WATER (CONT)

PIPES; SWMH1 – MH4 (CONT)

<u>Precast conc. pipes</u>
<u>& fittings (cont)</u>

Nom. bore 150mm;
in tr. dep. 2-2.5m
(I 214)

0.51		G7 (gully connections)
0.14		G3
0.81		G8
0.22		G4
	1.68	

Nom. bore 150mm;
in tr. dep. 2.5-3m
(I 215)

0.42		G8 (gully connections)
0.12		G4
	0.54	

Road

SURFACE WATER (CONT)

MANHOLES AND
PIPEWORK ANCILLARIES

Precast conc. man-
holes with in-situ
surr; as dwg no.
RD/GJN/3; cover &
fr. to BS 497 grade
A ref MA 55

$$\frac{dep}{\begin{array}{r} 75275 \\ 73825 \\ \hline 1450 \\ 100 \\ \hline 1550 \end{array}}$$

dep. 1.5 – 2m; SWMH1
(K152)

1	
	1

$$\frac{dep}{\begin{array}{r} 75400 \\ 72925 \\ \hline 2475 \\ 100 \\ \hline 2575 \end{array}}$$

dep. 2.5 – 3m; SWMH2
(K154)

1	
	1

$$\frac{dep}{\begin{array}{r} 75550 \\ 72575 \\ \hline 2975 \\ 100 \\ \hline 3075 \end{array}}$$

dep. 3 – 3.5m; SWMH3
(K155)

1	
	1

Explanatory Notes

SURFACE WATER (CONT)

MHS. AND. PIPEWORK
ANCILL (CONT)

Precast' conc. mhs (cont)

	dep
	75625
	72225
	3400
	100
	3500

dep. 3-3.5m; SWMH4

(K155)

1

1

Gullies

Precast' conc. trapped;   The type and loading duty
as drwg. no. RD/GJN/3;   of gully covers must be
cover & ft. to   stated in item descriptions
B.S. 497 grade A
ref. GA 1450

(K360)

(all gullies)

8/ 1

8

			SURFACE WATER (CONT)	Explanatory Notes

			MHS AND PIPEWK ANCILL (CONT)
			<u>Crossings</u>
			Electric main; pipe bore n.e. 300mm
			(K681·1)
1			
	1		

Note that the surface water sewer passes above the gas main and does not have to be measured as a crossing.

Also note that the lowest invert of the pipe including its bed (72075) or manhole to the underside of its base (SWMH4 at 71925) are both higher than the level of rock and hence there will not be any rock excavation for surface water drainage.

			PIPEWORK SUPPORTS
			<u>Beds 150dp in type B gran. mat; with Surr. in type A gran. mat.</u>
			pipe nom. bore 150mm (L331)
	88·40	88·40	(SWMH 1 – MH2)
4/	4·15	16·60	}(gully runs)
4/	1·15	4·60	
		109·60	

Note that the length of the bed and surround is measured along the pipe centre line (M11). Although not specifically stated, the length occupied by the gully is deducted as this is not classified as a fitting or valve.

467

<u>Road</u>

<u>Explanatory Notes</u>

SURFACE WATER (CONT)

PIPEWORK SUPPORTS (CONT)

<u>Beds 150dp in
type B gram mat (cont)</u>

pipe nom. bore
225 mm
                    (L332·1)

| 33·40 | | (SWMH 2- MH3) |

| | 33·40 | |

pipe nom. bore
300mm
                    (L332·2)

| 33·40 | | (SWMH 3- MH4) |

| | 33·40 | |

		SERVICE TRENCH FOR STATUTORY AUTHORITIES	Explanatory Notes
		PIPEWORK – MANHOLES	
		Tr. for pipes or cables not laid by contractor; backfilling above bed & surr. with selected Exc. Mal. after install. of services by others	Although not specifically required, details of backfilling requirements (if any) should always be given in item descriptions for service trenches
		Cross-sect. area 0·5– 0·75m² (K483)	
210·00	210·00		
		PIPEWORK SUPPORTS	This is a rogue item as it is not classified strictly in accordance with the Third Division, and is therefore coded 9. Also, the size of the bed and surround has been stated because it is not particularly related to the bore of the services or pipes.
		Sand bed & surr. 750 × 300mm around services	
210·00	210·00	(L3919)	

Road

ROADS AND PAVINGS

Sub·bases, flexible
road bases and
surfacing.

Gran. mat.
D.Tp. specified
type 1 ; dep.
350 mm

(R117)

	Wid
	7000
Kerb fndn 2/300	600
	7600

210·00
7·60

(subbase)

1596·00

Dense tarmacadam;
D.Tp. Clause 810;
dep. 100mm

(R253)

210·00
7·00

(road
base) $

1470·00

Rolled asphalt;
D.Tp. clause 902;
28mm nom. size
agg.; dep. 60mm

(base
course) $

(R322)

Rolled asphalt;
D.Tp. clause 907;
14mm nom. size
agg.; dep. 40mm

(wearing
course)

(R322·1)

M1 states the width of each course to be measured at the top surface. In the case of the sub·base this is taken as extending under the kerb foundation. It is not necessary to state the depth in bands as the actual depth is given (Paragraph 3.10)

It is not necessary to state the size of aggregate as there there is only one size in the specification. No deductions are made for the areas of manholes and gullies as their plan size does not exceed 1m² (M1).

It is necessary to state the aggregate size for both the base course and wearing course as there are several sizes in the specification

Road

ROADS AND PAVINGS (CONT)

Kerbs, channels
and edgings

			Precast. conc. Kerb to B.S.7263:Part1,Fig1(F) Straight; incl. fndn & haunch as drwg no. RD/GJN/1 (R631)	It is not necessary to state the dimensions in the item description in accordance with A7 as all the relevant information is contained on the drawing.
2/	210·00			
		420·00	&	
			Precast conc. Edging to B.S.7263 Part1,Fig(m)straight; incl. fndn & haunch. (R661)	The length of the edging is exactly the same as the kerb, hence the use of the ampersand.

Explanatory Notes

ROADS AND PAVINGS (CONT)

Light duty
pavements

Hardcore base;
dep. 150mm

(R724)

$\dfrac{wid}{1800}$

Kerb haunch
300-150 = 150
Edging haunch
$\dfrac{150-50}{2}$ = 50 $\dfrac{200}{1600}$

2/210.00			(footpath)
1.60			
		672.00	

As the width is defined as
the width along the top
surface. the haunch to
both the kerb and edging
must be deducted.

			Sand base
			(R712)
2/210.00			(blinding)
1.80			&
		756.00	

Precast conc.
flags as spec.
Clause 22.12
(R782)

Although the bottom of the
sand layer is the same as
the hardcore width the
top surface is wider and
it is this width upon
which the measurement
is based.

APPENDIX E

REINFORCEMENT

MEMBER	BAR MARK	TYPE & SIZE	No. OF MBRS.	No. IN EACH	TOTAL No.	LENGTH OF EACH BAR	SHAPE CODE	A	B	C	D	E/r
COMPONENT 'A'	24	T16	2	24	48	1450	37	1290				
	25	T10	1	86	86	1675	37	1205				
	26	T10	3	108	324	5075	20	STRAIGHT				
	27	T12	1	8	8	3000	38	1370	330			
COMPONENT 'B'	21	T10	2	20	40	5550	37	5385				
	22	T10	1	72	72	5425	37	5260				
	23	T12	6	36	216	2150	38	1000	215			
	24a	T12	1	4	4	2675	37	2515				
	24b	T12	10	4	40	2175	37	2010				
	24c	T10	5	8	40	1925	37	1755				
	24d	T16	1	8	8	1750	37	1580				
	24e	T16	8	8	64	1625	37	1455				
	24f	T10	1	24	24	1450	37	1290				
	25a	T10	10	78	780	1675	37	1205				
	26a	T20	1	100	100	5075	20	STRAIGHT				
	27a	T20	4	4	16	3000	38	1370	330			
	28	T10	4	8	32	2600	20	STRAIGHT				
	29	T10	4	8	32	1225	37	1055				
	30	T12	6	32	192	4350	38	2050	330			
	31a	T16	2	4	8	2075	38	920	330			

ALL BENDING DIMENSIONS ARE IN ACCORDANCE WITH BS 4466

<u>Reinforcement</u>

				<u>Explanatory Notes</u>
			Bar bending schedule BBS-21	Reinforcement quantities may be prepared by one of three methods.

1) By direct taking off from the drawings.

2) From bar bending schedules

3) From average weights of steel per m^3 of concrete, divided into various diameters. This method should only be used for the tender quantities and never for the final account.

The following is an example of the calculation of reinforcement weights from a typical bar bending schedule.
The schedule gives the total number and girth of each bar as calculated in accordance with BS4449.
The measurement involves abstracting the bar lengths according to their diameter and type, multiplying them by the total number of each and applying the conversion factor to arrive at the gross weight.
The schedules are normally prepared by the Engineer.

Explanatory Notes

Deformed High Yield
Steel bars to
BS. 4449

10mm nominal
size
(bar mark No) (G 523)

86/	1.68	144.48	(25
324/	5.08	1645.92	(26
40/	5.55	222.00	(21
72/	5.43	390.96	(22
40/	1.93	77.20	(24b
24/	1.45	34.80	(24f
780/	1.68	1310.40	(25
32/	2.60	83.20	(28
32/	1.23	39.36	(29
		3948.32	
		×0.616	Kg/m
		2432.17	÷1000 = 2.43217T

12mm nominal size
(bar mark No) (G 524)

8/	3.00	24.00	(27
216/	2.15	464.40	(23
4/	2.68	10.72	(24a
40/	2.18	87.20	(24b
192/	4.35	835.20	(30
		1421.52	
		×0.888	kg/m
		1262.31	÷1000 = 1.26231T

The weights of reinforcement per mm² of cross-sectional area are given in kg/m in the British Standard. The total weight of reinforcement must therefore be divided by 1000 to convert it to tonnes.

			16mm nominal size
			(bar mark (G 525)
			No.)
48/	1·45	69·60	(24
8/	1·75	14·00	(24d
64/	1·63	104·32	(24e
8/	2·08	16·64	(31a
		204·56	
		×1·579	kg/m
		323·00	÷ 1000 = 0·3230T

			20mm nominal size
			(bar mark (G 526)
			No.)
100/	5·08	508·00	(26a
16/	3·00	48·00	(27a
		556·00	
		×2·466	kg/m
		1371·10	÷ 1000 = 1·3711T

APPENDIX F

GATEHOUSE

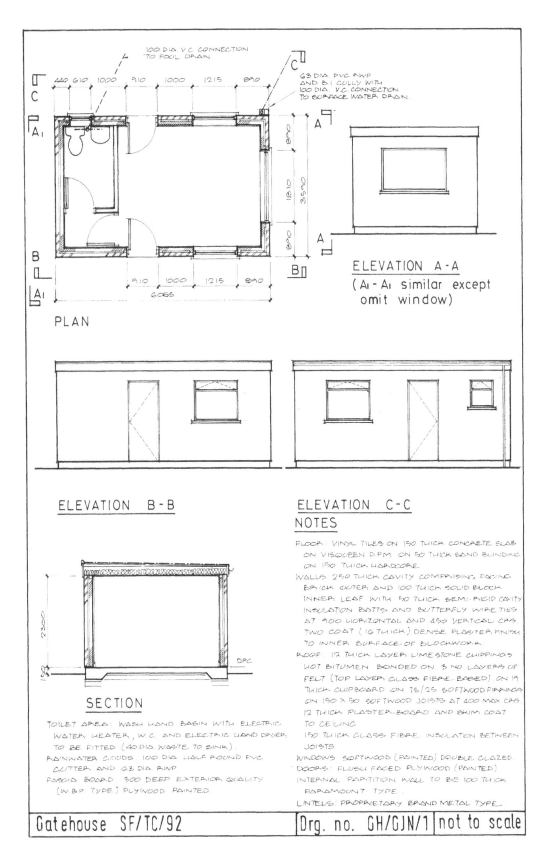

PLAN

100 DIA V.C. CONNECTION TO FOUL DRAIN

440 610 1000 910 1000 1215 890

63 DIA PVC RWP AND B I GULLY WITH 100 DIA V.C. CONNECTION TO SURFACE WATER DRAIN

890 3590 1810 890

910 1000 1215 890
6065

ELEVATION A-A
(A₁-A₁ similar except omit window)

ELEVATION B-B

ELEVATION C-C

2300
150
DPC

SECTION

NOTES

FLOOR: VINYL TILES ON 150 THICK CONCRETE SLAB ON VISQUEEN D.P.M ON 50 THICK SAND BLINDING ON 150 THICK HARDCORE.

WALLS: 250 THICK CAVITY COMPRISING FACING BRICK OUTER AND 100 THICK SOLID BLOCK INNER LEAF WITH 50 THICK SEMI-RIGID CAVITY INSULATION BATTS AND BUTTERFLY WIRE TIES AT 900 HORIZONTAL AND 450 VERTICAL CRS. TWO COAT (16 THICK) DENSE PLASTER FINISH TO INNER SURFACE OF BLOCKWORK.

ROOF: 12 THICK LAYER LIMESTONE CHIPPINGS HOT BITUMEN BONDED ON 3 NO LAYERS OF FELT (TOP LAYER GLASS FIBRE BASED) ON 19 THICK CHIPBOARD ON 75/25 SOFTWOOD FIRRINGS ON 150 X 50 SOFTWOOD JOISTS AT 400 MAX CRS. 12 THICK PLASTERBOARD AND SKIM COAT TO CEILING. 150 THICK GLASS FIBRE INSULATION BETWEEN JOISTS.

WINDOWS: SOFTWOOD (PAINTED) DOUBLE GLAZED.
DOORS: FLUSH FACED PLYWOOD (PAINTED).
INTERNAL PARTITION WALL TO BE 100 THICK PARAMOUNT TYPE.
LINTELS: PROPRIETARY BRAND METAL TYPE.

TOILET AREA: WASH HAND BASIN WITH ELECTRIC WATER HEATER, W.C. AND ELECTRIC HAND DRYER TO BE FITTED (40 DIA WASTE TO SINK).
RAINWATER GOODS: 100 DIA HALF ROUND PVC GUTTER AND 63 DIA RWP.
FASCIA BOARD: 300 DEEP EXTERIOR QUALITY (W.B.P TYPE) PLYWOOD PAINTED.

| Gatehouse SF/TC/92 | Drg. no. GH/GJN/1 | not to scale |

481

Drawing No. GH/GJN/1

Explanatory Notes.

The following example shows a typical simple building which is often found on a water or sewage treatment works and is 'incidental to civil engineering works' and should be measured in accordance with the provisions of Class Z.

It is assumed that the foundations have been measured under Classes E, F and G, the brickwork under classes U and W and the decorating under Class V.

CARPENTRY AND JOINERY

Structural and
carcassing timber;
flat roofs.

6.07
overhang 2/0.75 .15
400 |6.22
15.55 + 1

= 17

Divide the length of building by the joist centres to calculate the number of joists required.

3.59
overhang 2/0.75 .15
3.74

17/	3.74	63.58

Softwood joists;
150 × 50 mm
(Z.113.1)

&

Firring pieces; 75
to 25 × 50 mm
(Z.113.2)

These tapered pieces (maximum height 75mm, minimum 25mm) create the slope to the roof.

Sheet boarding; sloping upper surfaces.

	6.22		Chipboard 19mm
	3.74	23.26	thick.
			(2.132)

Miscellaneous joinery.

2/	6.22	12.44	Fascia board; WPB
2/	3.74	7.48	exterior quality plywood;
		19.92	300 x 25mm
			(2.159)

Skirtings; softwood.

2/	5.57	11.14	75 x 18mm
2/	3.09	6.18	(2.651) WC
2/	1.40	2.80	6.07
	2.00	2.00	.50
		22.12	wall thickness 5.57
			2/250
	Ddt.		3.59
4/	.91	3.64	doors .50
		18.48	2/.25 3.09

484

<u>Insulation.</u>

16/3.74			Glass fibre
	.40	23.93	quilt 150mm thick,
			(aid between joists)
			(Z.229)

Windows, doors
<u>and glazing.</u>

	1		Standard softwood
			window type 107v
			size 630 x 750 mm
			(Z.311.1)

2/	1	2	Standard softwood
			window type 210w
			size 1200 x 1050 mm
			(Z.311.2)

	1		Special observation
			window consisting
			of softwood members
			as specification
			clause 27.9; size
			1810 x 1050mm.
			(Z.311.3)

	1	1	Standard flush plywood faced internal flush door, 40mm thick, size 1981 x 762 mm (Z.313.1)
2/	1	2	Standard flush plywood faced external quality door, 44mm thick, size 1981 x 838 mm (Z.313.2)
	1	1	Standard softwood door lining size 27 x 94 mm for door size 1981 x 762 mm (Z.314.1)
2/	1	2	Standard softwood door frame size 33 x 64 mm for door size 1981 x 838 mm (Z.314.2)

2/	4.87	9.74	Miscellaneous
	4.95	4.95	joinery; softwood
	4.95	4.95	architraves once
		19.64	chamfered; size

75 × 18mm

     (2.152)

Intl. door 2/1.98 3.96

          2/.075   .15

                 .76

               4.87

Extt doors 2/1.98 3.96

          2/.075   .15

                 .84

               4.95

Ironmongery.

3/	1 pr	3 pr	Steel butt hinges;

100mm   (2.341)

Note deviation from CESMM. Hinges are usually bought and sold in pairs not enumerated. The change should be listed in the Preamble.

2/	1	2	Yale cylinder

night rim latch

     (2.343.1)

	1	1	SAA mortice latch

with lever handles

     (2.343.2)

SAA is an abbreviation for satin anodised aluminium.

	1	1	SAA wc indicating bolt	
			(z.343.3)	
2/	2	4	SAA Casement stay and fastener.	It is assumed that the windows are supplied without any ironmongery.
	1	1		
		5		
			(z.349)	

Glazing

2/	1.15	1.61	Standard plain glass; clear float; to wood frames in putty; 6mm thick.
	.70		
2/	1.15	.41	
	.18		
			(z.351.1)
	1.70	1.61	
	.95	3.94	

	.55	.22	Standard plain glass; white patterned; to wood frames in putty 6mm thick
	.40		
	.55	.09	
	.18	.31	
			(z.351.2)

488

Surface finishes, linings and partitions.

	5.57			
	3.09	17.21		

Insitu finishes, cement and sand (1:3) floor screed, steel trowelled finish, 25mm thick.

(2.411)

&

Vinyl 'Polyflex' floor tiles, 2mm thick, fixed with adhesive

(2.421)

The screed and floor tiling has been measured under the partition although it may not be done this way on site.

	17.32		
	2.30	39.83	
	17.55		
	.10	1.75	
		41.58	

Insitu finishes, one coat cement and sand (1:3) backing coat, one coat Thistle class B plaster to blockwork, steel trowelled finish 16mm thick.

(2.413)

The plaster is measured overall and the doors and windows are then deducted. CESMM does not mention work to reveals so these areas are added in.

		2/5.57	11.14
		2/3.09	6.18
			17.32

Ddt.

.63		
.75	0.47	
1.20		
1.05	2.52	
1.81		
1.05	1.90	
26.91		
2.00	3.64	8.53
		32.95

Reveals

Windows	2/.75	1.50
		.63
	2/2/1.05	4.20
	2/1.20	2.40
	2/1.05	2.10
		1.81

Doors.	2/2.00	4.00
		.91
		17.55

489

| | | 5.57 | | | Gypsum plasterboard fixing with nails to underside of softwood joists, 12.5 mm thick |
| | | 3.09 | 17.21 | |

Gypsum plasterboard
fixing with nails
to underside of
softwood joists,
12.5 mm thick

(2.434)

&

One coat Thistle
board finish,
steel trowelled, to
soffit, 3 mm thick.

(2.414)

Paramount partition
complete with
softwood floor,
wall and ceiling
battens as specification
clause 31.9, 57 mm
thick.

(2.479)

2.20
1.50
3.70

&

One coat Thistle
board finish,
steel trowelled
to walls 3 mm
thick.

(2.413)

Ddt.
.85
2.00

3.70
2.30   8.51

1.70
6.81

490

Piped building services.

Pipework, copper pipe to BS2871, Table X lead free pre-soldered capillary joints, 15 mm diameter.

It is unusual for the pipework for the plumbing services to be shown on a drawing and the take off usually has to prepare an isometric sketch for his own use.

1.00	1.00	to softwood skirting (Z.511.1)
1.00 1.75		to plastered wall (Z.511.2)
	2.75	

Extra over for

1	1	15 mm equal tee (Z.512.1)
4	4	15 mm elbow (Z.512.2)
1	1	tap connector (Z.512.3)
1	1	WC connector (Z.512.4)
1	1	water heater connector (Z.512.5)

491

1	1	Stopcock to BS1010, lead free pre-soldered capillary joints, gunmetal with brass headwork, 15mm diameter. (Z512.6)
		Equipment.
1	1	'Streamline' oversink water heater, 7 litre, 1kw, with spout and valve, fixed to plastered wall. (Z.529.1)
1	1	'Handidry' electric hand drier, 1.4kw, fixed to wall. (Z.529.2)
		Sanitary appliances and fittings.
1	1	Wash basin, vitreous china, complete with chromium plated waste, overflow with chain and plastic plug, polyprophylene P trap 40mm, 560x430mm white, wall mounted on brackets. (Z.530.1)

1	1	'Aztec' chromium plated pillar tap 15mm (2.512.7)
1	1	wc suite washdown type, vitreous china with plastic seat and cover, 9 litre cistern, ball valve, flush pipe, pan, connection to S trap. (2.530.2)
1.30	1.30	Ultra ABS marley waste system, solvent-welded joints, to clips, 40mm diameter. (2.511.3)
		Extra over for
1	1	bend (2.512.8)

6.07	6.07	UPVC Terrain system rainwater gutter, straight half round, joint bracket joints, to softwood fascia with support brackets at 1m maximum centres
		(x .331)
1	1	Stop end
		(x .332.1)
1	1	Stop end outlet
		(x.332.2)
2.55	2.55	UPVC Terrain system rainwater pipe, straight, connector joints, to brickwork with clips at 2m centres, plugging
		(x .333)
1	1	Shoe
		(x.334)

Roofing.

Three layers of bituminous felt roofing fibre based surface type 18 weighing 25 kg/ 10 m², limestone chipping to top surface.

	6.07		Upper surfaces inclined at an
	3.59	21.79	angle not exceeding 30° to the horizontal
			(W.341)

2/	6.07	12.14	Surfaces at
2/	3.59	7.18	width 100 mm
		19.32	

(W.347)

495

APPENDIX G

MENSURATION AND USEFUL DATA

FIGURE	AREA	PERIMETER

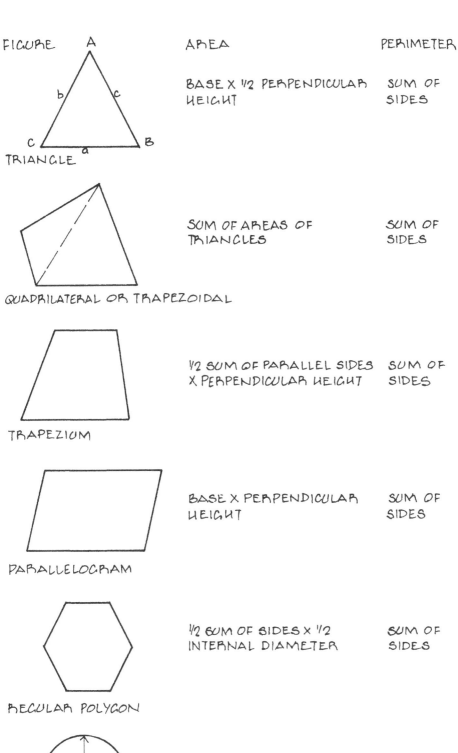

TRIANGLE

BASE X ½ PERPENDICULAR HEIGHT — SUM OF SIDES

QUADRILATERAL OR TRAPEZOIDAL

SUM OF AREAS OF TRIANGLES — SUM OF SIDES

TRAPEZIUM

½ SUM OF PARALLEL SIDES X PERPENDICULAR HEIGHT — SUM OF SIDES

PARALLELOGRAM

BASE X PERPENDICULAR HEIGHT — SUM OF SIDES

REGULAR POLYGON

½ SUM OF SIDES X ½ INTERNAL DIAMETER — SUM OF SIDES

CIRCLE

πr^2
$\pi = 3.1416$

πd OR $2\pi r$

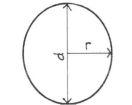

FIGURE	AREA	PERIMETER

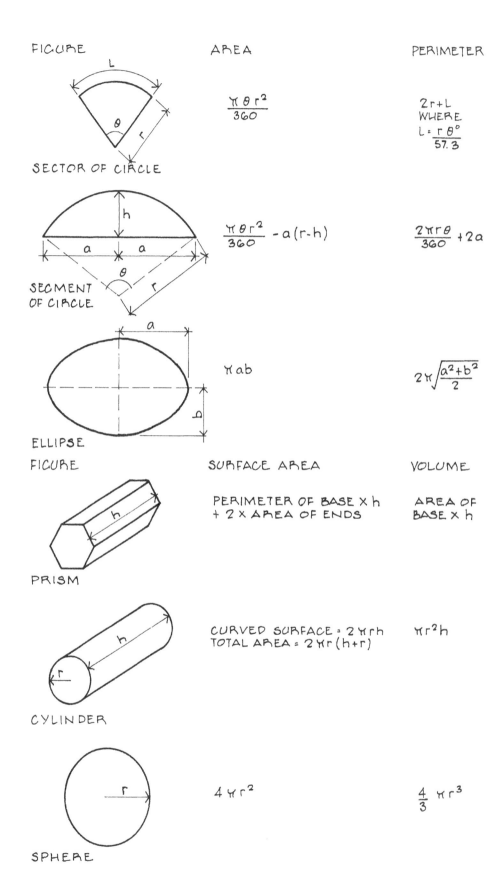

SECTOR OF CIRCLE

Area: $\dfrac{\pi \theta r^2}{360}$

Perimeter: $2r+L$ WHERE $L = \dfrac{r\theta°}{57.3}$

SEGMENT OF CIRCLE

Area: $\dfrac{\pi \theta r^2}{360} - a(r-h)$

Perimeter: $\dfrac{2\pi r \theta}{360} + 2a$

ELLIPSE

Area: πab

Perimeter: $2\pi \sqrt{\dfrac{a^2+b^2}{2}}$

FIGURE	SURFACE AREA	VOLUME

PRISM

Surface Area: PERIMETER OF BASE × h + 2 × AREA OF ENDS

Volume: AREA OF BASE × h

CYLINDER

Surface Area: CURVED SURFACE = $2\pi rh$ TOTAL AREA = $2\pi r(h+r)$

Volume: $\pi r^2 h$

SPHERE

Surface Area: $4\pi r^2$

Volume: $\dfrac{4}{3}\pi r^3$

FIGURE	SURFACE AREA	VOLUME

SEGMENT OF SPHERE

CURVED SURFACE = $\pi(r^2 + h^2)$

$\frac{\pi}{6} h(h^2 + 3r^2)$

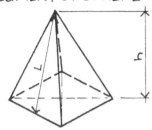

PYRAMID

½ PERIMETER OF BASE × L PLUS AREA OF BASE

⅓ BASE × h

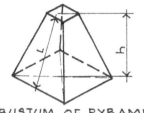

FRUSTUM OF PYRAMID

LATERAL AREA = ½ MEAN PERIMETER × L

$\frac{h}{3}(A + B + \sqrt{AB})$

WHERE A = AREA OF LARGE END AND B = AREA OF SMALL END

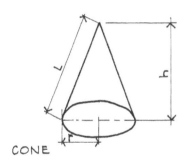

CONE

$\pi r(L + r)$

$\frac{1}{3} \pi r^2 h$

FRUSTUM OF CONE

CURVED SURFACE $\pi L(R + r)$

$\frac{\pi h}{3}(R^2 + r^2 + Rr)$

MENSURATION AND USEFUL DATA

The metric system

Linear

1 centimetre (cm)	= 10 millimetres (mm)
1 decimetre (dm)	= 10 centimetres (cm)
1 metre (m)	= 10 decimetres (dm)
1 kilometre (km)	= 1000 metres (m)

Area

100 sq millimetres (mm^2)	= 1 sq centimetre (cm^2)
100 sq centimetres(cm^2)	= 1 sq decimetre (dm^2)
100 sq decimetres(dm^2)	= 1 sq metre (m^2)

Capacity

1 millilitre (ml)	= 1 cubic centimetre (cm^3)
1 centilitre (cl)	= 10 millilitres (ml)
1 decilitre (dl)	= 10 centilitres (cl)
1 litre (l)	= 10 decilitres (dl)

Weight

1 centigram (cg)	= 10 milligrams (mg)
1 decigram (dg)	= 10 centigrams (cg)
1 gram (g)	= 10 decigrams (dg)
1 decagram (dag)	= 10 grams (g)
1 hectogram (hg)	= 10 decagrams (dag)

Imperial/Metric conversions

Linear

1 in = 25.4mm	1mm = 0.03937 in
1 ft = 304.8mm	1cm = 0.3937 in
1 yd = 914.4mm	1dm = 3.397 in
	1m = 39.37 in

Square

1 sq in = 645.16mm^2	1cm^2 = 0.155 sq in
1 sq ft = 0.0929m^2	1m^2 = 10.7639 sq ft
1 sq yd = 0.8361m^2	1m^2 = 1.196 sq yd

Cube

$$1 \text{ cu in} = 16.3871 \text{cm}^3 \qquad 1\text{cm}^3 = 0.061 \text{ cu in}$$
$$1 \text{ cu ft} = 0.0283 \text{m}^3 \qquad 1\text{m}^3 = 35.3148 \text{ cu ft}$$
$$1 \text{ cu yd} = 0.7646 \text{m}^3 \qquad 1\text{m}^3 = 1.307954 \text{ cu yd}$$

Capacity

$$1 \text{ fl oz} = 28.4 \text{ml} \qquad 1\text{ml} = 0.0352 \text{ fl oz}$$
$$1 \text{ pt } 0 = 0.568 \text{ l} \qquad 1\text{dl} = 3.52 \text{ fl oz}$$
$$1 \text{ gallon} = 4.546 \text{ l} \qquad 1\text{L} = 1.7598 \text{ pt}$$

Weight

$$1 \text{ oz} = 28.35 \text{g} \qquad 1\text{g} = 0.035 \text{ oz}$$
$$1 \text{ lb} = 0.4536 \text{kg} \qquad 1\text{kg} = 35.274 \text{ oz}$$
$$1 \text{ st} = 6.35 \text{kg} \qquad 1\text{t} = 2204.6 \text{ lb}$$
$$1 \text{ ton} = 1.016 \text{t} \qquad 1\text{t} = 0.9842 \text{ ton}$$

Temperature equivalents

In order to convert Fahrenheit to Celsius deduct 32 and multiply by 5/9. To convert Celsius to Fahrenheit multiply by 9/5 and add 32.

Fahrenheit	Centigrade
230	110.0
220	104.4
210	98.9
200	93.3
190	87.8
180	82.2
170	76.7
160	71.1
150	65.6
140	60.0
130	54.4
120	48.9
110	43.3
90	32.2
80	26.7
70	21.1
60	15.6
50	10.0
40	4.4
30	-1.1
20	-6.7
10	-12.2
0	-17.8

Bricks   Number of bricks per square metre in half brick thick wall in stretcher bond

50 x 102.5 x 215mm	74
65 x 102.5 x 215mm	59
75 x 102.5 x 215mm	52

Blocks   Number of blocks per square metre

450 x 225mm	10
450 x 300mm	7
600 x 225mm	7

Timber   1 standard = 4.67227 cubic metres

1 cubic metre = 35.3148 cubic feet

10 cubic metres = 2.140 standards

Number of slates/tiles per m2

Asbestos-free slates

Size (mm)	Lap (mm)	Nr of slates
400 x 200	70	30.0
400 x 200	76	30.9
400 x 200	90	32.3
400 x 240	80	26.1
500 x 250	90	19.5
500 x 250	80	19.1
500 x 250	70	18.6
500 x 250	76	18.9
500 x 250	90	19.5
500 x 250	106	20.5
500 x 250	100	20.0
600 x 300	106	13.6
600 x 300	100	13.4
600 x 300	90	13.1
600 x 300	80	12.9
600 x 300	70	12.7
600 x 350	100	11.5

Blue Welsh Slates

Size (mm)	Nr of slates
405 x 205 (16" x 8")	29.59
405 x 255 (16" x 10")	23.75
405 x 305 (16" x 12")	19.00
460 x 230 (18" x 9")	23.00

Blue Welsh Slates (cont'd)

460 x 255 (18" x 10")	20.37
460 x 305 (18" x 12")	17.00
510 x 255 (20" x 12")	18.02
510 x 305 (20" x 12")	15.00
560 x 280 (22" x 11")	14.81
560 x 305 (22" x 12")	14.00
610 x 305 (24" x 12")	12.27

Westmorland Green Slates

1 ton (Imperial) standard quality covers approximately 18-20m2

1 ton (Imperial) Peggies covers approximately 15-16m2

Marley Tiles (100mm gauge)

Type	Nr/m2
Plain	60.0
Feature	56.0
Ludlow Plus	17.4
Anglia Plus	17.3
Ludlow Major	10.7
Mendip	10.6
Double Roman	10.4
Modern	10.8
Wessex	11.0
Bold Roll	10.6
Monarch	14.5

Standard tiles

150 x 150mm	44	250 x 125mm	32
100 x 200mm	50	230 x 230mm	19
200 x 200mm	25		

Lead

Code	kg/m2
3	14.97
4	20.41
5	25.40
6	30.05
7	36.72
8	40.26

Coverage of plasters

Carlite premixed browning		m2/1000kg
	11mm floating coat	130-150
Metal lathing	11mm pricking up and floating	60-70
Bonding coat	8mm floating coat on concrete	145-155
	11mm floating coat on brickwork and blockwork	100-110
	8mm floating coat on plasterboard	150-165
Finish	2mm finishing coat on floating coat	410-500
Thistle finish	2mm finishing coat on sanded undercoat	350-450
Thistle board finish	5mm finishing in two coats	160-170
Thistle renovating plasters		
Thistle undercoat	11mm thick	120
Thistle finish	2mm thick	380-420
Sirapite B plasters		
Sirapite B	3mm finishing coat on sanded undercoat	250-270

Mortar mixes

Cement mortar	1:3	0.48 tonnes cement/m3 1.45 tonnes sand/m3
	1:4	0.36 tonnes cement/m3 1.45 tonnes sand/m3
Cement lime mortar	1:1:6	0.22 tonnes cement/m3 0.11 tonnes lime/m3 1.45 tonnes sand/m3

Mortar mixes (cont'd)

1:2:9   0.16 tonnes cement/m3
0.14 tonnes lime/m3
1.45 tonnes sand/m3

Velocity

Miles per hour into kilometres per hour	1.60934
Feet per second into metres per second	0.3048
Feet per minute into metres per second	0.00508
Feet per minute into metres per minute	0.3048
Inches per second into milimetres per second	25.4
Inches per minute into milimetres per second	0.423333
Inches per minute into centimetres per minute	2.54

Fuel Consumption

Gallons per mile into litres per kilometre	2.825
Miles per gallon into kilometres per litre	0.354

Density

Tons per cubic yard into kilogrammes per cubic metre	1328.94
Pounds per cubic foot into kilogrammes per cubic metre	16.0185
Pounds per cubic inch into grammes per cubic centimetre	27.6799
Pounds per gallon into kilogrammes per litre	0.09978

Average plant outputs (cubic metres per hour)

Bucket size (litres)	Soil	Sand	Heavy Clay	Soft Rock
Face Shovel				
200	11	12	7	5
300	18	20	12	9
400	24	26	17	13
600	42	45	28	23
Backactor				
200	8	8	6	4
300	12	13	9	7
400	17	18	11	10
600	28	30	19	15

Average plant outputs (cont'd)

Dragline

200	11	12	8	5
300	18	20	12	9
400	25	27	16	12
600	42	45	28	21

Bulkage of excavation

Multiply volume by %

Soil	25
Gravel	15
Sand	12.5
Chalk	50
Clay (Heavy)	30
Rock	30

Reinforcement mass

Hot Rolled Bars

Stainless Steel Bars

Size in mm	Mass per Metre in kg		Size in mm	Mass per Metre in kg
6	0.222		10	0.667
8	0.395		12	0.938
10	0.616		16	1.628
12	0.888		20	2.530
16	1.579		25	4.000
20	2.466		32	6.470
25	3.854			
32	6.313			
40	9.864			
50	15.413			

Mesh fabric	Mesh Size		Wire Size		
	Main	Cross	Main	Cross	
	mm	mm	mm	mm	kg
A 393	200	200	10	10	6.16
A 252	200	200	8	8	3.95

Reinforcement mass (cont'd)

	Mesh Size		Wire Size		
	Main	Cross	Main	Cross	
	mm	mm	mm	mm	kg
A 191	200	200	7	7	3.02
A 142	200	200	6	6	2.22
A 98	200	200	5	5	1.54
B1131	100	200	12	8	10.90
B 785	100	200	10	8	8.14
B 503	100	200	8	8	5.93
B 385	100	200	7	7	4.53
B 283	100	200	6	7	3.73
B 196	100	200	5	7	3.05
C 785	100	400	10	6	6.72
C 503	100	400	8	5	4.34
C 385	100	400	7	5	3.41
C 283	100	400	6	5	2.61
D 98	200	200	5	5	1.54
D 49	100	100	2.5	2.5	0.770

Standard wire gauge

SWG	mm
3	6.40
4	5.89
5	5.38
6	4.88
7	4.47
8	4.06
9	3.63
10	3.25
11	2.95
12	2.64
13	2.34
14	2.03
15	1.83
16	1.63
17	1.42
18	1.21
19	1.02
20	0.91
21	0.81
22	0.71

Standard wire gauge (cont'd)

SWG	mm
23	0.61
24	0.56
25	0.51
26	0.46

Paper sizes

Size	mm	inches
A0	841 x 1189	33.11 x 46.81
A1	594 x 841	23.39 x 33.11
A2	420 x 594	16.54 x 23.39
A3	297 x 420	11.69 x 16.54
A4	210 x 297	8.27 x 11.69
A5	148 x 210	5.83 x 8.27
A6	105 x 148	4.13 x 5.83
A7	74 x 105	2.91 x 4.13
A8	52 x 74	2.05 x 2.91
A9	37 x 52	1.46 x 2.05
A10	26 x 37	1.02 x 1.46

Average weights of materials

Material	tonnes per m3
Ashes	0.68
Aluminium	2.68
Asphalt	2.31
Brickwork - Engineering	2.24
Brickwork - Common	1.86
Bricks -Engineering	2.40
Bricks - Common	2.00
Cement - Portland	1.45
Cement - Rapid Hardening	1.34
Clay - dry	1.05
Clay - wet	1.75
Coal	0.90
Concrete	2.30
Concrete - reinforced	2.40
Earth - topsoil	1.60
Glass	2.60
Granite - solid	2.70
Gravel	1.76

Average weights of materials (cont'd)

Material	tonnes per m3
Iron	7.50
Lead	11.50
Limestone - crushed	1.75
Plaster	1.28
Sand	1.90
Slate	2.80
Tarmacadam	1.57
Timber - general construction	0.70
Water	1.00

INDEX

NOTES

NOTES

NOTES

NOTES

NOTES

NOTES

NOTES

NOTES

NOTES

NOTES

Printed and bound by CPI Group (UK) Ltd, Croydon, CR0 4YY

01/11/2024

01782605-0013